Lecture Notes in Computer Science

Edited by G. Goos and J. Hartmanis

Advisory Board: W. Brauer   D. Gries   J. Stoer

Lecture Notes in Computer Science

Edited by G. Goos and J. Hartmanis

Advisory Board: W. Brauer   D. Gries   J. Stoer

649

A. Pettorossi (Ed.)

Meta-Programming in Logic

Third International Workshop, META-92
Uppsala, Sweden, June 10-12, 1992
Proceedings

Springer-Verlag

Berlin Heidelberg New York
London Paris Tokyo
Hong Kong Barcelona
Budapest

Series Editors

Gerhard Goos
Universität Karlsruhe
Postfach 69 80
Vincenz-Priessnitz-Straße 1
W-7500 Karlsruhe, FRG

Juris Hartmanis
Cornell University
Department of Computer Science
4130 Upson Hall
Ithaca, NY 14853, USA

Volume Editor

Alberto Pettorossi
University of Rome Tor Vergata, Via della Ricerca Scientifica
I-00133 Roma, Italy

CR Subject Classification (1991): I.2.2-4, F.4.1

ISBN 3-540-56282-6 Springer-Verlag Berlin Heidelberg New York
ISBN 0-387-56282-6 Springer-Verlag New York Berlin Heidelberg

Typesetting: Camera ready by author/editor
Printing and binding: Druckhaus Beltz, Hemsbach/Bergstr.
45/3140-543210 - Printed on acid-free paper

Foreword

This volume contains the papers presented at the Third International Workshop on "Meta-Programming in Logic" held at the Department of Computer Science of the University of Uppsala (Sweden), 10-12 June, 1992. This workshop is the successor of the ones organized by John Lloyd in Bristol, U.K. (June 1988) and by Maurice Bruynooghe in Leuven, Belgium (April 1990). This volume also includes the invited lectures and the advanced tutorials from the workshop.

The Programme Committee received 35 papers, from which 18 were chosen for a long presentation and 6 for a short one. The papers give an interesting and stimulating view of the major topics under investigation on: i) the *foundations and applications of meta-programming and transformational programming*, ii) the *design and implementation of language facilities for meta-programming*, and iii) *knowledge representation and meta-programming*.

Unfortunately, we were not able to include in this volume the following 6 papers relative to the short presentations:
i) F. Giunchiglia and L. Serafini (IRST, Povo, Italy): "Hierarchical Meta-Logics (or: How We Can Do Without Modal Logics)", ii) K. Hinkelmann (DFKI, Kaiserslautern, Germany): "Forward Logic Evaluation: Developing a Compiler from a Partially Evaluated Meta-Interpreter", iii) R. M. Jones (Optismsoft, Wegberg, Germany): "Why Must the King Speak Out?", iv) H. Ohnishi and S. Akama (Asao-ku, Kawasaki-shi, Japan): "Indexed Knowledge in Epistemic Logic Programming", v) J. S. Santibáñez (A.I. Dept., Polytechic University Madrid, Spain): "A Formal Model for Temporal Knowledge Based Systems Verification", and vi) D. G. Schwartz (Dept. Computer Engineering, CWRU, Cleveland, Ohio): "Metaprograms: the Glue to Integrate and Control Blackboard Knowledge Sources".

Their abstracts have been published in a document of the Computing Science Department of the University of Uppsala (Sweden).

I would like to thank the members of the Programme Committee who carefully read the submitted papers with the help of external referees and reported on time their comments and opinions, and in particular those who attended the Programme Committee meeting held in Rome at the beginning of February 1992. Special thanks also go to Maurice Bruynooghe and Robert Kowalski who supported my work with useful advice and suggestions.

I acknowledge the dedicated, patient, and skilful cooperation of Maurizio Proietti.

I am grateful to Jonas Barklund and the members of the Organizing Committee for their efforts in making the workshop possible and preparing this volume. They were very good at overcoming the inevitable difficulties due to the distance between Rome and Uppsala, which sometimes made things not so easy.

On behalf of the participants and the Programme Committee I would like also to thank the invited speakers, who delivered very interesting and stimulating lectures.

Finally, I want to thank the Department of Electronic Engineering of the University of Rome 'Tor Vergata' (Italy), the IASI Institute of the Italian National Research Council in Rome (Italy), and the Department of Computer Science of Uppsala University (Sweden) for their financial support and for providing the necessary facilities.

Alberto Pettorossi

Acknowledgements

The Meta 92 Workshop has been made possible through generous financial support from several sponsors:
- the Swedish Board for Technical Development (NUTEK),
- Uppsala University,
- the Association for Logic Programming,
- Swedish Institute of Computer Science (SICS), and
- SUN Microsystems AB.
The help from Civildatalogförbundet is also appreciated.

Being the chairman for the workshop, I find this an appropriate opportunity also to thank the other members of the Organizing Committee for the considerable time they have spent for preparing the workshop. I hope our combined efforts have produced a stimulating setting for the workshop.

The success of this workshop is of course mostly dependent on the quality of the program. The members of the Programme Committee have made a good effort to judge each submission carefully and to include as many significant contributions as time could allow.

In particular, we owe much to the programme chairman for the workshop, Alberto Pettorossi, and to his colleague Maurizio Proietti for hosting the Programme Committee meeting and for careful editing of the contributions.

Finally, I would like to thank Danny De Schreye for sharing valuable experiences from the preceding workshop in this series.

Jonas Barklund

Program Committee

L. Aiello Carlucci (Rome, Italy)
J. Barklund (Uppsala, Sweden)
H. Blair (Syracuse, U.S.A.)
K. A. Bowen (Syracuse, U.S.A.)
M. Bruynooghe (Leuven, Belgium)
A. Bundy (Edinburgh, U.K.)
W. Drabent (Warsaw, Poland)
K. Furukawa (Tokyo, Japan)
J. Gallagher (Bristol, U.K.)
J. Komorowski (Trondheim, Norway)

R. A. Kowalski (London, U.K.)
G. A. Lanzarone (Milan, Italy)
W. Marek (Ithaca, U.S.A.)
D. Miller (Philadelphia, U.S.A.)
L. M. Pereira (Lisbon, Portugal)
A. Pettorossi (Rome, Italy), chairman
J. Staples (Queensland, Australia)
L. Sterling (Cleveland, USA)
S.-Å. Tärnlund (Uppsala, Sweden)
F. Turini (Pisa, Italy)

Organizing Committee

J. Barklund (Uppsala, Sweden), chairman
A. Hamfelt (Uppsala, Sweden)

T. Hjerpe (Uppsala, Sweden)
F. Möllerberg (Uppsala, Sweden)

List of Referees

L. Aiello, J. Alferes, J. Aparicio, R. Barbuti, J. Barklund, H. Blair, K. A. Bowen, A. Brogi, M. Bruynooghe, A. Bundy, M. Cialdea, S. Costantini, M. Danelutto, B. Demoen, D. De Schreye, W. Drabent, G. Filè, H. Fujita, K. Furukawa, J. Gallagher, E. Hainicz, Y. J. Jiang, T. Kawamura, J. Komorowski, R. A. Kowalski, G. Lanzarone, W. Lukaszewicz, W. Marek, B. Martens, D. Miller, D. Nardi, E. G. Omodeo, D. Pedreschi, L. M. Pereira, A. Pettorossi, M. Proietti, T. Shintani, J. Staples, L. Sterling, A. Takeuchi, S.-Å. Tärnlund, M. Temperini, F. Turini, K. Verschaetse, S. Wierzchon, A.Wrzos-Kaminska, J. Wrzos-Kaminski.

Preface

Meta-programming is a technique which is widely used in logic programming. It is also an important technique for other programming paradigms and it has been present in various areas of computer science throughout the history of its development.

Meta-programming can be understood as the treatment of programs as objects. Thus, the various methodologies which refer to program transformation, program analysis, and program manipulation for imperative, functional, and logic languages, are all included in the area of meta-programming.

It is assumed that the programs under consideration have an associated semantics. This hypothesis is essential, because program manipulations and program transformations are meaningful only if they are performed while preserving the semantic values.

In the case of logic languages where computations and programs can naturally be viewed as proofs (recall the 'programs as proofs' paradigm), one can consider the techniques for representing and manipulating proofs to be an essential part of the meta-programming idea.

In particular, one may refer to proofs by using the 'demo' predicate, which can be decorated with suitable arguments for explicitly describing the relevant properties of the proofs in hand.

Proofs may belong to different logical theories. One may stratify those theories in a hierarchy or one may amalgamate them into a unique theory. There are advantages and disadvantages for either choice: the best solution very much depends on the application domain.

In both cases one may refer to a proof in a given logical theory by a name (or a term). The reader will find in this book some papers which address this 'naming problem' and propose some solutions. The origin of the naming problem is linked to Gödel's incompleteness theorem, which uses a naming technique for encoding logical formulas.

A particularly interesting field where logical theories are used is knowledge representation, where it is often required to represent what agents know about their own knowledge and the knowledge of other agents. This 'reflexive' power can easily be encoded into suitable logic programs, provided that the language includes some meta-programming features.

It is not required to use modal theories to represent these situations in a natural way. Some of these issues have been considered in the invited lectures and in other papers of these proceedings.

This book also includes papers on: i) logical foundations of meta-programming, ii) model-theoretic and proof-theoretic problems, iii) analysis and transformation of logic programs, iv) use of meta-programming for deductive databases, and v) implementation aspects related to meta-programming, like modularization, compiler optimization, process communication, and object-orientation.

Meta-programming is not a special feature of logic programming. As we already said, it has been part of computer science since its very beginning. Indeed, if meta-programming is considered to be manipulation of programs as objects, one can say that it was already required when writing programs for the von Neumann computer. For instance, if we want to compute the sum of k integers which are assumed to be stored from memory location i to memory location i+k−1, and we assume that the values of k and i are known at run time only, it is necessary to write a program whose instructions manipulate the program itself (if indirect addressing and index registers are not available). Some of the instructions which will be executed at run time, are themselves obtained as the result of a computation.

Also interpretation or compilation can be viewed as instances of meta-programming. In both cases the executable program in machine language is the result of a computation over the source program given as input.

Abstraction mechanisms present in some programming languages, like i) making a procedure, say '$\lambda x.x+1$', out of an expression, say '$x+1$', or ii) making a block, say 'begin S1; S2 end', out of a sequence of statements, say 'S1; S2', can be viewed as meta-programming techniques.

We do not have here the space to thoroughly explore many other programming techniques (such as partial evaluation) and examine their relationship to meta-programming. However, we want to remark that meta-programming is also present in functional languages, in particular if they allow a higher-order type discipline. In that case, in fact, one may obtain a function to be used as the result of the application of a higher-order function to an argument which itself is a function. Obviously, higher-order functions are extremely useful for conciseness and clarity. A standard example is the function composition operator, which is very often used in functional programming.

Rome, Italy
September 1992

Alberto Pettorossi

Table of Contents

Invited Papers

P. Gärdenfors
Belief Revision: A Vademecum 1

L. Fariñas Del Cerro and A. Herzig
Meta-Programming Through Intensional Deduction: Some Examples 11

K. Konolige
An Autoepistemic Analysis of Metalevel Reasoning in Logic Programming ... 26

Advanced Tutorials

J. Komorowski
An Introduction to Partial Deduction 49

D. De Schreye and K. Verschaetse
Tutorial on Termination of Logic Programs 70

Languages and Applications I

F. van Harmelen
Definable Naming Relations in Metalevel Systems 89

A. Brogi, P. Mancarella, D. Pedreschi, and F. Turini
Meta for Modularizing Logic Programming 105

S. K. Debray
Compiler Optimizations for Lowlevel Redundancy Elimination:
An Application of Metalevel Prolog Primitives 120

S. Costantini, P. Dell'Acqua, and G. Lanzarone
Reflective Agents in Metalogic Programming 135

Languages and Applications II

I. Cervesato and G.F. Rossi
Logic Meta-Programming Facilities in 'Log 148

R. Bahgat
The Pandora Deadlock Handler Metalevel Relation 162

K. Benkerimi and P.M. Hill
Object-Oriented Programming in Gödel: An Experiment 177

Logical Foundations

D. De Schreye and B. Martens
A Sensible Least Herbrand Semantics for Untyped Vanilla Meta-Programming
and its Extension to a Limited Form of Amalgamation 192

H. Christiansen
A Complete Resolution Method for Logical Meta-Programming Languages ... 205

P. Bonatti
Model Theoretic Semantics for Demo.. 220

F. Giunchiglia, L. Serafini, and A. Simpson
Hierarchical Meta-Logics: Intuitions, Proof-Theory, and Semantics 235

Transformation and Analysis I

G.A. Wiggins
Negation and Control in Automatically Generated Logic Programs 250

A. Bossi, N. Cocco, and S. Etalle
Transforming Normal Programs by Replacement 265

Transformation and Analysis II

J.L. Träff and S.D. Prestwich
Meta-Programming for Reordering Literals in Deductive Databases 280

M. Bruynooghe and G. Janssens
Propagation: A New Operation in a Framework for Abstract Interpretation of
Logic Programs 294

F. Mesnard and J.-G. Ganascia
CLP(Q) for Proving Interargument Relations 308

Knowledge Representation

A. Hamfelt and Å. Hansson
Representation of Fragmentary Multilayered Knowledge 321

J. Grabowski
Metaprograms for Change, Assumptions, Objects, and Inheritance 336

Author Index 353

Belief revision: A vade-mecum

Peter Gärdenfors

Lund University Cognitive Science,
Kungshuset, Lundagård,
S–223 50 LUND, Sweden

Abstract. This paper contains a brief survey of the area of belief revision and its relation to updating of logical databases as it has developed during the last years. A set of rationality postulates for revisions are presented. A couple of semi-algorithmic models of revision and updating are introduced. The models are connected to the postulates via some representation theorems. Finally, the relation between belief revision and nonmonotonic logic is explored.

1. Introduction

Suppose you have a database that contains, among other things, the following pieces of information:

A: Gold can only be stained by aqua regia.
B: The acid in the bottle is sulphuric acid.
C: Sulphuric acid is not aqua regia.
D: My wedding ring is made of gold.

The following fact is derivable from A – D:

E: My wedding ring will not be stained by the acid in the bottle.

Now suppose that, *as a matter of fact*, the ring is indeed stained by the acid. This means that you want to add the negation of E to the database. But then the database becomes *inconsistent*. If you want to keep the database consistent you need to *revise* it. You don't want to give up all of the beliefs since this would be an unnecessary loss of valuable information. So you have to *choose* between retracting A, B, C or D.

When trying to handle belief revisions in a computational setting, there are three main methodological questions to settle:

(1)   How are the beliefs in the database represented?

(2)   What is the relation between the elements explicitly represented in the database and the beliefs that may be *derived* from these elements?

(3)   How are the choices concerning what to retract made?

When beliefs are represented by *sentences* in a belief system K, one can distinguish three main kinds of belief changes:

(i) *Expansion*: A new sentence together with its logical consequences is added to a belief system K. The belief system that results from expanding K by a sentence A will be denoted K+A.

(ii) *Revision*: A new sentence that is inconsistent with a belief system K is added, but in order that the resulting belief system be consistent some of the old sentences in K are deleted. The result of revising K by a sentence A will be denoted K*A.

(iii) *Contraction*: Some sentence in K is retracted without adding any new facts. In order that the resulting system be consistent some other sentences from K may have to be given up. The result of contracting K with respect to the sentence A will be denoted K–A.

Expansions of belief systems can be handled comparatively easily. K+A can simply be defined as the logical closure of K together with A:

(Def +) K+A = {B: K ∪ {A} ⊢ B}

It is not possible to give a similar explicit definition of revisions and contractions in logical and set-theoretical notions only. When tackling the problem of belief revisions (and contractions) there are two general strategies to follow, namely to present explicit *constructions* of the revision process and to formulate *postulates* for such constructions. After presenting postulates and proposals for revision methods, the two approaches will be connected via a number of *representation theorems*.

2 . Preliminaries

I shall work with a language L which is based on first order logic. It will be assumed that the underlying logic includes *classical propositional logic* and that it is compact. If K logically entails A we will write this as K ⊢ A. Where K is a set of sentences we shall use the notation Cn(K) for the set of all logical consequences of K, i.e. Cn(K) = {A: K ⊢ A}.

The simplest way of modelling a belief state is to represent it by a *set* of sentences from L. Accordingly, we define a *belief set* as a set K of sentences in L which satisfies the following constraint:

(⊢) If K logically entails B, then B ∈ K.

If belief states are modelled by belief sets, all possible changes can be described as expansions, contractions and revisions, or defined by a series of such operations.

Other models of belief states with a richer structure may allow other types of belief changes. Some such models and the corresponding changes are presented in [4].

There is exactly one inconsistent belief set under our definition, namely the set of all sentences of L. We introduce the notation $K_\perp$ for this belief set.

Against modelling belief states as belief sets it has been argued that some of our beliefs have no independent standing but arise only as inferences from our more basic beliefs. Formally, this idea can be modelled by saying that B_K is a *base for a belief set* K iff B_K is a finite subset of K and $Cn(B_K) = K$. Then instead of introducing revision and contraction functions that are defined on belief sets it is assumed that these functions are defined on bases. Such functions will be called *base revisions* and *base contractions* respectively (see [2]). This approach introduces a more finegrained structure since we can have two bases B_K and C_K such that $Cn(B_K) = Cn(C_K)$ but $B_K \neq C_K$.

3. Rationality postulates for belief revisions

In the remainder of the paper, it will be assumed that belief sets are used as models of belief states. The goal is now to formulate postulates for rational revision and contraction functions defined over such belief sets. The following postulates are taken from [4]:

(K*1)   For any sentence A and any belief set K, K*A is a belief set.

(K*2) $A \in K*A$.

(K*3) $K*A \subseteq K+A$.

(K*4)   If $\neg A \notin K$, then $K+A \subseteq K*A$.

(K*5) $K*A = K_\perp$ if and only if $\vdash \neg A$.

(K*6)   If $\vdash A \leftrightarrow B$, then $K*A = K*B$.

The postulates (K*1) – (K*6) are elementary requirements that connect K, A and K*A. This set will be called the *basic* set of postulates. The final two conditions concern *composite* belief revisions.

(K*7) $K*A\&B \subseteq (K*A)+B$.

(K*8)   If $\neg B \notin K*A$, then $(K*A)+B \subseteq K*A\&B$.

The postulates for the contraction function '–' are to an even larger extent than for revisions motivated by the principle of informational economy.

(K–1)   For any sentence A and any belief set K, K–A is a belief set.

(K–2) $K–A \subseteq K$.

(K–3)   If $A \notin K$, then $K–A = K$.

(K–4)   If not ⊢ A, then A ∉ K–A.

(K–5)   If A ∈ K, then K ⊆ (K–A)+A.

(K–6)   If ⊢ A ↔ B, then K–A = K–B.

Postulates (K–1) – (K–6) are called the *basic set* of postulates for contractions. Again, two further postulates for contractions with respect to conjunctions will be added.

(K–7)   K–A ∩ K–B ⊆ K–A&B.

(K–8)   If A ∉ K–A&B, then K–A&B ⊆ K–B.

We next turn the connections between revision and contraction functions. A revision of a knowledge set can be seen as a composition of a contraction and an expansion:

(Def *)   K*A = (K–¬A)+A

Theorem 1: If a contraction function '–' satisfies (K–1) to (K–4) and (K–6), then the revision function '*' obtained from (Def *) satisfies (K*1) – (K*6). Furthermore, if (K–7) also is satisfied, (K*7) will be satisfied for the defined revision function; and if (K–8) also is satisfied, (K*8) will be satisfied for the defined revision function.

Conversely, contractions can be defined in terms of revisions.

(Def –)   K–A = K ∩ K*¬A.

Theorem 2: If a revision function '*' satisfies (K*-1) to (K*6), then the contraction function '–' obtained from (Def –) satisfies (K–1) – (K–6). Furthermore, if (K*7) is satisfied, (K–7) will be satisfied for the defined contraction function; and if (K*8) is satisfied, (K–8) will be satisfied for the defined contraction function.

4. Models and representation results

The problem in focus of this section is how to define the contraction K–A with respect to a belief set K and a proposition A. Via (Def *) this will then give us a definition of a revision function. A general idea is to start from K and then give some recipe for choosing which propositions to delete from K so that K–A does not contain A as a logical consequence. We should look at as large a subset of K as possible.

The following notion is useful: A belief set K' is a *maximal subset of K that fails to imply A* if and only if (i) K' ⊆ K, (ii) A ∉ K', and (iii) for any sentence B that is in K but not in K', B → A is in K'. The set of all belief sets that fail to imply A will be denoted K⊥A.

A first tentative solution to the problem of constructing a contraction function is to identify K–A with one of the maximal subsets in K⊥A. Technically, this can be done with the aid of a *selection function* S that picks out an element S(K⊥A) of K⊥A for any K and any A whenever K⊥A is nonempty. We then define K–A by the following rule:

(Maxichoice)   K–A = S(K⊥A) when not ⊢A, and K–A = K otherwise.

Contraction functions determined by some such selection function were called *maxichoice contraction functions* in [1]. A first test for this construction is whether it has the desirable properties. It is easy to show that any maxichoice contraction function satisfies (K–1) – (K–6). But they also satisfy the following *fullness* condition:

(K–F)   If B ∈ K and B ∉ K–A, then B → A ∈ K–A for any belief set K.

Theorem 3: Any contraction function that satisfies (K–1) – (K–6) and (K–F) can be generated by a maxichoice contraction function.

However, in a sense maxichoice contraction functions in general produce contractions that are *too large*. Let us say that a belief set K is *maximal* iff for every sentence B, either B ∈ K or ¬B ∈ K.

Theorem 4: If a revision function '*' is defined from a maxichoice contraction function '–' by means of the Levi identity, then, for any A such that ¬A ∈ K, K*A will be maximal.

A second tentative idea is to assume that K–A contains only the propositions that are *common to all* of the maximal subsets in K⊥A:

(*Meet*)   K–A = ∩(K⊥A) whenever K⊥A is nonempty and K–A = K otherwise.

This kind of function was called *full meet contraction function* in [1]. Again, it is easy to show that any full meet contraction function satisfies (K–1) – (K–6). They also satisfy the following *intersection* condition:

(K–I)   For all A and B, K–A&B = K–A ∩ K–B.

Theorem 5: A contraction function satisfies (K–1) – (K–6) and (K–I) iff it can be generated as a full meet contraction function.

The drawback of full meet contraction is that it results in contracted belief sets that are far *too small*.

Theorem 6: If a revision function '*' is defined from a full meet contraction function '–' by means of the Levi identity, then, for any A such that ¬A ∈ K, K*A = Cn({A}).

Given these results it is natural to investigate the consequences of using only *some* of the maximal subsets in K⊥A when defining K–A. Technically, a *selection function* S can be used to pick out a nonempty *subset* S(K⊥A) of K⊥A, if the latter is nonempty, and that puts S(K⊥A) = K in the limiting case when K⊥A is empty. The contraction function can then be defined as follows:

(*Partial meet*):   K–A = ∩S(K⊥A).

Such a contraction function was called a *partial meet contraction function* in [1].

Theorem 7: For every belief set K '–' is a partial meet contraction function iff '–' satisfies postulates (K–1) – (K–6).

The idea of S picking out the 'best' elements of K⊥A can be made more precise by assuming that there is an *ordering* of the maximal subsets in K⊥A that can be used to pick out the top elements. Technically, we do this by introducing the notation M(K) for the *union* of the family of all the sets K⊥A, where A is any proposition in K that is not logically valid. Then it is assumed that there exists a *transitive and reflexive* ordering relation ≤ on M(K). When K⊥A is nonempty, this relation can be used to *define* a selection function that picks out the top elements in the ordering:

(Def S) S(K⊥A) = {K' ∈ K⊥A: K" ≤ K' for all K" ∈ K⊥A}

A contraction function that is determined from ≤ via the selection function S given by (Def S) will be called a *transitively relational partial meet contraction function*.

Theorem 8: For any belief set K, '−' satisfies (K−1) – (K−8) iff '−' is a transitively relational partial meet contraction function.

Thus we have found a way of connecting the rationality postulates with a general way of modelling contraction functions. The drawback of the partial meet construction is that the *computational costs* involved in determining what is in the relevant maximal subsets of a belief set K are so overwhelming that other solutions to the problem of constructing belief revisions and contractions should be considered.

As a generalization of the AGM postulates several authors have suggested postulates for revisions and contractions of *bases* for belief sets rather than the belief sets themselves [2].

A second way of modelling contractions is based on the idea that some sentences in a belief system have a higher degree of *epistemic entrenchment* than others. The guiding idea for the construction of a contraction function is that when a belief set K is revised or contracted, the sentences in K that are given up are those having the lowest degrees of epistemic entrenchment. If A and B are sentences in L, the notation A ≤ B will be used as a shorthand for "B is at least as epistemically entrenched as A". We assume the following postulates for epistemic entrenchment:

(EE1)	If A ≤ B and B ≤ C, then A ≤ C	(transitivity)
(EE2)	If A ⊢ B, then A ≤ B	(dominance)
(EE3)	For any A and B, A ≤ A&B or B ≤ A&B	(conjunctiveness)
(EE4)	When K ≠ K⊥, A ∉ K iff A ≤ B, for all B	(minimality)
(EE5)	If B ≤ A for all B, then ⊢ A	(maximality)

We now turn to the connections between orderings of epistemic entrenchment and the contraction and revision functions presented above. We will accomplish this by providing two conditions, one of which determines an ordering of epistemic entrenchment assuming a contraction function and a belief set as given, and the other of which determines a contraction function assuming an ordering of epistemic entrenchment and a belief set as given:

(C≤) A ≤ B if and only if A ∉ K−A&B or ⊢ A&B.

(C–)   B ∈ K–A if and only if B ∈ K and either A < A ∨ B or ⊢A.

Theorem 9: If an ordering ≤ satisfies (EE1) – (EE5), then the contraction function which is uniquely determined by (C–) satisfies (K–1) – (K–8) as well as the condition (C≤).

Theorem 10: If a contraction function '–' satisfies (K–1) – (K–8), then the ordering ≤ that is uniquely determined by (C≤) satisfies (EE1) – (EE5) as well as the condition (C–).

These results suggest that the problem of constructing appropriate contraction and revision functions can be *reduced* to the problem of providing an appropriate ordering of epistemic entrenchment. Furthermore, condition (C–) gives an explicit answer to which sentences are included in the contracted belief set, given the initial belief set and an ordering of epistemic entrenchment. From a computational point of view, applying (C–) is trivial, once the ordering ≤ of the elements of K is given. However, in order to determine what is contained in various belief sets, any implementation of the belief revision processes considered here presumes that efficient theorem provers are available .

5. Connections with nonmonotonic logic

Belief revision and nonmonotonic logic are motivated by quite different ideas. The theory of belief revision deals with the dynamics of belief states, that is, it aims at modelling how an agent or a computer system updates its state of belief as a result of receiving new information. Nonmonotonic logic, on the other hand, is concerned with a systematic study of how we jump to conclusions from what we believe. By using default assumptions, generalizations etc. we tend to believe in things that do not follow from our knowledge by the classical rules of logic. Within the research on nonmonotonic logic one can identify several systems which have been put forward as potential tools for handling nonmonotonic reasoning within artificial intelligence. Among the principal approaches one may mention the default systems of Reiter [12], Shoham's [13] preferential models, Poole's [11] maxiconsistent constructions, and probabilistic default reasoning [10].

Despite the differences in motivation for the theories of belief revision and nonmonotonic logic, it can be shown that the formal structures of the two theory areas, as they have developed, are surprisingly similar. The aim of this section is to show that it is possible to translate concepts, models, and results from one area to the other. This translation was first given in [9].

In the same way as for the theory of belief revision one can discuss postulates for nonmonotonic reasoning. It was Gabbay [3] who initiated this kind of investigation by focussing on the formal properties of a nonmonotonic inference relation ⊢~ so that A ⊢~ B hold between two propositions A and B, if B follows nonmonotonically given A as a premise. The classical inference relation ⊢ is monotonic in the sense that if A ⊢ B, then A&C ⊢ B, for any C. By definition, nonmonotonic inferences do not satisfy this condition, but Gabbay proposes that such an inference relation ⊢ should at least satisfy *cautious monotony* which is the condition that if A ⊢ B and A ⊢ C, then A&C ⊢ B.

Gärdenfors and Makinson [9] suggest a method of translating postulates for belief revision into postulates for nonmonotonic logic, and vice versa. The key idea for the translation from belief revision to nonmonotonic logic is that a statement of the form B ∈ K*A is seen as a nonmonotonic inference from A to B given the set K of sentences as background (default) information. So the statement B ∈ K*A for belief revision is translated into the statement A ⊢ B for nonmonotonic logic. Conversely, a statement of the form A ⊢ B for nonmonotonic logic is translated in to a statement of the form B ∈ K*A for belief revision, where K is introduced as a fixed belief set.

Using this recipe it is possible to translate all the postulates (K*1) – (K*8) for belief revision into conditions for nonmonotonic logic. It turns out that every postulate translates into a condition on ⊢ that is valid in some kinds of nonmonotonic inferences in the literature. In particular, most of them hold in the 'classical stoppered ranked preferential model structures' as presented in [8].

Conversely, every postulate on ⊢ that is known in the literature translates into a condition on belief revision that is a consequence of (K*1) – (K*8). For example, cautious monotony translates into "if B ∈ K*A and C ∈ K*A, then B ∈ K*A&C" which follows from (K*1) – (K*8). To sum up, using the proposed translation it can been shown that there is a very tight connection between postulates for belief revision and nonmonotonic logic.

Turning finally to the connection between *models* for belief revision and models for nonmonotonic inferences, there are several constructions in the two areas that seem to be closely related. For example, Poole's [12] 'maxiconsistent' construction for nonmonotonic inference is very similar to the 'full meet revision function' studied by Alchourrón, Gärdenfors, and Makinson [1] and Gärdenfors [4]. Furthermore, Shoham's [13] preferential entailment approach, using an ordering among models to determine those that are minimal, is reminiscent of the orderings used in belief revision to determine transitively relational partial meet revision functions.

More detailed investigations into the connections between the two kinds of models have begun recently (see [6]). The guiding idea is that when we try to find out whether C follows nonmontonically from B, the background information that we use for the inference does not only contain what we firmly believe, but also information about what we *expect* in the given situation. If we denote the set of sentences which represent expectations by Δ, then the key idea can be put informally as follows:

B nonmonotonically entails C iff C follows logically from B together with as many as possible of the sentences in Δ that are compatible with B.

The gist of the analysis can be illustrated by an example. Let the language L contain the following predicates:

Sx: x is a Swedish citizen
Ix: x has Italian parents
Px: x is a protestant

Assume that the set Δ of expectations contains Sx → Px and Sx & Ix → ¬Px. If we assume that Δ is closed under logical consequences, it also contains Sx → ¬Ix and, of course, the logical truth Sx & Ix → Sx. If we now learn that a is a Swedish citizen, that is Sa, this piece of information is consistent with Δ and thus we can conclude that Sa |~ Pa according to the recipe above.

On the other hand, if we learn both that a is a Swedish citizen and has Italian parents, that is, Sa & Ia, then this information is *inconsistent* with Δ and so we cannot use all expectations in Δ when determining which inferences can be drawn from Sa & Ia. The most natural expedient is to give up the expectation Sx → Px and the consequence Sx → ¬Ix. The contracted set of expectations, which contains Sx & Ix → ¬Px and its logical consequences contains as many sentences as possible in Δ that are compatible with Sa & Ia. So, by the general rule above, we have Sa & Ia |~ ¬Pa. This shows that |~ is indeed a nonmonotonic inference operation.

In order to make this idea more precise, one may use an ordering of epistemic entrenchment of the sentences as presented above, now interpreted as an ordering of *expectations*. It turns out that in this context only the properties (EE1) – (EE3) are required. Let us call an ordering ≤ satisfying (EE1) – (EE3) an *expectation ordering*.

An expectation ordering ≤ can be used to determine whether B nonmonotonically implies C by saying that C follows from B together with all the propositions that are 'sufficiently well' expected in the light of B. How well is 'sufficiently well'? A natural idea is to require that the added sentences be strictly more expected than ¬B in the ordering. This motivates the following:

Definition: |~ is a *comparative expectation* inference relation iff there is an expectation ordering ≤ such that the following condition holds:

(C|~) B |~ C iff C ∈ Cn({B} ∪ {D: ¬B < D})

This definition can now be used together with the translations of the revision postulates into the nonmonotonic formalism to give a representation theorem (proved in [6]).

Theorem 11: Let ≤ be an expectation ordering over L. Then the inference relation |~$_≤$ that it determines by (C|~) satisfies all the translations of postulates (K*1) – (K*8).

Theorem 12: Let |~ be any inference relation on L that satisfies all the translations of postulates (K*1) – (K*8). Then |~ is a comparative expectation inference relation, i.e., there exists an expectation ordering ≤ over L such that |~ = |~$_≤$.

References

1. Alchourrón, C.E., P. Gärdenfors, and D. Makinson (1985): "On the logic of theory change: Partial meet contraction and revision functions", *The Journal of Symbolic Logic 50*, 510–530.

2. Fuhrmann, A. (1991): "Theory contraction through base contraction" *Journal of Philosophical Logic 20*, 175–203.

3. Gabbay, D. (1985): "Theoretical foundations for nonmonotonic reasoning in expert systems", in *Logic and Models of Concurrent Systems*, K. Apt ed. (Berlin: Springer-Verlag).

4. Gärdenfors P. (1988): *Knowledge in Flux: Modeling the Dynamics of Epistemic States* (Cambridge, MA: The MIT Press, Bradford Books).

5. Gärdenfors, P. and D. Makinson. (1988): "Revisions of knowledge systems using epistemic entrenchment", in *Proceedings of the Second Conference on Theoretical Aspects of Reasoning about Knowledge*, M. Vardi ed. (Los Altos, CA: Morgan Kaufmann).

6. Gärdenfors, P. and D. Makinson. (1992): "Nonmonotonic inferences based on expectations", manuscript, Cognitive Science, Lund University.

7. Kraus, S., D. Lehmann, and M. Magidor, (1990): "Nonmonotonic reasoning, preferential models and cumulative logics", *Artificial Intelligence 44*, 167–207.

8. Makinson, D. (1991): "General patterns in nonmonotonic reasoning", to appear as chapter 2 of *Handbook of Logic in Artificial Intelligence and Logic Programming, Volume II: Non-Monotonic and Uncertain Reasoning*. (Oxford: Oxford University Press).

9. Makinson, D. and P. Gärdenfors (1991): "Relations between the logic of theory change and nonmonotonic logic", *The Logic of Theory Change*, ed. by A. Fuhrmann and M. Morreau (Berlin: Springer-Verlag), 185–205.

10. Pearl, J. (1988): *Probabilistic Reasoning in Intelligent Systems* (Los Altos, CA: Morgan Kaufmann).

11. Poole, D. (1988): "A logical framework for default reasoning", *Artificial Intelligence 36*, 27–47.

12. Reiter R. (1980): "A logic for default reasoning", *Artificial Intelligence 13*, 81–132.

13. Shoham, Y. (1988): *Reasoning about Change*. (Cambridge: Cambridge University Press).

METAPROGRAMMING THROUGH INTENSIONAL DEDUCTION: SOME EXAMPLES

Luis Fariñas del Cerro, Andreas Herzig

IRIT, Université Paul Sabatier
118 route de Narbonne, F-31062 Toulouse Cedex, France
email: {farinas,herzig}@irit.fr, fax: (33) 61 55 62 58

Abstract. Intensional logics have become a comprehensive framework for many domains of programming. In this paper we argue that metaprogramming is a natural application area of intensional systems, in the sense that intensional languages allow to capture many basic metalogical concepts.

1   INTRODUCTION

Intensional concepts are at the base of languages that allow to qualify expressions and to define contexts in which these expressions must be evaluated. Syntactically, such contexts are materialized by intensional operators which qualify formulas. Traditionally they are designated by the box and diamond operators [R] and <R>, where R is an argument representing the context.

In this paper we wish to illustrate the power and the scope of such a device by three examples.

- the modal logic of provability possesses an intensional operator [] such that []A is intended to mean "A is provable in Peano Arithmetic".
- In hypothetical logic, the intensional operator depends on formulas. The formula [A]B means "assuming that A holds B must hold".
- In dynamic logic of programs, each intensional operator is parametrized by a program. Then [p]A means "A must hold after every execution of program p".

In the last case the programs may be logic programs, too. This shows already that there is a strong connection between intensional logics and metaprogramming: The intensional language can be seen as a metalanguage, and deducing in the intensional logic then becomes executing a metaprogram.

Kowalski and Bowen (Kowalski Bowen 1982) inaugurated logic metaprogramming using a predicate of provability Demo with two arguments, T and G, where T represents a theory T and G a goal G. Demo(T,G) is true if G is derivable from T. This predicate allows an elegant implementation of Prolog in itself.

Since then metalogic programming has become a large field of research in which many classical domains are at home. Examples of it are compilation, partial evaluation, debugging, tracing, interfaces, and expert system shells. We think that there are other domains where metalogic programming techniques should be fruitful.

Program transformations have not interested too much the logic programming community up to now, but should play an important role in the future, given their applicability to other domains of metaprograming such as partial evaluation. Knowledge base updating is a hot domain in databases and in artificial intelligence for which many important results have been obtained in the last few years. Other examples include new features of the logic programming paradigm such as introducing explanation facilities or non monotonic reasoning.

Following Montague (Montague 1963), we may consider the concept "Demo" as a *metapredicate* which verifies the following natural properties:

. $Demo(\rightarrow(A,B)) \rightarrow (Demo(A) \rightarrow Demo(B))$

. $Demo(A) \rightarrow A$

. If A is a theorem then $Demo(A)$ is a theorem

where $\rightarrow$ represents implication in the object language. If we consider the above as a first-order theory then it can be shown that the system thus defined is inconsistent. This gives some hints about the logical difficulties of metaprogramming. On the other hand, if the concept Demo is an intensional operator then the above properties define a well-known intensional logic, for which a natural semantics in terms of Kripke models can be defined.

More generally speaking when computer systems and in especially logic programming systems try to get closer to human reasoning, they have to deal with various concepts such as time, belief, knowledge, contexts, etc. Prolog is just what is needed to handle the Horn clause fragment of first order logic, but what about these intensional logics? Just suppose we want to represent in Prolog time, knowledge, hypotheses, or two of them at the same time; or to organize our program in modules, to have equational theories, to treat fuzzy predicates or clauses. In all these cases we need particular ways of deducing, which in logic programming means computing a new goal from an existing one.

In this note we present some case studies in metaprogramming, using intensional logics, concerning the notion of provability and hypothetical reasoning. The domains where these notions apply are:

- negation as failure in Prolog
- the dynamics of logic programs (like the definition of the update operations)
- modularity in logic programming (which can be viewed as a particular form of dynamics)

Finally we present an abstract machine enabling us to deduce, in logic programming style, with intensional concepts. This machine can be considered as a general engine for implementing metainterpreters.

2   INTENSIONAL LOGICS: SYNTAX AND SEMANTICS

In the present section we describe a scheme of intensional languages with a family of intensional operators determined by *intensional expressions*. Our intensional language extends that of first-order logic by operators [.] and <.> which quantify formulas and may take expressions as arguments. Such expressions may be constants, terms, predicate names, or formulas, or sets of the latter. We shall abreviate the set of expressions by IOP. Any particular intensional language that we introduce in this paper is obtained from this scheme by specifying sets of variables, constants, and predicates, and the type of expressions intensional operators may take as arguments.

Given a formula F and an intensional expression R, then [R]F and <R>F are formulas. Every particular intensional language will be an instance of that scheme. For example, in epistemic logics, the arguments of the context operators are constants representing agents. [a]P means that agent a knows that P, and <a>P means that it is compatible with a's knowledge that P.

By a *model* of an intensional language we mean a quadruple of the form:

$$M = <ST,D,\{r_R: R \in IOP\},m>$$

where ST is a nonempty set whose elements are calles states, D is a set called domain of discourse, for each constant $R \in IOP$, r_R is binary relation on set ST, called accessibility relation, m is a meaning function such that

$$m(p) \in ST \times D^n \text{ for each n-ary predicate name p}$$

In model M a state s *satisfies* formula F (M s $\models$ F) as usual, in particular:

M s $\models$ [R]F   iff   for all t∈ ST if (s,t)∈ r_R then M t $\models$ F

M s $\models$ <R>F   iff   there exists t∈ ST such that (s,t)∈ r_R and M t $\models$ F

A formula F is *true in a model* M iff for all s∈ ST we have M s $\models$ F. F is *satisfiable* iff there are M and s such that M s $\models$F. F is *valid* or is a tautology ($\models$ F) iff it is true in all models. Each intensional logic is defined semantically by some conditions in the accessibility relation.

We give a deductive structure to intensional logics by specifying *axioms* and *inference rules*. All the logics that we shall present in this paper are normal, i.e. they contain

- Classical logic
- the normality axiom $[R](F \rightarrow G) \rightarrow ([R]F \rightarrow [R]G)$
- the necessitation rule   from F infer [R]F

A *proof* in a intensional logic is a finite sequence of formulas each of which is either an axiom, or else is obtainable from earlier formulas by a rule of inference. A formula F is a *theorem* or F is provable ($\vdash$ F) iff it is a last line of a proof.

In the next sections we present three examples of formalization, viz. of modules in logic programing, of the dynamics of knowledge bases, and of negation by failure in Prolog.

3 MODULES

Modules are a useful concept to structure programs, to make them more readable, to make debugging easier. Several similar but different propositions have been made, for example (Miller 1986, Giordano et al. 1988, 1992, Monteiro Porto 1987, Lamma et al. 1990).

In a first approach, one may partition the logic program into modules, without any relation between them. In order to give more power to these modules, a first thing to do is to add the concept of *static import*: E.g. module M1 imports everything defined in M2. A more interesting concept is that of *dynamic import*: When a goal is asked, it must be said which modules can be used to solve it. Now if there is more than one module to load, one must choose: One may decide to put all the contents of the loaded modules together. Inspired by the deduction theorem for classical logic, this has been proposed in (Miller 1986), and also in terms of hypothetical reasoning in (Gabbay Reyle 1986). The second possibility is to load the modules in sequence. Again there are several choices concerning the effect of loading first module M1 and then module M2: M2 may use the definitions in M1 (but not the converse), or M2 may overwrite the definitions in M1 (where we must make precise what "overwrite" means). In (Giordano Martelli 1988), it has been shown that modal logic S4 is the right logic to handle such modules. In (Monteiro Porto 1987, Lamma et al. 1990) there have proposed modules or units for logic programming. Goals are resolved in the context of units. The central concept is that of context extension, which permits to solve goals in modules in a dynamic way. This is a quite powerful approach. In this spirit we shall give in the sequel a module logic possessing an intensional semantics in terms of Kripke models. The logic presented here has been introduced in (Bieber et al. 1988), where more details can be found.

The basic hypothesis is that to a module we may attach a set of predicate names M whose definitions are contained in it. Hence from a logical point of view a module is a theory which describes the extension of the elements of M. One of the principles of modularity is that things defined in some module are a priori invisible for every other module. So in our case the extension of a predicate may vary according to its definition (or "non-definition") in a module. Hence we consider a module to be the context of a formula F, and we may say that F is true in the context of that module. We express contexts by intensional operators qualifying formulas. Following the ideas in (Monteiro Porto 1987) and (Lamma Mello 1990), a *context* is a sequence of modules. It may be *extended* by a further module M in the way that we keep those definitions which do not appear in M, and adopt those of M otherwise.

Language. Intensional operators are associated with modules. More precisely, the arguments of intensional operators are finite sets of predicate names. This is a rather general way of defining modular structures, where no restriction is made neither on the form of the formulas nor on the relation between the latter and the set of defined predicate names. Thus, [M]F means that F is true after loading module M (i.e. after redefinition of the predicate names in M). We need also a particular intensional operator [] permitting to say that a formula F is true in every context. Examples of formulas are [M1](p←q), [][M1](p ← [M2]q).

Semantics. In the models, the accessibility relation is as follows.
- R = {$R_{[M]}$: [M] ∈ IOP} is a set of binary relations on ST all of which are serial (i.e. for every state s there is at least one state s' such that sRs')
- $R_{[]}$ is a reflexive and transitive relation on ST containing every $R_{[M]}$ in R, i.e. ∪{$R_{[M]}$, $R_{[M]}$ ∈ R} ⊆ $R_{[]}$,
- m(p,s) = m(p,t) if s $R_{[M]}$ t and p∉ M

What we are interested in is whether P →[M]G is valid for a given program P and goal ←[M]G. For example, let

$$P = \{ \quad [][q]q←,$$
$$[][p](p(a)←q),$$
$$[][p,r]p(b)← \quad \}.$$

Then P → [p,r][p][q]p(b) is not valid. This can be seen by the following countermodel

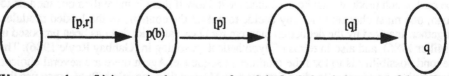

We can see that p(b) is true in the context of module [p,r], but it becomes false after extending this context by [p] and [q].

Axiomatics. We give the following axiom and inference schemes.

Barcan[M]	$\forall x[M]A \rightarrow [M]\forall xA$
Barcan[]	$\forall x[]A \rightarrow []\forall xA$
persistence[M]	$p(t1,...,tn) \rightarrow [M]p(t1,...,tn)$ if $p \notin M$
D[M]	$[M]A \rightarrow \neg[M]\neg A$
T[]	$[]A \rightarrow A$
[]-[M]	$[]A \rightarrow [][M]A$

This axiomatics is sound and complete with respect to the semantics.

4 DYNAMICS OF KNOWLEDGE BASES

In computer science, in the field of *database updates* (Winslett 1990) two paradigms have been given. The first one might be called global. In this case updates are interpreted as a means to obtain a new state of the database. Modifications of the data base produce a sequence of states ordered temporally. The second approach can be called local: We do not modify the database itself, but rather the reasoning which can be done on the base of information collected in it, under the hypothesis that the update was done. In other words we reason about updates. It is local with respect to the database, because it is concerned only with a fragment of the database. Moreover, updates are organized without explicit reference to time. These two dichotomic approaches can be found in many studies of database updates.

In the philosophical *theory of change* (Alchurron et al. 1985), three kinds of changes have been considered:

- expansion: a piece of information is added without any modification of the database.
- revision: a piece of information is introduced and requested modifications of database are made, for example any piece of information which is inconsistent with the new information is eliminated.
- contraction: a piece of information is withdrawn (retracted) from the database.

In complete databases any piece of information which does not occur in the database is treated as false information. Hence the fact of retracting a piece of information from a complete database is interpreted as addition of its negation. Consequently, contraction and revision coincide, and the only operation needed to make updates is revision.

The main issue in formalization of updates is: Which information from the database has to be brought to a new state after an update is made? The classical answer to this question is: A new state after an update is the state obtained from the given state by the least change which consistent with the update. In several applications some additional assumption concerning the new state is admitted in the form of some preservation criterion saying which formulas can be brought from one state to another. A system which involves minimal changes and a preservation criterion can be inconsistent. Thus a balance must be found between minimality and preservation.

In the rest of this section we consider an intensional logic, ASSUME logic (Fariñas Herzig 1988, 1992), that is able to support some kind of updates.

Language. Here we consider that hypotheses are contexts of formulas. The hypotheses being itself formulas it is possible that hypotheses are nested. The intended meaning of a formula [p]q is "assuming p, q holds". More formally, we suppose that our intensional expressions are formulas of a propositional language build on a set of propositional variables PVAR. In the formula [A]B, A is called the *change formula*. We call a formula A a *clausal change formula* if the change formulas appearing in A are all clauses, and we call A a *literal change formula* if the change formulas appearing in A are restricted to literals. We shall use a function *pvar* which associates to every formula the set of propositional variables appearing in it. For example, pvar([p](q∧r)) = {p,q,r}.

Semantics. Usually, semantics of conditional logics is given in terms of possible world models. In our case, we might as well do this, but it is simpler to only refer to a single classical interpretation (which we suppose to be a subset of the set of propositional variables PVAR). In the sequel an interpretation satisfying A will be called an *A-interpretation*.

Given two interpretations I and J, the *distance* between I and J is defined as the symmetric difference $I \div J = I \setminus J \cup J \setminus I$. Hence the distance between two interpretations is the set of propositional variables on which they disagree. Our notion of distance is that of (Winslett 1988).

We are going to say first what the change operation on complete knowledge bases is like: The set I*A is the set of A-interpretations whose distance to I is minimal. Then we are able to define the satisfaction relation. Let I a classical interpretation and A a formula. I*A is defined by

I*A = {J: J is an A-interpretation, and there is no A-interpretation J'

such that $I \div J \supset I \div J'$}

Now we give the satisfaction relation for the conditional operator:

I ⊨[A]C if for every J ∈ I*A, J ⊨ C

An example of a model is the following:

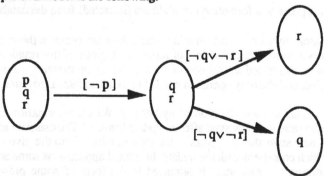

We can see that after updating by ¬p, q and r remain true, and that updating then by ¬q∨¬r leads to two states: One where q becomes false and r remains true, and vice versa.

Let I an interpretation and A a formula. Then we have the following properties:
(1) If A is a literal then I*A interprets all propositional variables as I except pvar(A).
(2) If A is a clause and I ⊨ A then I*A = {I}.
(3) If A is a clause of the form B∨C and I⊭B∨C then I*B∨C = I*B ∪ I*C.

Axiomatics. In the style of (Chellas 1975) we give the following inference rules and axiom schemes for the conditional logic ASSUME, based on some axiomatization of classical propositional logic.

RCEA
$$\frac{A \leftrightarrow B}{[A]C \leftrightarrow [B]C}$$

D	[A]A
CA	([A]C ∧ [B]C) → [A∨B]C
MP (weak centering)	[A]C → (A→C)
CS (strong centering)	(A∧C) → [A]C
atomic consistency	¬[A]¬A if A is a literal
non-interference	C → [A]C if A and C do not interfere
disjunctive non-interference	¬A → ([A∨B]C→[B]C) if A and B do not interfere

where we say that two formulas A and B *interfere* (Nute 1980) if they have some propositional variables in common, i.e. pvar(A) ∩ pvar(B) ≠ Ø.

All the axioms are standard, except the last two ones. Thus, ASSUME logic is build on the system CK (Nute 1980, Chellas 1985) which contains just the inference rules (RCEA) and (RCK). The last two axioms are supposed to approximate the requirements of the non-interference postulate. (atomic consistency) expresses that propositional variables correspond to consistent statements. (Note that false is not a propositional formula.) The reading of (non-interference) is clear. (disjunctive non-interference) can be read as follows: If A does not hold in the current knowledge base, and A and B do not interfere, to change by B has at most the consequences of changing by A∨B.

Thus, the axiomatics is given for change formulas of any kind. It is sound, but it is complete only for update formulas which are clauses, i.e. disjunctions of literals. This is due to the fact that (disjunctive non-interference) is apparently too weak, and that there is no axiom for conjunctive changes.

A main property of ASSUME logic is that each assume formula is equivalent to a classical formula (see Fariñas Herzig 1992). An example of a normalization theorem is
[A∨B]C ↔ (((A∨B)∧C) ∨ ([A]C ∧ [B]C))

It is easy to verify that the axioms of ASSUME are valid, and that the inference rules preserve validity. For clausal change formulas, completeness can be proved using the fact that every ASSUME formula is equivalent to a classical formula.

The normalization furnishes a tool for proving completeness and also for performing automated deduction in ASSUME logic. Here a spectrum of possibilities appears, depending how the normalization theorem is used. It can be used globally: In a first step we transform the formula KB → [A]B into an equivalent classical one, and then a classical proof method can be used. As normalization is exponential this is not very efficient. In the other extremity, we use a lazy normalization, in other words we use the normalizsation theorem only when it is necessary to go on with the deduction.

Given a knowledge base KB and a clausal change formula A, we are now able to turn the membership problem "Is a formula B in the revision of KB by A?" into the deduction problem: "Is KB → [A]B is a theorem of ASSUME logic?" Thus we have defined a change (update) operation from a conditional logic.

Example. In (Winslett 1988) there is given the following example. Let two knowledge bases KB1 = {p, q} and KB2 = {p, p→q}. KB1 and KB2 are equivalent, and updating by ¬q should have the same effect on KB1 as on KB2. Indeed, we have ⊢KB1→[¬q](p∧¬q) as well as ⊢KB2 → [¬q](p∧¬q).

A main hypothesis in this section concerns the fact that different propositional variables do not interfere. The following example shows that this is not realistic: Let p and q propositional variables with the respective readings "The traffic light is red" and "The traffic light is green". As p and q are different propositional variables, it follows from our non-interference postulate that p → [q](p∧q) is a theorem of ASSUME logic. But it is absurd that after the traffic light has become green it continues to be red! What we need is some extra information able to tell us that although the propositional variables are different, p and q do not interfere.

Two ways out are possible: The first approach has been followed e.g. in (Ginsberg Smith 1988) or (Seguin 1992), where the notion of a constraint has been employed. We are currently working in another direction, where the notion of interference is extended: We suppose given a relation of similarity between formulas, which is a reflexive and symmetric relation between propositional variables. We say that two formulas interfere if they are similar. Thus we exploit metalinguistical information about formulas having something to do with each other. Formally, we suppose given a relation ~ between formulas which is reflexive and symmetric (but not transitive). Furthermore, we suppose that the whole relation on formulas can be constructed from its part concerning only propositional variables for example in the following way:

A ~ B iff there are p ∈ pvar(A) and q ∈ pvar(B) such that p ~ q

Now the axiom (non-interference) becomes

C → [A]C if not (A ~ C)

With this new axiom, the formula p → [q](p∧q) is no longer a theorem as soon as we have p ~ q.

5   PROVABILITY IN LOGIC PROGRAMMING

In (Gabbay 1991) and (Balbiani 1991) it has been established that provability and unprovability in logic programs with respect to SLD-resolution and SLDNF-resolution can be characterized in terms of intensional validity.

We suppose an intensional language with two intensional operators to characterize provability (using intensional operator [+]) and unprovability (using intensional operator [-]) in logic programming. The relation R^+ is transitive and the relation R- is transitive and reverse well-founded. It has been shown that this corresponds to two well-known modal logics (Chellas 1975, Boolos 1979): [+] defines the logic K4 and [-] the logic of provability Pr.

The intensional completion formula scheme associated to the program B→A is the expression:

$$[+]([+]B \to A), [-]([-]B \to A) \text{ and } [-]B.$$

This intensional completion is based on Clark's idea that a predicate is totally defined by the bodies of the clauses defining it in some logic program.

To be more precise, let $\forall x_1 ... \forall x_n (E_1 \lor .. \lor E_L \leftrightarrow p(x_1,...,x_n))$ be Clark's completed definition of p in the program P. In the first order predicate calculus, this formula is logically equivalent to the conjunction of the formulas (i) and (ii):

(i) $\forall x_1 ... \forall x_n (E_1 \lor ... \lor E_L \to p(x_1,...,x_n))$.

(ii) $\forall x_1 ... \forall x_n (\neg E_1 \land ... \land \neg E_L \to \neg p(x_1,...,x_n))$.

Each E_1 is of the form $\exists y_1 ... \exists y_k (x_1=t_1 \land ... \land x_n=t_n \land B_1 \land ... \land B_m)$ and each negation $\neg E_1$ of a E_1 is of the form $\forall y_1 ... \forall y_k (x_1=t_1 \land ... \land x_n=t_n \to \neg B_1 \lor ... \lor \neg B_m)$. The *intensional completion* is obtained by the following steps

1) In a first step, let us prefix with the intensional operator [+] every positive literal and let us replace every occurrence of the negation symbol by the intensional operator [-]. Thus we obtain the formulas :

(i.1) $\forall x_1 ... \forall x_n (E_1^+ \lor ... \lor E_L^+ \to [+]p(x_1,...,x_n))$

(ii.1) $\forall x_1 ... \forall x_n (E_1^- \land ... \land E_L^- \to [-]p(x_1,...,x_n))$

where E_1^+ and E_1^- are the following intensional formulas :

$$\exists y_1 ... \exists y_k (x_1=t_1 \land ... \land x_n=t_n \land [+]B_1 \land ... \land [+]B_m)$$
$$\forall y_1 ... \forall y_k (x_1=t_1 \land ... \land x_n=t_n \to [-]B_1 \lor ... \lor [-]B_m)$$

2) Then we syntactically introduce in this transformed completed definition of p the idea of a relation of cause and effect between $E_1^+, ..., E_L^+$ and [+]p on one hand and $E_1^-, ..., E_L^-$ and [-]p on the other hand. To do so we move the intensional operators prefixing p in (i.1) and (ii.1). We then obtain the formulas :

(i.2) $\forall x_1 ... \forall x_n [+](E_1^+ \lor ... \lor E_L^+ \to p(x_1,...,x_n))$

(ii.2) $\forall x_1 ... \forall x_n [-](E_1^- \land ... \land E_L^- \to p(x_1,...,x_n))$

which constitute the *intensional completed definition* of p. We note P^+ the set of the "positive" completed definitions (i") and P^- the set of the "negative" completed definitions (ii.2) of the predicate symbols of the language.

For example, let P={ $A(s(x)) \to A(x)$, $C(x) \to B(x)$, $C(0)$ }. The intensional completion formula of P consists in the two formulas P^+ and P^- defined as follows:

P^+ : $\forall x([+]([+]A(s(x)) \to A(x)) \land [+]([+]C(x) \to B(x)) \land [+](x=0 \to C(x)))$

P^- : $\forall x([-]([-]A(s(x)) \to A(x)) \land [-]([-]C(x) \to B(x)) \land [-](\neg(x=0) \to C(x)))$

We have $P^+ \models_{K4} [+]C(0)$ and $P^+ \models_{K4} [+]B(0)$. Moreover, $P^- \models_{K4} [-]C(s^{n+1}(0))$, and $P^- \models_{K4} [-]B(s^{n+1}(0))$. Finally, for every integer n, corresponding to the fact that there is no SLD-refutation of $P \cup \{A(s^n(0)) \to \}$, we have that $P^- \models_{HPr} [-]A(s^n(0))$, where $P^- \models_{HS} [-]A$ means that A is valid with respect to the S-validity for the Herbrand models and the intensional system S and program P^-.

The formal relation between the classical completeness of SLD-resolution and negation as failure was proved in (Balbiani 1991):

Theorem. For every definite program P and for every ground atom A:

(1) There is an SLD-refutation of P∪{A→} iff $P^+ \models_{K4} [+]A$ iff $P^+ \models_{HK4} [+]A$.

(2) A belongs to the SLD-finite failure set of P iff $P^- \models_{K4} [-]A$.

This theorem allows to gives an intensional semantics of negation by failure in Prolog.

6 TARSKI: AN INFERENCE MACHINE FOR METAPROGRAMMING

The main problem becomes now to make inferences with intensional logics. Practically speaking, inference engines for intensional logics must reckon for the particular behaviour of the intensional operators.

In order to solve this problem, based on work in (Fariñas 1986) we have defined at IRIT a general inference machine, called TARSKI (Toulouse Abstract Reasoning System for Knowledge Inference) for building a large class of metainterpreters (Balbiani et al. 1991, Alliot et al. 1992). In particular, this machine is suitable for implementing extensions of Prolog with intensional logics. The particular properties of the intensional operators will be handled by built-in features of the inference engine.

It is clear that in every case it is possible to write specific meta-interpreters in Prolog that implement these intensional logics. But there are disadvantages of a meta-interpreter: lower speed and compilation notoriously inefficient. If we want to go a step further, and to write proper extensions of Prolog, then the problem is that costs for that are relatively high (because for each case we will lead to write a new extension), and we are bound to specific domains: we can only do temporal reasoning, but not reasoning about knowledge (and what if we want to add modules to all that?).

The aim is to define a framework wherein a superuser can create easily his extension of Prolog. This framework should be as general as possible and thus provide a general methodology to implement intensional logics. It is built on four basic assumptions:

1. to keep as a base the fundamental logic programming mechanisms that are backward chaining, depth first strategy, backtracking, and unification
2. to parametrize the inference step: it is the superuser who specifies how to compute the new goal from a given one, and he specifies it in a logic form.
3. to be able to rewrite goals
4. to select clauses "by hand"

Points (2) and (3) postulate a more flexible way of computing goals than that of Prolog, where first a clause is selected from the program, then the unification algorithm is applied to the clause and the head of the goal, and finally a new goal is produced. Point (4) introduces a further flexibility: the superuser may select clauses that do not unify exactly with the current goal, but just resemble it in some sense. Even more, if the current goal contains enough information to produce the next goal, or if we just want to simplify a goal or to reorder literals we don't need to select a fact clause at all. The assumptions (1) and (2) were at base of the development of a meta-level inference system called MOLOG (Fariñas 1986).

In the rest of the section we give the definition of intensional Horn clauses and of the inference rules by which the superuser can specify his metainterpreter. As an example we show how modules can be specified easily in terms of inference rules.

Intensional Horn clauses. The *conditio sine qua non* for logic programming languages is that they possess an implicational symbol to which a procedural sense can be given. To define a logic programming language it's less important if this is material implication or not, but it is rather the dynamic aspect of implication that makes the execution of a logic program possible. That is why the language is built around some arrow-like symbol. Hence the base of the language is that of Prolog. That language can (but need not) be enriched with context operators if one wants to mechanize intensional logics. Intuitively, an *intensional Horn Clause* is a formula built with the above connectors, such that dropping the context we may get a classical Horn clause. For example, [m1](p←q) and ←[m1][m2]p are intensional Horn clauses.

The general mechanism is as follows: Just as in Prolog, to decide whether a given goal follows from the database essentially means to compute step by step new subgoals from given ones. In our case, the computation of the new subgoal is specified by the superuser. There are five steps:

- Clause selection: We select a clause to solve the first sub-goal of the question.
- Rule selection: We select a rule to be applied to the current clause and the current question.
- Rule execution: The execution of the rule "modifies" the current clause and the current question and builds a resolvent.
- Rewriting of the resolvent: When we reach a termination rule, we rewrite the resolvent into a new question.
- End of resolution: A resolution is completed when we reach a final form, which is the goal clause ←*true*.

This system is doubly non determinist, because we have both a clause selection (as in standard Prolog) and a rule selection. In the sequel we are going to explain how this mechanism can be implemented.

Inference Rules. An inference rule is of the form A, ←B ⊢ ←C where A is a definite clause and B and C are goal clauses. It can be read: If the current goal clause unifies with B and the selected database clause unifies with A then a new goal can be inferred that is unified with C. In the style of Gentzen's sequent calculus, inference rules can be defined recursively as follows:

$$\frac{A, \leftarrow B \vdash \leftarrow C}{A', \leftarrow B' \vdash \leftarrow C'}$$

where A and A' are definite clauses and ←B, ←B', ←C and ←C' are goal clauses. As usual in metaprogramming, objects of the object language are represented by variables of the metalanguage. Partial termination rules are written A, ←B ⊢ ←C and end the recursivity in resolution. For example, the Prolog partial termination rule is

$$p\leftarrow, \leftarrow p \vdash \leftarrow true.$$

(Note that here we make use of unification.) The Prolog rule for goal conjunctions is

$$A, \leftarrow B \wedge C \vdash \leftarrow D \wedge C$$
$$\overline{A, \leftarrow B \vdash \leftarrow D}$$

and the Prolog rule for implications in database clauses:

$$A \leftarrow B, \leftarrow C \vdash \leftarrow B \wedge D$$
$$\overline{A, \leftarrow C \vdash \leftarrow D}$$

These three rules are exactly what is needed to implement pure Prolog.

As soon as we have reached a partial termination rule, we rewrite the resolvent to create the new goal. Rewriting is useful not only in order to simplify goals, but also in order to eliminate the predicate "true" from the new goal clause. Rewrite rules are of the form $\leftarrow G1 \approx> \leftarrow G2$ and allow to transform a resolvent that is matched by $\leftarrow G1$ into the new goal $\leftarrow G2$. For example, the Prolog rewrite rule is $\leftarrow true \wedge G \approx> \leftarrow G$. Another example is the rule $\leftarrow [c][c]G \approx> \leftarrow [c]G$ which is a useful simplification in some logics.

To summarize, the execution of an inference rule modifies the current fact and the current question and constructs a resolvent. The resolvent has the same structure as the question or any other fact. Resolution is achieved in a branch when we reach a partial termination rule. How rules are selected is defined by the user. She can choose among a given set the way clauses are selected

Example. In the sequel we are going to show how to specify modules with dynamic import. Here, any module name such as M, M1, etc. is considered to be a context. We suppose that there are only intensional operators of type necessary [.] appearing in the clauses. The mecanism is as follows: A goal $\leftarrow [m1][m2]G$ succeeds if $\leftarrow G$ can be proved using clauses from the modules M1 and M2. The inference rules are that for Prolog, plus two supplementary rules to handle module operators:

$$[M]A, \leftarrow [M]B \vdash \leftarrow [M]C$$
$$\overline{A, \leftarrow B \vdash \leftarrow C}$$

$$A, \leftarrow [M]B \vdash \leftarrow [M]C$$
$$\overline{A, \leftarrow B \vdash \leftarrow C}$$

The first rule represents the case where a module M is used to compute a new goal, and the second where another module name occurring in B is used. The rewrite rules are:

$$\leftarrow true \wedge G \qquad \approx> \qquad \leftarrow G$$
$$\leftarrow [M]true \qquad \approx> \qquad \leftarrow true$$

Example. Consider the program
$$[M1](A \leftarrow [M2]B)$$
$$[M1][M2]B$$

and the goal ←[M1]A. Applying our inference rules recursively we first compute the resolvent ←[M1][M2](true∧B), which is rewritten to the subgoal ←[M1][M2]B. In a second step we compute ←[M1][M2]true which we rewrite to ←true.

Others types of modules such as modules with static import or with context extension can be specified by just adding a new inference rule. Further examples can be found in (Balbiani et al. 1991). It has been shown shown there how temporal logics, hypothetical reasoning and logics of knowledge and belief can be specified elegantly in our framework.

What the TARSKI abstract machine does is to bridge the gap between the description of inference rules in logical form as shown above, and the real implementation of the rule in an efficient programming language. Compared to the WAM, the TARSKI abstract machine deals with different objects, and has a quite different goal, but on the whole, principles are identical. TARSKI is also defined in terms of data, stacks, registers and instruction set (see (Alliot et al. 92)). The machine relies on classical structure sharing for unification, and on depth first search and backtracking. The development of an automatic translator from the logical shape of the rules to the abstract machine specifications suggests itself and is a subject of current work.

CONCLUSION

In this paper three examples of metaprogramming have been presented. The intensional character of the examples has been stressed and metaprogramming has been transformed into intensional deduction.

The main hypothesis of the paper is that intensional concepts are a natural way of speaking about logic programs: Given a logical program, the Kripke semantics represents the executions of the logical programs and the intensional language is a query language that allows to express properties on the Kripke structure (or on the executions of the program).

ACKNOWLEDGEMENTS

We would like to thank Ph. Balbiani, Ph Lamarre, A. Pettorossi, and L. Serafini for their comments on an earlier draft of this paper.

REFERENCES

Alchourron et al. 85   C. E. Alchourron, P. Gärdenfors and D. Makinson. On the Logic of Theory Change: Partial Meet Contractions and Revision Functions. The Journal of Symbolic Logic. 50. 2. pp 510-530, 1985.

Alliot et al. 92   Jean-Marc Alliot, Andreas Herzig, Mamede Lima Marques. Implementing Prolog Extensions : A Parallel Inference Machine. Proc. of the Int. Conf. on Fifth Generation Computer Systems (FGCS'92), 1992.

Balbiani 91   Ph. Balbiani. Modal Logic and Negation as Failure, Journal of Logic and Computation I(1991) pp. 331-356.

Balbiani et al. 88   Philippe Balbiani, Mamede Lima Marques, Andreas Herzig. Declarative Semantics for Modal Logic Programs. Proc. of the Int.

24

Workshop on Processing Declarative Knowledge (PDK'91). Springer Verlag, LNAI, 1991.

Balbiani et al. 88   Philippe Balbiani, Andreas Herzig, Mamede Lima Marques. TIM: The Toulouse Inference Machine for Nonclassical Logic Programming. Proc. of the Int. Conf. on Fifth Generation Computer Systems '88.

Bieber et al. 88   P. Bieber, L. Fariñas del Cerro and A. Herzig. MM: A modal logic for modules. Report LSI, Université Paul Sabatier Toulouse, 1988.

Bibel et al. 89   W. Bibel, L. Fariñas del Cerro, B. Fronhöfer and A. Herzig. Plan Generation by Linear Proofs:On Semantics. German Workshop on AI (GWAI 89). Springer Verlag 1989.

Boolos 79   G. Boolos. The Unprovability of Inconsistency. Cambridge University Press, Cambridge, 1979.

Boolos Sambin 92   G. Boolos and G. Sambin. Provability: the emergence of a mathematical modality, Studia Logica, to appear.

Bowen Kowalski 82   K. A. Bowen and R.K. Kowalski. Amalgamating language and metalanguage. IN K.L.Clark and S. A. Tarnlund, editors Logic and Programming , Academic Press, Londin 1982 pp 153-172

Chellas 75   B.F. Chellas. Basic conditional logic. J. of Philos. Logic, 4, 1975, pp 133-53.

Dalal 88   M. Dalal. Investigations into Theory of Knowledge Bases Revision: Preliminary Report. Proc. of the Seventh Nat. Conf. on AI, Minneapolis pp 475-479. 1988.

Fariñas 86   L. Fariñas del Cerro Molog: A system that extends Prolog with modal logic. New Generation Computer Journal 1986.

Fariñas Herzig 88   L. Fariñas del Cerro and A. Herzig. An automated modal logic for elementary changes. Non-Standard Logics for Automated Reasoning (ed. P Smets, A, Mandani, D. Dibois and H. Prade). Academic Press, 1988, pp 63-79.

Fariñas Herzig 88   L. Fariñas del Cerro and A. Herzig. Constructive minimal changes. Draft, IRIT, april 1992.

Fariñas Penton92   Intensional logics for Programming .L. Fariñas del Cerro and M. Pentonnen editors, Oxford University Press, 1992.

Gabbay 91   D. M. Gabbay. Intensional Provability Foundations for Negation by Failure. In : P. Schröder-Heister (Ed.), Extensions of Logic Programming. Springer-Verlag, 1991, pp. 179-222.

Gabbay Reyle 86   N-Prolog: An extension of Prolog with hypothetical implications. I. Journal of Logic Programming 4, 1984, 319-355.

Gärdenfors 88   P. Gärdenfors. Knowlege in Flux. MIT Press, 1988.

Ginsberg Smith 88   M. L. Ginsberg, D.E. Smith. Reasoning about action I: A possible world approach. Readings in Nonmonotonic Reasoning, M.L. Ginsberg ed., Morgan Kaufmann, 1987.

Giordano et al. 88 L. Giordano, A. Martelli, G.F. Rossi. Local definitions with static scope rules in logic languages. Proc. of the Int. Conf. on Fifth Generation Computer Systems. 1988.

Giordano et al. 92 L. Giordano, A. Martelli, G.F. Rossi. Extending horh clauses logic with implication goals. To appear in T.C.S.

Grahne 91   G. Grahne. Updates and Counterfactuals. Principles of Knowlege Representation and Reasoning. Morgan and Kaufmann, 1991.

Grove 86   A. Grove. Two modelings for Theory Change, Auckland Philosophy Papers 13, 1986.

Hegner 87   S. Hegner. Specification and Implementation of Programs for Updating Incomplete Information Databases. Proc. of the Sixth Symp. on Principles of Database Systems. San Diego, pp 146-158, 1987.

Hughes Cressw86   G. Hughes & M.J. Cresswell, A Companion to Modal Logic. Methuen & Co. Ltd., London, 1986.

Lamma et al. 90   E. Lamma, PK Mello, A. Natali, The design of an abstract machine for efficient implementation of contexts in logic programming. Proc. Sixth Int. Conf. on Logic Programming (ed. G. Levi, M. Martelli), The MIT Press, 1990, pp. 303-317.

Lewis 73   D. K. Lewis. Counterfactuals. Blackwell, Oxford, 1973.

Katsuno Mende 91   H. Katsuno and A.O. Mendelzon. On the Difference between Updating a Knowledge Base and Revising it. Principles of Knowledge Representation and Reasoning, Morgan Kaufmann 1991.

Kunen 87   K. Kunen. Negation in Logic Programming. Journal of Logic Programming 4 (1987) pp. 289-308.

Miller 86   D. A. Miller. A theory of modules for logic programming. in Proc. IEEE Symp. on Logic Programming 1986; pp 106-114.

Montague 63   Montague, Syntactical treatements of modality with corollaries on reflexion principle and finite axiomatizability. Acta Philosophica Fennica, vol. 16, 1963.

Monteiro Porto 87   L.Monteiro, A.Porto, Contextual Logic Programming. Proc. 6th Int. Conf. on Logic Programming, 1989, pp 284-299.

Nute 80   Donald Nute. Topics in Conditional Logic. D. Reidel Publishing Company, 1980.

Satoh 88   K. Satoh. Nonmonotonic Reasoning by Minimal Belief Revision. Proc. of the Int. Conf. on Fifth Generation Computer Systems. 1988, pp 455-462.

Segerberg 86   Krister Segerberg. On the logic of small changes in theories, I. Auckland Philos. Papers, 1986.

Seguin 92   Christel Seguin. De l'action à l'intention : Une caractérisation formelle des agents. Phd. thesis, Universié Paul Sabatier, Toulouse, march 1992.

Shepherdson 88   J. Shepherdson. Negation in Logic Programming. In : J. Minker (Ed.), Foundations of Deductive Databases and Logic Programming. Morgan Kaufmann, Los Altos, 1988, pp. 19-88.

Winslett 88   M. Winslett. Reasoning about actions. Proc. AAAI 1988, pp 89-93.

Winslett 90   M. Winslett. Updating Logical Databases. Cambridge Tracts in Theoretical Computer Science. Cambridge University Press 1990.

An Autoepistemic Analysis of Metalevel Reasoning in Logic Programming

Kurt Konolige

Artificial Intelligence Center, SRI International
Menlo Park, CA 94025

Abstract. An introspective agent can represent the connection between her own beliefs and the state of the world, using introspection principles to reason about this connection. An ideal introspective agent is one who makes the fewest possible assumptions about her beliefs. There is a strong formal connection between ideal introspection and metalogical systems in Logic Programming. We trace this connection in two cases: Reflective Prolog and negation as failure. The main contribution of this paper is to show how principles of soundness, completeness, and groundedness can be carried over from introspective to metalogic systems.

1 Introduction

Autoepistemic (AE) logic is the study of how an agent can reflect on her own beliefs, and use her knowledge of the connection between beliefs and the world to come to further conclusions. An example (from [20]) is the following reasoning:

> Do I have an older sister?
> Well, if I did have an older sister, I would have the belief that I have
> such a sister, by virtue of having seen her, fought with her, etc.
> I do not have such a belief.
> Therefore, I must not have an older sister.

Note that this reasoning involves introspection by an agent about the state of her own knowledge. From the lack of belief in a certain proposition (*I do not believe I have an older sister*), the agent comes to the belief in the negation of the proposition (*I do not have an older sister*).

Autoepistemic logic characterizes agents that have complete introspective access to their own beliefs. It relies on a fixed point definition that has two significant parts. The first part is a set of assumptions or hypotheses about the contents of the fixed point. The second part is a set of reflection principles that link sentences with statements about their provability. In recent research [13], we characterize a family of ideal AE reasoners in terms of the minimal hypotheses that they can make, and the weakest and strongest reflection principles that they can have, while still maintaining the interpretation of AE logic as self-belief.

These results can help in an analysis of metalevel constructions in logic programming (LP). The point of view we take is that a logic program is meant as a representation of the world, the belief set of some software agent. The metalevel of a logic program then functions as a means of describing the belief set and its relation to the world. Then, we can use results from AE logic to guide us in the selection of reflection principles for the metalevel.

We apply these results in two ways. First, we look at the Reflective Prolog system of Costantini and Lanzarone [4, 5], which embodies a semantic account of metalevel reflection in LP. We can show that there is an exact correspondence to AE logic, and that they introduce an extra reflection principle in order to guarantee the existence of fixed-points. We argue that this reflection principle is unsound in general but may be useful in certain cases.

Second, we examine some of the semantics proposed for negation-as-failure, namely the stable model semantics [2, 7] and the well-founded semantics of Bonatti (this volume). Here we can show that, if negation is treated as nonbelief, both of these semantics are problematic in terms of metalevel reflection. The stable semantics is too strong in allowing unsound reflection principles, while the well-founded semantics is too weak in not admitting valid ones.

2 Issues and Language

The first question we must ask is what purpose a metalevel serves in LP. Without a clear answer to this question, it is impossible to support arguments for or against the multitude of introspection principles. There are three basic reasons for wanting a metalevel.

1. Procedural control. If the proof process of LP is described by a metalevel axiomatization, then it is possible to change the evaluation of LP's by changing the axioms.

2. Language extensions. Facts about relations can be incorporated into the metalevel, thereby extending the expressive power of the language. For example, we might want to characterize a relation as being symmetric or transitive, and use this fact to derive more consequences from particular instances of the relation. Further, we can characterize the class of valid LP's by giving integrity constraints at the metalevel, e.g., DEMO('a') ∨ DEMO('b') says that a LP must either prove a or b in order to be valid. (For arguments as to the need for a metalevel in expressing integrity constraints, see [3]).

3. Connection to the world. The LP designer may have information about how the LP will track the actual state of affairs. For example, in developing a database of airline flight information, she may know that the database is incomplete with respect to the actual flights available. In this case,

incorporating that knowledge into the LP can help in answering queries, for example:

Q: How many flights are there to Rome from Stockholm?
A: I know of three, but there may be more.

Of these, we will be concerned with the nonprocedural aspects of representation in (2) and (3). From our point of view, a LP is a database with information about the world. The metalevel characterizes the internal connections among the LP predicates, as well as their relation to the actual state of affairs.

We have the choice as to whether to keep the metalevel separate from the LP, or to try to incorporate it into an amalgamated LP. The former case is much easier analytically, and from our point of view would be characterized by the monotonic epistemic logics, e.g., [9, 10]. But since metalevel reasoning can lead to new conclusions about the world, as in the example of the older sister, it can be reasonably argued that they should also become the beliefs of the agent, and so the metalevel must itself be part of the LP. The study of such amalgamated systems has been made in the context of nonmonotonic epistemic logics [8, 11, 13, 15, 16, 20]. We will use the system of ideal introspective logic proposed recently by this author [13] as the starting point for the exploration of metalevel principles in LP.

2.1 Language

There are three choices for language at the metalevel:

1. Reified FOL: $Holds(p, x, y)$

2. Gödel terms: $Demo('P(x, y)')$

3. Modal: $\Box P(x, y)$

Reified FOL uses standard first-order language, and contains terms (p) that refer to properties, i.e., the properties are "reified" as terms. One can quantify over property names here, but not over formulas in general. This is not a very expressive language, but has the advantage of simplicity and consistency.

Gödel term languages are first-order, but introduce the complication of an explicit naming relation between formulas of the language and terms in the language, usually with some quotation device. Thus the terms '$P(x, y)$' stands for the formula $P(x, y)$. This language is much more expressive than the reified one, since one can quantify over formulas and parts of formulas, and also nest names, e.g., $Demo('Demo('P(x, y)')')$. It is possible to formulate self-canceling sentences in this language, given a reasonable substitutional capability.

Modal languages use a modal operator, with one propositional argument, to express an intensional mode of holding that proposition: that it is necessary, or provable, or believed, for example. There is no quantification over propositions or formulas, only over their arguments. Modal languages are intermediate between reified and Gödel term languages. They were the language of choice for axiomatization of intensional concepts, especially because of the results of Montague [19] showing how natural axiomatizations of, e.g., belief in Gödel term languages were inconsistent. In fact, modal languages with an adequate substitutional framework suffer from the same inconsistency problems as Gödel term languages, so in either case substitution must be restricted.

In this paper we will use a modal language, in keeping with the work on AE logic. In fact a Gödel term language would also be appropriate, although the notation is a little harder. The important part is not the language itself, but whether the modal operator or *Demo* predicate is interpreted specially in the semantics. When an appropriate semantics is specified, it makes the logic more readily understood. Having a modal language forces one to do this, since the ordinary semantics of FOL does not apply (for examples of interpreted predicates in a Gödel term FOL, see Bonatti (this volume) and [4, 12]). Here we take the interpretation of the modal operator to be self-belief in a belief system, described in the next section.

The language $\mathcal{L}$ is a first-order language $\mathcal{L}_0$, with the addition of modal atoms of the form $L\phi$, with ϕ a sentence (closed formula) of $\mathcal{L}_0$. The treatment of quantification in AE logic is difficult (see [14, 16]), and we will not attempt to deal with it here, nor is it really necessary, since expressions of the form $\forall x.\alpha(x)$ can be considered substitutionally as the set of all ground instances. Similarly, we restrict $\mathcal{L}$ to only nonnested modal atoms, as we will not need any recursion of metalevels.

3 Ideal Introspection

We take the point of view that a LP represents the state of the world, that its main role is the set of beliefs of an agent about the world. Given this viewpoint, the natural question to ask is, What kind of introspective capability can we expect an ideal agent to have? This question is not easily answered, since it depends on what kind of model we take for the agent's representation of her own beliefs. Let us start with a simple situation, in which we clearly distinguish between the agent's beliefs about the world, and the agent's beliefs about her own beliefs. The diagram of Figure 1 is a model of a reflective agent.

The box labeled "BELIEFS" contains the initial beliefs of the agent about the world, together with any consequences she can derive from those. We represent these beliefs as propositions in $\mathcal{L}_0$; the arrow indicates that the intended semantics of the beliefs is given by the real world, e.g., the belief p is a true belief just in case it holds in the real world. Of course an agent's beliefs may be false; the term "Facts" is used with a neutral meaning, as being any proposition that the agent believes, whether it is true or not.

An ideally rational agent with a set of propositional beliefs about the world should be able to infer all of the tautological consequences of those beliefs, e.g., from p and $p \supset q$ he should derive q. We will assume the following rule:

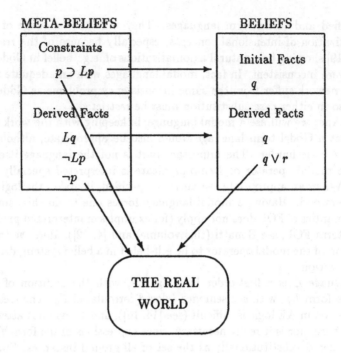

Figure 1: A reflective agent

BELIEF Rule Taut.

If Γ is a finite subset of BELIEFS, and ϕ is a tautological consequence of Γ, then add ϕ to BELIEFS.

In addition to beliefs about the world, an agent may also consider beliefs about her own beliefs, and constraints between her beliefs and the state of affairs in the real world. For example, the agent might have beliefs about her own perceptual abilities: she believes that she will perceive the condition p should it occur, and so believes that $p \supset Lp$. The intended meaning of Lp is that the agent believes p: more specifically, that p is part of the agent's BELIEFS (the arrow leading from Lp to BELIEFS indicates this intended semantics). The nonmodal proposition p refers to the state of the world, so that the constraint is a *meta*-belief about the correlation between the agent's simple beliefs and the world.

From the constraints and from the set BELIEFS the agent may derive further facts (again, in the neutral sense of what the agent takes to be true) about her own beliefs or about the world. These are in META-BELIEFS. Again, since the agent is an ideal propositional reasoner, she should be able to infer all the tautological consequences of her META-BELIEFS.

META-BELIEF Rule Taut.

If Γ is a finite subset of META-BELIEFS, and ϕ is a tautological consequence of Γ, then add ϕ to META-BELIEFS.

The connection between BELIEFS and META-BELIEFS is given by the intended semantics of the L operator, and by the recognition that BELIEFS and META-BELIEFS are beliefs of the *same* agent. Let us examine this relationship more closely. Given the semantics of the L operator, for the META-BELIEFS to be correct and complete we want to have the following relationships hold:

DEFINITION 3.1 (REFLECTIVE SEMANTICS)

$$\phi \in BELIEFS \rightarrow \quad L\phi \in META\text{-}BELIEFS$$
$$\phi \notin BELIEFS \rightarrow \quad \neg L\phi \in META\text{-}BELIEFS$$

The first condition can be easily met by adding another rule to META-BELIEFS, so that whenever ϕ is contained in BELIEFS, we add the corresponding $L\phi$ to META-BELIEFS.

META-BELIEF Rule Up.
If ϕ is in BELIEFS, then add $L\phi$ to META-BELIEFS.

It is not so clear how to implement the second semantic condition. If we assume that BELIEFS contains the beliefs of some *other* agent (say, BELIEFS are the beliefs of the agent Hermione and META-BELIEFS those of Ralph), and that the initial facts are all of Hermione's beliefs, then the correct rule for Ralph would be something like the following.

META-BELIEF Rule Neg.
If ϕ is not a tautological consequence of the initial facts of BELIEFS, then add $\neg L\phi$ to META-BELIEFS.

This rule formed the basis for one investigation into how an agent could reason about what another agent did not believe, given full knowledge of initial beliefs [11]. However, it is not adequate for treating a reflective agent reasoning about her own beliefs, because the contents of META-BELIEFS can have an effect on BELIEFS. To see how this is so, consider the situation described in Figure 1, and assume that Rule Neg has been used to infer both $\neg Lp$ and $\neg L\neg p$, since neither of these are consequences of the initial facts in BELIEFS. By META-BELIEF Rule Taut, $\neg p$ can be added to META-BELIEFS. This is a simple fact about the world, and if we consider META-BELIEFS and BELIEFS to be attached to the same agent, then $\neg p$ should also appear in BELIEFS, which is the repository for such facts. We call this linkage the introspective hypothesis.

Introspective Hypothesis
If BELIEFS and META-BELIEFS belong to the same agent, then the conclusions reached in META-BELIEFS should also be BELIEFS of the agent.

In a formal way, the introspective hypothesis can be implemented by the following rule.

BELIEF Rule Down.

If ϕ is a nonmodal proposition in META-BELIEFS, then add it to BE-LIEFS.

This rule is only appropriate under the introspective hypothesis, when the agent has privileged introspective access to her beliefs. It makes no sense, for example, to say that all of Hermione's beliefs should also be Ralph's; but if Hermione comes to the conclusion on the basis of META-BELIEFS that condition $\neg p$ holds in the world, then that can also be part of her BELIEFS.

The presence of Down complicates the implementation of the semantic rules in 3.1, since now the contents of META-BELIEFS as well as BELIEFS must be taken into account when deciding that a proposition ϕ will not appear in BELIEFS. Rule Neg can be contradictory, giving $\neg L \neg p$ while Rule Up yields $L \neg p$ in the example of Figure 1. To get a coherent belief set, it is necessary to use a fixed point definition. First we define a derivation operator for BELIEFS that uses all of the introduced rules:

DEFINITION 3.2 (R1 DERIVATION)

Let A be a set of sentences of $\mathcal{L}_1$, and B a set from $\mathcal{L}_0$. The operator $A, B \vdash_{R1} \phi$ means that the nonmodal sentence ϕ can be derived in BE-LIEFS using the rules Taut, Up, and Down, given that A is the set of constraints of META-BELIEFS, and B the set of initial BELIEFS.

Now suppose our agent starts with an initial set of BELIEFS B and META-BELIEFS A. Using the operator just defined, we can calculate the set of consequences

$$S = \{\phi \mid A, B \vdash_{R1} \phi\}.$$

As we just noticed, adding $\neg L \psi$ to A for every ψ not occurring in S may cause further consequences to be derived, with the possibility of inconsistency. Instead, we could first try to guess at the possible outcome of S, given that we initially add a set of modal atoms $\neg L \psi$ to A. If this turns out to give an S that contains all but the initial ψ's we chose as not derivable, then we have succeeded in satisfying the semantics of L. The idea of guessing the set S can be described by a fixed point of the operator $A, B \vdash_{R1}$. We define the assumptions of nonderivability by:

$$\neg L \overline{S} = \{\neg L \psi \mid \psi \notin S \text{ and } \psi \in \mathcal{L}_0\}.$$

Then the fixed point is defined as follows.

DEFINITION 3.3 (R1 EXTENSION)

Any set S satisfying

$$S = \{\phi \mid A \cup \neg L \overline{S}, B \vdash_{R1} \phi\}.$$

is an R1 extension of A.

The agent of Figure 1 serves as a good example of the fixed point construction. Assume initially that $S = \text{Cn}(q, \neg p)$, where $\text{Cn}(\Gamma)$ is the set of tautological

META-BELIEFS   BELIEFS

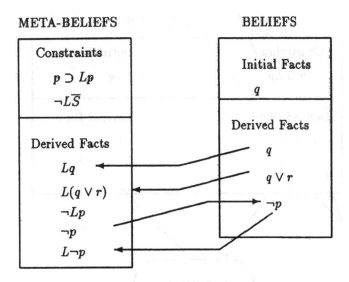

Figure 2: Example of reflection

consequences of Γ. Then $\neg L\overline{S}$ is the set of initial assumptions that we add to META-BELIEFS; among them is $\neg Lp$, because $p \notin \text{Cn}(q, \neg p)$. This situation is described in Figure 2.

Since both q and $q \vee r$ are consequences of the initial facts in BELIEFS, the Up rule can be used to add Lq and $L(q \vee r)$ to META-BELIEFS. From the initial constraints and the assumptions $\neg L\overline{S}$, the fact $\neg p$ can be derived in META-BELIEFS. Now the Down rule adds $\neg p$ to BELIEFS, and further use of Up adds $L\neg p$ to META-BELIEFS. By now the only further additions to BELIEFS will come from tautological consequences of $\neg p$ and q, so that the fixed point definition is satisfied by S.

The rules Up and Down are *reflection principles* [6, 23] that relate the sentences in BELIEFS and META-BELIEFS. Together with the Taut rules, they constitute the a set of inferences for the fixed point definition that will enforce the semantic rules for L, and at the same time make the set of nonmodal META-BELIEFS part of BELIEFS.

PROPOSITION 3.1 *The equations of Reflective Semantics are satisfied by every R1 extension of A and B.*

Proof. Since BELIEFS is closed under tautological consequence, every extension will be. The first semantic condition holds because of rule Up. The second condition holds since any proposition ϕ not derivable in BELIEFS will not be in S, and so part of the assumptions $\neg L\overline{S}$.

The reflection rules can be shown to be minimal, in that any belief set sat-

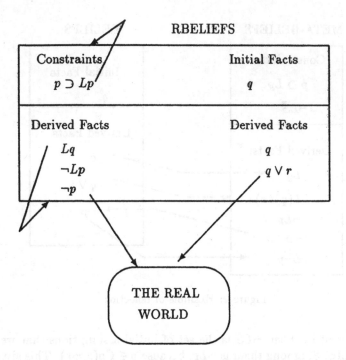

Figure 3: An amalgamated reflective agent

isfying the semantic rules admits them as valid (see [13]).

3.1 Amalgamation

So far we have kept BELIEFS and META-BELIEFS as separate boxes, using reflection principles to transfer sentences between them. But since these both contain beliefs of the agent, ideally we would like to merge or amalgamate them into one set. The only change is that META-BELIEFS, instead of containing just a subset of all the facts about the world, will now contain a full set. As it turns out, this makes no difference for the extensions of an agent's initial beliefs.

To accomplish the amalgamation, we consider just one set RBELIEFS containing both constraints and initial facts. The initial facts and constraints are put together; facts are still sentences of $\mathcal{L}_0$, and constraints contain at least one modal operator. The amalgamated belief set is diagrammed in Figure 3. Note that the semantics for L is the amalgamated set RBELIEFS, so it is now self-referential.

With this amalgamation, the semantic rules are changed to refer to RBELIEFS.

DEFINITION 3.4 (AMALGAMATED SEMANTICS)

$$\phi \in RBELIEFS \text{ and } \phi \in \mathcal{L}_0 \rightarrow L\phi \in RBELIEFS$$
$$\phi \notin RBELIEFS \text{ and } \phi \in \mathcal{L}_0 \rightarrow \neg L\phi \in RBELIEFS$$

Any set from $\mathcal{L}$ that satisfies these conditions, contains A, and is closed under tautological consequence, will be called a *stable set for A*.[1] The kernel of a stable set S is $S \cap \mathcal{L}_0$. A stable set for A is minimal if there is no other stable set for A with a smaller kernel.

The rules must be modified to apply to just the set RBELIEFS.

Rule Taut.

If Γ is a finite subset of RBELIEFS, and ϕ is a tautological consequence of Γ, then add ϕ to RBELIEFS.

Rule Reflective Up.

If ϕ is a sentence from $\mathcal{L}_0$ in RBELIEFS, then add $L\phi$ to RBELIEFS.

Only one tautological rule is required. No Down rule is necessary to transfer sentences from BELIEFS to META-BELIEFS, because these are now the same set. The Up rule is modified to apply only to facts, i.e., sentences from $\mathcal{L}_0$.

Because of the amalgamation, there is only one premise set for derivations, and we define the following derivation operator.

DEFINITION 3.5 (RN DERIVATION)

Let A be a set of sentences of $\mathcal{L}_1$. The operator $A \vdash_{RN} \phi$ means that the nonmodal sentence ϕ can be derived in RBELIEFS from the initial set A (containing both constraints and facts) using the rules Taut and Reflective Up.

Finally, the fixed point equation is modified.

DEFINITION 3.6 (RN EXTENSION)

Any set S satisfying

$$S = \{\phi \mid A \cup \neg L\overline{S} \vdash_{RN} \phi\}.$$

is an RN extension of A.

The amalgamated belief system produces exactly the same extensions as the original definition.

PROPOSITION 3.2 *S is an R1 extension of A and B iff it is the kernel of an RN extension of $A \cup B$.*

Proof. The consequence operator $A, B \vdash_{R1}$ is the same as $A \cup B \vdash_{RN}$ restricted to $\mathcal{L}_0$. Since $\neg L\overline{S}$ only considers nonmodal sentences from S, the fixed points are equivalent.

As a consequence, the amalgamated extensions satisfy the semantic rules, that is, they are stable sets.

COROLLARY 3.3 *Every RN extension of A is a stable set containing A.*

---

[1] The definition and term "stable set" is from Stalnaker [22], although he considered a more expressive language in which nested modal operators were allowed.

The amalgamated system simplifies the consequence operator, and makes the rule Down superfluous, but is not very different from the separated system with the introspective hypothesis. It is this hypothesis, the identification of the META-BELIEF consequences with the BELIEFS of the agent, that is important, and which the amalgamated system embodies.

As with R1 extensions, in RN extensions the reflection rules can be shown to be minimal, in that any amalgamated belief set satisfying the semantic rules admits them as valid (see [13]).

3.2 Reflection Principles

The rules RN are minimal ones that might be used by an agent reasoning about its own beliefs. But are there other reflective reasoning principles that could be incorporated? In this section we will give a partial answer to this question by examining several standard modal axiomatic schemata, and showing how some of them are appropriate as general reasoning principles, while others must be regarded as specific assumptions about the relation of beliefs to the world.

The most well-known modal schemata (without nested modal operators) are the following.

$$K. \quad L(\phi \supset \psi) \supset (L\phi \supset L\psi)$$
$$T. \quad L\phi \supset \phi \tag{1}$$
$$D. \quad L\phi \supset \neg L\neg\phi$$

Which of these schemata are sound with respect to the semantics of amalgamated belief sets? It should be clear that K is sound, since if its antecedents are true of a stable set, then so is its consequent, and stable sets are closed under tautological consequence. The schema D is true only of stable sets that are consistent: they do not contain both a sentence and its negation. This might be a valid assumption for some applications.

The schema T, on the other hand, is not semantically valid, as can be shown with the following counterexample. Let an agent believe ϕ, so that $L\phi$ is true, but suppose that in the actual world ϕ is in fact false. Then $L\phi \supset \phi$ is false. Asserting T for a particular fact p says something about the agent's knowledge of how her belief in p is related to its truth in the world.

Here is a short example of how the sentence $Lp \supset p$ could be used by an agent. Consider the propositions:

$$p = \text{The copier repairman has arrived}$$
$$q = \text{The copier is ok}$$

Suppose an agent believes that if she has no knowledge that the repairman has arrived, the copier must be ok. Further she believes that the copier is broken. We represent this as:

$$A = \{\neg q, \neg Lp \supset q\} . \tag{2}$$

The premises A do not have any RN extension, because while Lp is derivable, p is not. One solution is to give the agent confidence in her own beliefs, e.g.,

$$A' = \{\neg q, \neg Lp \supset q, Lp \supset p\} . \tag{3}$$

Now there is an RN extension in which p is true, since from Lp the agent can derive p. It is as if the agent says, "I believe that p, therefore p must be the case."

Although one might not want to use this type of reasoning in a particular agent design, the point is that T sanctions a certain type of reasoning about the connection of beliefs to the world, and is thus a "nonlogical" axiom, similar to $\neg Lp \supset q$.

A weaker form of T is the following rule:

Rule Strip.
   If $L\phi$ is a proposition in RBELIEFS, then add ϕ to RBELIEFS.

Strip is the converse of Reflective Up. Like T, it is unsound with respect to the semantics of amalgamated beliefs.

Adding reflective axioms can only increase the number of extensions. For example, the theory $A = \{L(a \wedge a) \supset a, \neg La \supset a\}$ has no RN extensions, but adding K to RN yields one extension whose kernel is $Cn(a)$. If we take a proposition ϕ to be a consequence of a theory A when it is in *every* extension, then we have the curious case that adding reflection principles can only decrease the consequences of A.

3.3   Groundedness

An ideal agent is one that makes as few assumptions about the world or her own beliefs as possible, consistent with maintaining the amalgamated semantics. For example, an agent who is allowed to make the assumption ϕ for any nonmodal proposition ϕ can come to believe anything about the world, no matter what her initial beliefs. We state the following principle:

Principle of Groundedness
   An agent should believe only those propositions that follow from her initial beliefs, together with a minimal set of assumptions about what she does not believe.

RN extensions were formulated with this principle in mind. To give an idea of groundedness for some AE theories, consider the following table.

Theory	RN extensions	AE extensions	min stable sets
$La \supset a$	$Cn()$	$Cn(), Cn(a)$	$Cn()$
$\neg La \supset b$ $La \supset a$	$Cn(b)$	$Cn(b), Cn(a)$	$Cn(b), Cn(a)$
$\neg La \supset a$ $La \supset a$	$Cn(a)$	$Cn(a)$	$Cn(a)$
La	none	none	$Cn(a)$

AE extensions are reviewed in the next subsection. In comparison to RN extensions, they permit the derivation of a from the sentence $La \supset a$ and the assumption of La, and so are less grounded. Note that minimal stable sets always exist, but are not necessarily grounded.

The system RN is almost equivalent to default logic [21]. It is not quite as strongly grounded as the latter; for while there exists a translation from default logic to RN that preserves extensions, the inverse translation fails in a few cases. For a complete comparison of RN and its relatives with AE logic, modal nonmonotonic logics [17], and default logics, see [13].

3.4   AE Semantics

We will use some semantic concepts from Moore's original definition of AE logic [20]. An interpretation I of $\mathcal{L}$ assigns true or false to every sentence of $\mathcal{L}$, following the normal truth-recursion rules for the boolean connectives. Each modal atom $L\phi$ is treated as a separate predicate, so that it is possible to have $I \models La \wedge \neg a$.

DEFINITION 3.7 (AE MODEL) *A model I respects a set of sentences S if whenever $\phi \in S$, $I \models \phi$. An AE model I of a theory S is a model of S that respects S.*

This semantics can be used to define notions of soundness and completeness for AE extensions. A theory $T \subseteq \mathcal{L}$ is said to be *sound* with respect to a premise set A if every AE model of T is a model of A. It is *complete* if it contains every sentence true in all of its AE models.

An AE extension of A is defined by the following fixed point equation.

$$T = \{\phi \mid A \cup \neg L\overline{T} \cup LT \vdash \phi\}. \tag{4}$$

where $LT = \{L\phi \mid \phi \in \mathcal{L}_0 \text{ and } \phi \in T\}$. Moore showed that AE extensions are exactly the class of all theories T that are sound and complete with respect to A.

RN extensions are also AE extensions, and so satisfy AE semantics. But they exclude some ungrounded AE extensions, because they do not have the assumptions LT.

3.5   Special Forms

In relating AE theories to LP, we take advantage of the special form of LP theories. The primary simplification is that the arguments to the modal operator are always atomic: $L(a \vee b)$, $L(a \wedge b)$, and $L(\neg a)$ are not allowed. We define:

DEFINITION 3.8 $\mathcal{L}^+$ *is the subset of $\mathcal{L}$ consisting of those sentences without modal atoms of the form $L\phi$ for nonatomic ϕ.*

Of the reflection schemata in Section 3.2, only T is expressible in $\mathcal{L}^+$. This schema is important in our analysis of LP metatheory.

Herbrand models for $A \subseteq \mathcal{L}^+$ are formed by taking subsets of atoms($\mathcal{L}^+$). Herbrand interpretations do not contain negated modal literals, but we can define a natural extension that does.

DEFINITION 3.9 (ASSOCIATED STABLE SET) *Let $I \subseteq$ atoms$(\mathcal{L}^+)$ be a Herbrand interpretation. Its associated set is the set $I \cup \{\neg L\phi \mid \phi \notin I\}$.*

We will often misuse terminology and speak of I as stable, meaning its associated set is stable.

In logic programming theories sentences are restricted to clauses of the form $a_1 \wedge a_2 \wedge \cdots \wedge a_n \supset c$. Depending on whether negation is present in the a_i, we distinguish two cases.

1. Definite clauses: c and each a_i are atoms. This case is related to Reflective Prolog.

2. Normal clauses: c is an atom and all a_i are either atoms or negative modal literals. This case is related to negation-as-failure semantics.

For AE theories that are programs, there is a close relationship between extensions of A and its Herbrand models. One can view Herbrand interpretations either as models, or as belief sets. This leads to the notion of a Herbrand interpretation respecting itself.

DEFINITION 3.10 *A Herbrand model I of $\mathcal{L}$ is self-respecting if it respects I.*

For normal programs A, every extension of A contains a self-respecting Herbrand model of A.

PROPOSITION 3.4 *If the set T is an extension of a normal program A, then atoms$(T \cap \mathcal{L}^+)$ is a self-respecting model of A, and is stable.*

> *Proof.* Let $I = $ atoms$(T \cap \mathcal{L}^+)$. For each clause $a_1 \wedge a_2 \wedge \cdots \wedge a_n \supset c$ of A, if all of a_i are in T, then c must be also. Since a_i are either nonmodal atoms or $\neg L\phi$, the same is true of I. And since T is a stable set, if I contains a nonmodal atom c, it contains Lc, and so is self-respecting. I is stable because T is.

4   Reflective Prolog

Reflective Prolog (RP) is an extension to logic programming proposed by Costantini and Lanzarone [4, 5] (hereafter C+L). The basic purpose is to extend the language of LP to represent metalevel relations among predicates. The metalevel language is a Gödel term FOL with an interpreted predicate SOLVE of one argument.

As an example of the use of RP, consider a dyadic predicate *Brother* that is symmetric. An RP program expressing this fact, together with some particular instances of the relation, would be:

> *Brother*(*ralph, harry*)
> SOLVE('$p(x, y)$') $\wedge$ *Symmetric*('p') $\supset$ SOLVE('$p(y, x)$')
> *Symmetric*('p')

When given the query $Brother(x, y)$, the program should answer with the tuples $ralph, harry$ and $harry, ralph$.

Much of design effort in RP has gone into careful construction of the naming relation that connects quoted forms such as '$p(x, y)$' and the object-level predications $p(x, y)$. We ignore this here, and instead concentrate on the reflection principles that are present, by translating RP into the modal language $\mathcal{L}$. This is possible because an RP program is considered to be a collection of its instances. For any instance SOLVE(α), the term α is always mapped into the same ground atom in all interpretations; this atom is denoted by $\downarrow\alpha$. We can therefore take any RP program P (which is a definite program) and translate it into a corresponding definite clause program P' in the language $\mathcal{L}$, using

$$\text{SOLVE}(\alpha) \quad \leftrightarrow \quad L\downarrow\alpha . \tag{5}$$

We will show that the semantics of P' given by self-respecting models is exactly the same as the RP semantics of P.

DEFINITION 4.1 (EXTENDED HERBRAND INTERPRETATION) *Let P be an RP program. An extended Herbrand interpretation I of P is an interpretation of P in which SOLVE(α) is true if $\downarrow\alpha$ is contained in I.*

According to C+L, the extended Herbrand models of a program P enjoy the model intersection property, so that if I and I' are two extended Herbrand models of P, so is $I \cap I'$. Thus there is a least extended Herbrand model I^* of every program.

Extended Herbrand models do not force SOLVE(α) to be true if and *only* if $\downarrow\alpha$ is true. For example, the program $P = \{\text{SOLVE}('a')\}$ has a minimal Herbrand model $\{\text{SOLVE}('a')\}$, which violates the intended interpretation of SOLVE. A further definition of reflective models is necessary to guarantee the presence of a.

DEFINITION 4.2 (REFLECTIVE MODEL) *A reflective model of an RP program P is an extended Herbrand model of $P \cup \{\text{SOLVE}(\alpha) \supset \downarrow\alpha\}$.*

The addition of the schema SOLVE(α) $\supset \downarrow\alpha$ satisfies the "only if" connection between SOLVE(α) and $\downarrow\alpha$.

These definitions have a straightforward interpretation in terms of the AE semantics of Section 3.4. As with programs, we can translate any Herbrand interpretation of an RP program into an interpretation of $\mathcal{L}$ by the substitution (5). The reflection condition satisfied by extended Herbrand interpretations is exactly that of self-respecting models, giving the following result.

PROPOSITION 4.1 *I is an extended Herbrand interpretation satisfying an RP program P if and only if its transform I' is a self-respecting model of P'.*

From the results of C+L, we know that there is a least self-respecting model $I^\#$ of P'. Since P' is a definite clause program in $\mathcal{L}$, the extensions of P' will be self-respecting models of P', by Proposition 3.4. But, as with RP programs, the least

Herbrand model is not always an extension. Consider the case of $P' = \{La\}$: the least model is $\{La\}$, which is not stable. However, we can show that if P' has an extension, then it is given by the least model.

PROPOSITION 4.2 (LEAST MODEL) *The following three statements are equivalent:*

> *i. $I^\#$ is a stable set for P'*
> *ii. $I^\#$ is the unique AE extension of P'*
> *iii. $I^\#$ is the unique RN extension of P'*

Proof. Recall that we refer to the associated set of $I^\#$. We need two facts about $I^\#$.

(a) If $P' \cup \neg L\overline{I^\#}$ is RN-consistent, then

$$I^\# = \{\phi \in \text{atoms}(\mathcal{L}^+) \mid P' \vdash_{RN} \phi\}$$
$$= \{\phi \in \text{atoms}(\mathcal{L}^+) \mid P' \cup \neg L\overline{I^\#} \vdash_{RN} \phi\} \, .$$

(b) If $I^\#$ is stable, then $P' \cup \neg L\overline{I^\#}$ is RN-consistent.

(a) holds because P' cannot use negative modal literals to deduce any new atoms; and (b) holds because a program P' is consistent with any of its stable sets.

From fact (a) the direction $i \Rightarrow ii$ follows, since $I^\#$ is a stable set for P' satisfying the conditions of RN extensions. The converse direction follows from Proposition 3.4 and fact (a).

For $ii \Leftrightarrow iii$, we need the following facts about AE extensions.

(a) $\{\phi \mid P' \cup \neg L\overline{T} \cup LT \vdash \phi\}$ is equal to $\{\phi \mid P' \cup \neg L\overline{T} \cup LT \vdash_{RN} \phi\}$.

(b) $\{\phi \in \text{atoms}(\mathcal{L}^+) \mid P' \cup LT \vdash \phi\}$ is equal to
$\{\phi \in \text{atoms}(\mathcal{L}^+) \mid P' \cup \neg L\overline{T} \cup LT \vdash \phi\}$.

(c) $\{\phi \in \text{atoms}(\mathcal{L}^+) \mid P' \vdash \phi\}$ is a subset of
$\{\phi \in \text{atoms}(\mathcal{L}^+) \mid P' \cup LT \vdash \phi\}$ for every extension T.

From (b) and (c) we know that there is a unique minimal AE extension of P', containing the atoms $\{\phi \in \text{atoms}(\mathcal{L}^+) \mid P' \vdash \phi\}$. From (a), this is equivalent to the unique RN extension of P', if it exists.

So in those cases in which P' has an RN extension, the minimal Herbrand model is that extension, and it is also the unique AE extension. In this case, for an RP program one could take the minimal extended Herbrand interpretation as the correct introspective semantics. In C+L, the reflective models of RP add the T schema $L\phi \supset \phi$. Is there any discrepancy between RP models and RN extensions? If P' has an extension, the answer is no. If $I^\#$ is not stable, however, then P' has no extension, because there are atoms $L\phi$ derivable from P' for which ϕ is not derivable. But, we can show that adding the schema T creates a unique extension for P'. We define $t(X) = X \cup \{L\phi \supset \phi\}$.

PROPOSITION 4.3 *The least model $I^\#$ of $t(P')$ is stable. Further, if $I^\#(P')$ is stable, then*

$$I^\#(P') = I^\#(t(P')) \, .$$

Proof. Since $I^\#(t(P'))$ is a model of $\{L\phi \supset \phi\}$, it must contain ϕ exactly when it contains $L\phi$, and so is stable. Further, if $I^\#(P')$ is stable, then it is a model for $\{L\phi \supset \phi\}$, and so is a model for $I^\#(t(P'))$.

So we have the following situation. The definite program $t(P')$ has the same unique extension as P' when the latter exists; and in those cases when P' has no extension, $t(P')$ still has exactly one extension, which is a stable set for P'. Thus the motivation for the reflective models of RP is clear: they always give a canonical model, and that model implements ideal introspection whenever a program has an RN extension.

How can the extensions of $t(P')$ be computed? Since this is a definite program, none of the assumptions $\neg L\psi$ are necessary in computing nonmodal consequences, and so the rules RN are sufficient. Further, the T schema can be eliminated if the rule Strip is added to RN. This gives a system with both upwards (Reflective Up) and downwards (Strip) reflection. The RSLD resolution procedure of C+L is complete, and contains just these rules for the SOLVE predicate.

5 Negation as Failure

The assumptions $\neg L\phi$ play no role in definite programs. In normal programs, with negative modal literals in the antecedent of clauses, the situation becomes more complicated. We distinguish two types of normal programs.

DEFINITION 5.1 (NORMAL MODAL PROGRAM) *A normal modal program contains clauses from $\mathcal{L}^+$ of the form*

$$l_1 \wedge \cdots \wedge l_n \supset c \, ,$$

with c atomic (modal or nonmodal) and l_i either atomic (modal or nonmodal) or a negative modal literal.

In normal modal programs, the only negative literals in the antecedent are modal. So negation as failure takes place only through modal literals. For example, the following is a normal modal program clause:

$$La \wedge \neg Lb \wedge e \supset Lc \, .$$

If we allow modal literals to occur only as negative antecedents, then we have restricted normal programs.

DEFINITION 5.2 (RESTRICTED NORMAL MODAL PROGRAM) *A restricted normal program is a normal program in which c is nonmodal and every atomic l_i is nonmodal.*

Every restricted normal modal program clause has the form:

$$a_i \wedge \neg L b_i \supset c \,.$$

Unlike definite programs, normal modal programs can have multiple extensions, e.g.,

$$\neg La \supset b$$
$$\neg Lb \supset a$$

There are two RN (and AE) extensions, whose kernels are $Cn(a)$ and $Cn(b)$. If multiple extensions are permitted or desired, as in abductive methods, then the RN extensions may be used directly. But if the desired semantics is a single canonical model, some additional effort is necessary. There are two strategies that have been taken.

- Change the definition of extensions.

- Restrict the language.

The first approach was taken by Bonatti (see his paper in this volume) in defining a kind of well-founded semantics for normal modal programs. We will briefly comment on the introspective nature of his formulation in the next subsection.

A second method is to put additional syntactic constraints on normal modal programs so that they have at most one extension. Stratification [1] is the standard technique: we will examine ideal introspection for this case.

5.1 Three-valued Semantics

Bonatti's revision of the fixed-point semantics of AE logic weakens the negative belief assumptions that are made. We give a proof-theoretic version of his generalized stable expansions, and compare it to the fixed-point equation (4).

Let $D = (D^+, D^-)$ be a tuple of sets, where D^+ (the "believed" sentences), and D^- (the "disbelieved") have no elements in common. Define a consequence operator $A \vdash_D c$ as equivalent to $A \cup \{L\phi \mid \phi \in D^+\} \cup \{\neg L\phi \mid \phi \in D^-\} \vdash c$.

DEFINITION 5.3 (GENERALIZED STABLE EXPANSIONS) *A pair of sentence sets*
$D = (D^+, D^-)$ is a generalized stable expansion of a theory $A \subseteq \mathcal{L}$ if it
satisfies the equations:

$$\begin{aligned} D^+ &= \{\phi \mid A \vdash_D \phi\} \\ D^- &= \{\phi \mid \forall G \supseteq D, \; A \vdash_G \phi\} \end{aligned}$$

From the point of view of ideal introspection, generalized stable expansions suffer from the same fault as AE extensions: they allow the assumption of positive self-belief. Sentences such as $La \supset a$ will allow the derivation of a, whereas ideal introspection will not allow it to be grounded. To bring generalized stable semantics more into line with ideal introspection, the fixed point could be modified to eliminate the positive belief assumptions, and use the rules RN. This would produce a much more cautious introspective believer, one who makes even fewer assumptions about what she does not believe.

5.2 Stratified Theories

We give a short review of results on stratified normal programs over $\mathcal{L}_0$. A stratified program can be divided into a totally ordered set of predicates $P^1 < P^2 \cdots < P^n$, such that if $\neg a$ appears in a clause in P^j, a does not appear in the head of any clause of P^i for $i \geq j$ (see [1]).

The stable model semantics [7, 2] of such programs is given by the following construction.

DEFINITION 5.4 (STABLE MODEL)   *Let P be a normal program over $\mathcal{L}_0$ and I a Herbrand interpretation of $\mathcal{L}_0$. Let $P^N = \{rule^N \mid rule \in P\}$, where*

$$(\neg c_1 \wedge \cdots \neg c_n \wedge b_1 \wedge \cdots \wedge b_m \supset a)^N = \left\{ \begin{array}{ll} b_1 \wedge \cdots \wedge b_m \supset a & \text{if } \forall j : c_j \notin I \\ true & \text{otherwise} \end{array} \right.$$

I is a stable model of P iff I is the minimal Herbrand model of P^N.

There is a unique stable model for a stratified program P, which is also the canonical *supported* Herbrand model of P [1]. The canonical supported model I^{supp} of P is constructed by iterating the RN consequence operator over the topology defined by the stratification, adding negative modal atoms whenever required by the clauses.

One translation from normal logic programs to $\mathcal{L}$ uses the restricted normal modal programs:

DEFINITION 5.5 (GELFOND AND LIFSCHITZ TRANSLATION)   *The transform of a program P to P^G is given by:*

$$\neg c_1 \wedge \cdots \wedge \neg c_n \wedge b_1 \wedge \cdots \wedge b_m \supset a \quad \Rightarrow \quad \neg L c_1 \wedge \cdots \neg L c_n \wedge b_1 \wedge \cdots \wedge b_m \supset a \,.$$

There is a unique AE and RN extension for the translation P^G of every stratified normal program P.

The restricted modal programs are not rich enough to express introspective concepts such as the schema T; normal modal programs are more interesting from an introspective point of view. Normal modal programs can be produced from normal programs either by considering the translation of [18]:

DEFINITION 5.6 (MAREK AND TRUSCZYNSKI TRANSLATION)   *The transform of a program P to P^M is given by:*

$$\neg c_1 \wedge \cdots \neg c_n \wedge b_1 \wedge \cdots \wedge b_m \supset a \quad \Rightarrow \quad \neg L c_1 \wedge \cdots \neg L c_n \wedge L b_1 \wedge \cdots \wedge L b_m \supset a \,.$$

Or, one might want to work directly within $\mathcal{L}$ itself, since some normal modal programs cannot be represented as translations, e.g.,

$$a \wedge L b \wedge \neg L c \supset d \,.$$

For stratified normal modal programs (La are treated as a in the definition of stratification), we have the following facts.

- Extensions may not exist, e.g., $P = \{La\}$.

- Because of the stratification condition, there is at most one RN extension, but there could exist multiple AE extensions, e.g., $P = \{La \supset a\}$ has two extensions.

- AE and RN extensions diverge, even if we consider only minimal AE extensions, e.g., $P = \{\neg La \supset b, \; La \supset a\}$ (see Section 3.3).

> *Proof.* The only one of these facts which is not obvious is the existence of at most one RN extension. But this follows in the same manner as the construction of the canonical supported model. The choice of assumptions $\neg L\phi$ are fixed by the stratification condition: at stratum j, if $\neg La$ is in the antecedent of a clause, and a has not been derived at this point, then $\neg La$ must be an assumption. For AE extensions this construction does not work because La can be an assumption, and La acts like a in the stratification, not $\neg La$. We could modify the stratification condition so that La was treated the same as $\neg La$, but this would exclude all the interesting cases, e.g., $La \supset a$. It would also mean that the Marek and Trusczynski translation of a stratified program would not necessarily be stratified in $\mathcal{L}$.

The advantage of groundedness means that RN extensions are preferable to AE semantics in this case. Further, we can use techniques similar to that of C+L to remedy the problem of having no extensions for some normal modal programs.

PROPOSITION 5.1 *If an RN extension of a normal modal program P exists, it is also the extension of $t(P)$.*

> *Proof.* By Proposition 3.4, $I = \mathrm{atoms}(T \cap \mathcal{L}^+)$ is a stable self-respecting model of P. Since I is stable, it also is a model of $\{L\phi \supset \phi\}$, and hence of $t(P)$. So, adding the schema $\{L\phi \supset \phi\}$ to P cannot increase the number of atoms deduced by $P \cup \neg L\overline{T}$, and hence T also satisfies the fixed point equation for $t(P)$.

Just as in the case of definite programs, when P has an extension, $t(P)$ has the same extension. But even when P has no extensions, $t(P)$ has one.

PROPOSITION 5.2 *For every normal modal program P, $t(P)$ has an extension.*

> *Proof.* The extensions is found by construction of a Herbrand model I, based on the stratification of $t(P)$. At each stratum j, all the atoms deducible from I^j and the clauses of all strata $i \leq j$ are added to form I^{j+1}. I^∞ is a self-respecting model of $t(P)$, and is stable because of $\{L\phi \supset \phi\}$. Further, by construction the stable set associated with I^∞ satisfies the fixed point equation for RN extensions.

6 Conclusion

We have tried to understand the reflection principles present in the metalogic of Logic Programming by considering logic programs as the belief sets of an agent. This approach leads naturally to identifying reflection principles with ideal introspection for the agent. Introspection is ideal when the agent makes as few assumptions as possible about the world or her own beliefs, consistent with maintaining complete introspective knowledge of her own beliefs. In formal terms, the beliefs of an agent are identified with RN extensions of the premises. Because of ideal introspection, RN extensions are more strongly grounded in the premises than the autoepistemic logic of Moore.

In the belief set approach, we investigated several introspective principles that an agent might use in reasoning about her own beliefs. Of these, the rule Reflective Up, which adds to an agent's knowledge information about what beliefs have been derived, was shown to be sound. The converse of this rule, Strip, introduces a belief about the world on the basis of introspective information; this rule was shown to be unsound, as was the stronger schema T: $L\phi \supset \phi$.

There is a strong formal connection between ideal introspection and metalogical systems in Logic Programming. We traced this connection in two cases: Reflective Prolog and negation as failure. The main results are the following.

- Reflective Prolog, at least the reflective part, can be understood as an ideal introspective system in which the schema T is added to ensure that a unique extension always exists. Although this schema is in general unsound, in the case of definite programs it makes no difference when an extension exists. If an extension does not exist, it makes the agent arrogant enough to derive an ungrounded conclusion about the world, based on introspective knowledge of her own beliefs.

- For *restricted* normal modal programs, RN extensions give the stable model semantics of the Gelfond/Lifschitz translation of logic program to $\mathcal{L}$. There is always a unique stable model and RN extension. However, this case is uninteresting from an introspective point of view, because the language is too limited to express introspective principles.

- For general normal model programs, RN extensions are a well-motivated implementation of reflection. Using the same techniques as in Reflective Prolog, we show that there is always a unique RN extension of $t(P)$ when P is stratified.

Acknowledgments

I thank Stefania Costantini and Elio Lanzarone for introducing me to Reflective Prolog, and for helpful discussions about reflection principles.

The research reported in this paper was supported by the Office of Naval Research under Contract No. N00014-89-C-0095.

References

[1] K. Apt, H. Blair, and A. Walker, Towards a theory of declarative knowledge, in: J. Minker, ed., *Foundations of Deductive Databases*, volume Chapter 2 (Morgan Kaufmann, 1988) 89–148.

[2] N. Bidoit and C. Froidevaux, Negation by default and nonstratifiable logic programs, Technical Report 437, Université Paris XI (1988).

[3] K. A. Bowen and R. A. Kowalski, Amalgamating language and metalanguage, in: K. L. Clark and S.-A. Tarnlund, eds., *Logic Programming* (Academic Press, 1982) 153–172.

[4] S. Costantini, Semantics of a metalogic programming language, *International Journal of Foundations of Computer Science* 1 (3) (1990).

[5] S. Costantini and G. A. Lanzarone, A metalogic programming language, in: G. Levi and M. Martelli, eds., *Proceedings of the Sixth International Conference on Logic Programming* (MIT Press, 1989) 218–233.

[6] S. Feferman, Transfinite recursive progressions of axiomatic theories, *Journal of Symbolic Logic* 27 (1962) 259–317.

[7] M. Gelfond and V. Lifschitz, The stable model semantics for logic programming, in: *Proceedings of the International Conference on Logic Programming* (1988).

[8] J. Y. Halpern and Y. O. Moses, Towards a theory of knowledge and ignorance: Preliminary report, in: *Nonmonotonic Reasoning Workshop*, New Paltz, New York (1984).

[9] J. Y. Halpern and Y. O. Moses, A guide to the modal logics of knowledge and belief, in: *Proceedings of the International Joint Conference on Artificial Intelligence*, Los Angeles (1985) 50–61.

[10] J. Hintikka, *Knowledge and Belief* (Cornell University Press, Ithaca, New York, 1962).

[11] K. Konolige, Circumscriptive ignorance, in: *Proceedings of the Conference of the American Association of Artificial Intelligence*, Pittsburgh, PA (1982).

[12] K. Konolige, *A Deduction Model of Belief* (Pitman Research Notes in Artificial Intelligence, 1986).

[13] K. Konolige, Ideal introspective belief, in: *Proceedings of the Conference of the American Association of Artificial Intelligence*, San Jose, CA (1992).

[14] K. Konolige, Quantification in autoepistemic logic, *Fundamenta Informaticae* (1992).

[15] H. J. Levesque, Foundations of a functional approach to knowledge representation, *Artificial Intelligence* **23** (2) (July 1984).

[16] H. J. Levesque, A study in autoepistemic logic, *Artificial Intelligence* **42** (2–3) (March 1990).

[17] W. Marek, G. F. Schwarz, and M. Truszczynski, Modal nonmonotonic logics: ranges, characterization, computation, in: *Proceedings of the Second International Conference on Principles of Knowledge Representation and Reasoning*, Cambridge, MA (1991).

[18] W. Marek and M. Truszczynski, Stable semantics for logic programs and default theories, Technical report, Department of Computer Science, University of Kentucky (1989).

[19] R. Montague, Syntactical treatments of modality, with corollaries on reflexion principles and finite axiomatizations, *Acta Philosophica Fennica* **16** (1963) 153–167.

[20] R. C. Moore, Semantical considerations on nonmonotonic logic, *Artificial Intelligence* **25** (1) (1985).

[21] R. Reiter, A logic for default reasoning, *Artificial Intelligence* **13** (1–2) (1980).

[22] R. C. Stalnaker, A note on nonmonotonic modal logic, Department of Philosophy, Cornell University, (1980).

[23] R. Weyhrauch, Prolegomena to a theory of mechanized formal reasoning, *Artificial Intelligence* **13** (1980).

An Introduction to Partial Deduction

Jan Komorowski

Department of Computer Science and Electrical Engineering
The Norwegian Institute of Technology, The University of Trondheim
N-7034 TRONDHEIM, NORWAY

Abstract

After several years of neglect, the importance of *partial deduction* (previously, partial evaluation in logic programming) as an omnipresent principle in Logic Programming, and to a certain degree in computing in general, is being recognized. This article provides a systematic introduction to partial deduction, its applications and open problems. Starting from an informal and intuitive presentation, the fundamental notions such as correctness and completeness are discussed. A selection of applications is presented to illustrate partial deduction in different contexts.

1 Introduction

In recent years a remarkable growth of interest in partial deduction (a.k.a. partial evaluation in logic programming) has occurred. The original motivation for partial deduction [25] came from partial evaluation in functional programming but Partial Deduction (hence, abbreviation PD) is now an independent and rich domain of study with its own set of methods and results. One possible explanation for the interest in PD is that it occurs in rather many different computational contexts, such as synthesis of software, refinement calculi for specifications, automatic generation of compilers from interpreters, optimization of deductive database queries, intensional answers to queries, machine learning, knowledge and inheritance compilation, planning, etc. This is probably not surprising, since logic is fundamental to all these fields of computing.

In response to the interest in partial deduction, a Special Issue of the Journal of Logic Programming on Partial Deduction will appear in 1992. It is worth noticing that partial deduction is given substantial attention in the Special Issue of the Communications of ACM on Logic Programming [46].

Aim of this paper Since much of the results on partial deduction are scattered throughout conference and journal articles, this paper aims at a systematic, self-contained and reasonably complete presentation of intuitions and formal foundations of the principle of Partial Deduction. Several examples illustrate the most representative applications.

50

It is assumed that the reader is acquainted with the fundamental notions of logic programming and has some experience with the Prolog programming language. An excellent introductory text to Logic Programming and Prolog is [36]. No background in partial evaluation (in functional programming) is assumed.

The paper is organized as follows: an informal and intuitive introduction with simple examples; formal definitions and the partial deduction theorem; a discussion of the relative value of partial deduction; a selection of applications of partial deduction; and open research issues.

2 Partial Deduction

Having originated from partial evaluation, which in turn can be traced back to Kleene's S-m-n theorem [24], and being related to in-line compilation, partial deduction inherits from these paradigms and extends them. Unification-based parameter propagation contributes to the relative power of partial deduction.

2.1 Informal presentation

The purpose of partial deduction is usually defined to be a synthesis of a specialized (and possibly more efficient) *residual* program P' given an *initial* program P and a *with respect to* atom A (hence abbreviation, wrt). It is required that the residual program will preserve its semantics for the atom. This is illustrated with the following logic program defining family relations.

EXAMPLE 1 *Family, initial program*
```
grandparent(X, Z) ← grandfather(X, Z).
grandparent(X, Z) ← grandmother(X, Z).
grandfather(A, C) ← father(A, B), parent(B, C).
parent(A, B) ← father(A, B).
parent(A, B) ← mother(A, B).
father(george, henry).
father(victor, helena).
father(henry, jan).
```

Assume we are now interested in finding a residual definition for grandfathers, that is, the wrt atom is set to be *grandfather(Who, GChild)*. A possibility is to replace *grandfather*/2 by a new definition where the condition of parenthood has been refined to fatherhood or motherhood. A reader with a compiler background will immediately recognize some similarity of partial deduction to macro opening.

EXAMPLE 2 *A residual family program*
```
grandparent(X, Z) ← grandfather(X, Z).
grandparent(X, Z) ← grandmother(X, Z).
grandfather(A, C) ← father(A, B), father(B, C).
grandfather(A, C) ← father(A, B), mother(B, C).
```

```
parent(A, B) ← father(A, B).
parent(A, B) ← mother(A, B).
father(george, henry).
father(victor, helena).
father(henry, jan).
```

As the next example will show, partial deduction is more complicated than macro-opening.

Partial deduction, as well as partial evaluation, is defined operationally. In the case of logic programming it is a proof-theoretic semantics. Recall that this semantics employs, among others, the well-known concepts of selection rules, derivations and derivation trees. The main task of partial deduction is to construct a *partial* derivation tree for the wrt atom, that is, a tree whose root is the wrt atom. Derivation trees are partial since they can end unresolved with a non-empty goal. (Unlike standard derivations which are either successful, failed, or infinite.) A so called *resultant* is formed from each finite and non-failing derivation. A resultant is a clause whose head is the substituted wrt atom and whose body is the unresolved goal at the end of derivation. The variables in the wrt atom and in the unresolved atoms are appropriately substituted. An intuitive interpretation of a resultant is that it summarizes a partial proof for the wrt atom. If the unresolved atoms of the goal are eventually proven then the conclusion (i.e. the head of the resultant) will be proven as well. Another intuition is that a resultant is a conditional answer for the wrt atom with the body expressing the enabling condition.

Let us consider the wrt atom to be *grandfather(Who, jan)*, that is, consider a task of finding a residual program that defines Jan's grandfathers. Let the selection rule be "select the leftmost atom, provided it is not *mother/2*". This particular selection rule is obviously dictated by the fact that the example program lacks a definition of the *mother/2* predicate. Given this rule, a partial derivation tree with the root *grandfather(Who, jan)* is constructed (see Figure 1). The first branch ends with a success, finding that George is Jan's grandfather. Hence, the first resultant is *grandfather(george, jan)*. The remaining derivations either end with an unresolved atom for predicate *mother/2* or are failing.

EXAMPLE 3 *Another residual family program*
```
grandparent(X, Z) ← grandfather(X, Z).
grandparent(X, Z) ← grandmother(X, Z).
grandfather(george, jan).
grandfather(george, jan)← mother(henry, jan).
grandfather(victor, jan)← mother(helena, jan).
grandfather(henry, jan)← mother(jan, jan).
parent(A, B) ← father(A, B).
parent(A, B) ← mother(A, B).
father(george, henry).
father(victor, helena).
father(henry, jan).
```

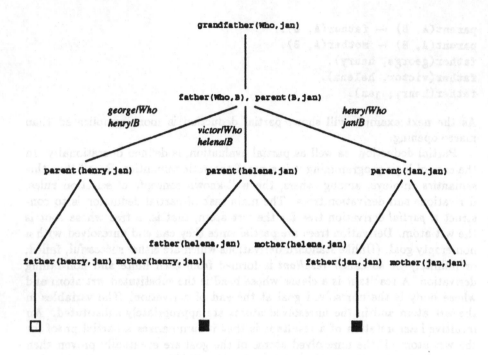

Figure 1: A partial derivation tree.

Intuitively, resultants are conditional answers. For instance, George is Jan's grandfather if Henry is Jan's mother(!); Victor is Jan's grandfather provided that Helena is Jan's mother, etc. These resultants would be called *intensional answers* in the context of deductive databases.

It is instructive to augment the definition of the predicate mother with, for instance, *mother(helena, jan)*. and *mother(helena, stan)*. and to construct derivations for *grandfather(Who, jan)* with the initial program augmented with this definition and the residual program also augmented with the same definition. The reader is encouraged to verify the derivations. Clearly, George and Victor are grandfathers and are the only grandfathers in both cases. This is known as *correctness* and *completeness* of partial deduction. More precise conditions under which these properties hold will be defined in the sequel.

2.2 Formalization of Partial Deduction

The notions that were informally introduced in the preceding section will be now made more precise. Among other issues, it is necessary to make it clear which semantics of the initial program should be preserved by the residue (for the wrt atom). Not surprisingly, a requirement that every semantics (such as model-theoretic and proof-theoretic) of a logic program should be preserved is not possible to achieve. In what follows, a foundation for proving correctness and completeness of partial deduction are laid down. Correctness of a residual program P' with respect to the initial program P and goal G for procedural (respectively, declarative), semantics means that computed (correct), answers for G in P' are computed (correct), answers for G in P. The converse of this is completeness.

Following [25] and [34], it appears that for definite logic programs partial deduction is always correct, but not necessarily complete [34]. For normal programs and goals, partial deduction is in general not correct. However, the same syntactic condition that guarantees completeness for definite programs ensures correctness in the normal case. An additional condition is necessary to guarantee completeness (for the procedural semantics) in the presence of negation. Due to the incompleteness of SLDNF-resolution, the results for declarative semantics are weaker.

A standard technique for defining partial evaluation (and partial deduction) is to base it on a suitably modified operational semantics. In the case of logic programming this operational semantics is proof-theoretic. The construction is as follows. Partial (or "unfinished") derivations for G in P (and partial derivation trees) are constructed first. Then, P' is obtained as an appropriate selection of instantiated clauses that label the derivation trees. To this end, the notions of goal, SLD-derivation, and SLD-tree are properly generalized. The generalizations allow us to extend the notion of derivation to an "unfinished" derivation, that is, to a finite derivation which can be neither a success nor a failure. The reader is referred to [33] or [36] for the standard notions of SLDNF-derivation, SLDNF-refutation and SLDNF-tree. With some exceptions, our formalization of partial deduction follows [34].

DEFINITION 1 (SLD-DERIVATION AND TREE) *Let P be a normal program and G a normal goal. An SLD-derivation (respectively, SLD-tree) of $P \cup \{G\}$ is an SLDNF-derivation (SLDNF-tree) of $P \cup \{G\}$ such that only positive literals are selected.*

DEFINITION 2 (RESULTANT) *A resultant is a formula $A_1, \ldots, A_m \leftarrow B_1, \ldots, B_n$ where $m > 0, n \geq 0$ and $A_1, \ldots, A_m$ and $B_1, \ldots, B_n$ are atoms.*

DEFINITION 3 (SLD-DERIVATION OF A RESULTANT) *Let C be a fixed computation rule. An SLD-derivation of a resultant R_0 is a finite or infinite sequence of resultants: $R_0 \Rightarrow_C R_1 \Rightarrow_C R_2 \Rightarrow_C \ldots$ where for each i*
R_i is of the form

$$A_1, \ldots, A_m \leftarrow B_1, \ldots, B_{i-1}, B_i, B_{i+1}, \ldots, B_n$$

R_{i+1} (if any) is of the form:

$$(A_1, \ldots, A_m \leftarrow B_1, \ldots, B_{i-1}, C_1, \ldots, C_j, B_{i+1}, \ldots, B_n)\theta$$

if (1) the C-selected atom in R_i is B_i, (2) there is a standardized apart program clause $H \leftarrow C_1, \ldots, C_j$ and (3) B_i and H have a most general unifier θ.

DEFINITION 4 (PARTIAL SLD-TREES) *The (partial) SLD-tree of a resultant R_0 under C is a rooted tree such that:*

  (i) its root is R_0, and

  (ii) each node R_i is either a leaf or its children are $\{R_{i+1} \mid R_i \Rightarrow_C R_{i+1}\}$.

DEFINITION 5 (PARTIAL DEDUCTION OF AN ATOM) *A partial deduction of an atom A in a program P is the set of all non-failing leaves of an SLD-tree of $A \leftarrow A$. A failing leaf is a selected atom which does not unify with any clause of the program.*

The assumption that A is an atom can be relaxed to allow a finite set of atoms, $\mathbf{A} = \{A_1, \ldots, A_m\}$, and a partial deduction of $\mathbf{A}$ in P is the union of partial deductions of $\{A_1, \ldots, A_m\}$ in P. These partial deductions are called *residual predicates*.

DEFINITION 6 (PARTIAL DEDUCTION OF A PROGRAM) *A partial deduction of P with respect to $\mathbf{A}$ is a program P' (also called a residual program) obtained from P by replacing the set of clauses in P whose heads contain one of the predicate symbols appearing in $\mathbf{A}$ with a partial deduction of $\mathbf{A}$ in P.*

A condition restricting arbitrary partial deductions to those which intuitively are "specializations" has to be imposed to guarantee completeness and, for the case of normal programs, even soundness of partial deduction.

DEFINITION 7 (CLOSEDNESS) *Let S be a set of first order formulae and $\mathbf{A}$ a finite set of atoms. S is called $\mathbf{A}$-closed if each atom in S containing a predicate symbol occurring in an atom in $\mathbf{A}$ is an instance of an atom in $\mathbf{A}$.*

2.3 Correctness and Completeness of Partial Deduction

For the procedural semantics partial deduction is correct and complete. These concepts are defined as follows. Let P be a definite program, G_0 a definite atom, and P' a partial deduction of P wrt to G_0.

DEFINITION 8 (CORRECTNESS) $P \cup \{G\}$ *has an SLD-refutation with computed answer θ, if $P' \cup \{G\}$ does.*

Completeness is the converse:

DEFINITION 9 (COMPLETENESS) $P' \cup \{G\}$ *has an SLD-refutation with computed answer θ, if $P \cup \{G\}$ does.*

The partial deduction theorem is formulated using the closedness condition.

THEOREM 1 *Let P be a program, G a goal, $\mathbf{A}$ a finite set of atoms, and P' a partial deduction of P wrt (with respect to) $\mathbf{A}$ (using SLD-trees):*

  i) P' is sound wrt P and G

ii) If $P' \cup \{G\}$ is A-closed, then P' is complete wrt P and G

A version of the lifting lemma that generalizes the standard lifting theorem for logic programming to partial derivation trees is used in the proof of the theorem.

LEMMA 1 (LIFTING) *Let A be an atom and ψ a substitution. Let R be a resultant obtained from an SLD-derivation D for the goal $\leftarrow A$. If there is a corresponding derivation D' for $\leftarrow A\psi$ then its resultant R' is an instance of R.*

LEMMA 2 (CORRESPONDING DERIVATIONS) *Let A and A' be atoms, P a program and $\leftarrow A$ a goal. Let R be the resultant associated with a derivation D for the goal $\leftarrow A$. Finally, let $\leftarrow A'$ be a goal such that A' unifies with the head of the resultant R with θ being the corresponding mgu. We put $R'' = R\theta$.*

i) There exists a derivation D' for A' corresponding to D that selects at each step a corresponding atom and that uses variants of the same clauses as derivation D. Moreover, if R' is the resultant associated with D' then R' is an instance of R''.

ii) If the atom A' is an instance of A, then there exists a corresponding SLD-derivation and its resultant $R' = R''$.

Given these lemmata, the proof that partial deduction is correct follows almost immediately. The proof of completeness is similar, but it additionally needs a version of the switching lemma. The interested readers are referred to [34].

Dropping the closedness condition leads to incompleteness. Consider the following program consisting of two clauses: $P = \{p \leftarrow q(a)., q(X).\}$ and the wrt set $\mathbf{A} = \{q(b)\}$. A residual program that is not A-closed is $P'_{q(b)} = \{p \leftarrow q(a)., q(b).\}$. Now, $P' \cup \{\leftarrow p\}$ does not have a refutation, while $P \cup \{\leftarrow p\}$ does.

2.4 Results for normal programs

An incorporation of the condition of *independence* leads to a version of Theorem 1 for normal programs.

DEFINITION 10 (INDEPENDENCE) *Let $\mathbf{A}$ be a finite set of atoms. $\mathbf{A}$ is independent if no pair of atoms in $\mathbf{A}$ have a common instance.*

THEOREM 2 *Let P be a normal program, G a normal goal, $\mathbf{A}$ a finite, independent set of atoms and P' a partial deduction of P wrt (with respect to) $\mathbf{A}$ using SLD-trees such that $P' \cup \{G\}$ is A-closed. Then the following hold.*

i) $P' \cup \{G\}$ has an SLD-refutation with computed answer θ iff $P \cup \{G\}$ does.

ii) $P' \cup \{G\}$ has a finitely failed SLD-tree iff $P \cup \{G\}$ does.

The results for normal programs can be extended to derivations using SLDNF-trees.

THEOREM 3 *Let P be a normal program, G a normal goal, **A** a finite, independent set of atoms, and P' a partial deduction of P wrt (with respect to) **A** using SLD-trees such that P' ∪ {G} is **A**-closed. Then the following hold.*

  i) P' ∪ {G} has an SLDNF-refutation with computed answer θ iff P ∪ {G} does.

  ii) P' ∪ {G} has a finitely failed SLDNF-tree iff P ∪ {G} does.

The independence condition cannot be omitted. Let $P = \{p(X) \leftarrow \neg q(X)\}$ and $\mathbf{A} = \{p(X), p(a)\}$, and let $P'_{\mathbf{A}} = \{p(X) \leftarrow \neg q(X)., p(a).\}$ be a residual program wrt **A**. Then $P' \cup \{\leftarrow p(X)\}$ is **A**-closed and has a refutation with computed answer $\{a/X\}$. The derivation of $P \cup \{\leftarrow p(X)\}$ flounders, however.

2.5 Results for declarative semantics

Results for declarative (model-theoretic) semantics are weaker. Let $comp(P)$ denote Clark's completion.

THEOREM 4 *Let P be a normal program, W a closed first order formula, **A** a finite set of atoms, and P' a partial deduction of P wrt **A** such that P' ∪ {W} is **A**-closed. If W is a logical consequence of comp(P'), then W is a logical consequence of comp(P).*

This theorem has an immediate consequence:

THEOREM 5 *Let P be a normal program, G a normal goal, **A** a finite set of atoms, and P' a partial deduction of P wrt **A** such that P' ∪ {G} is **A**-closed. Then the following hold.*

  i) If θ is a correct answer for comp(P') ∪ {G} then θ is a correct answer for comp(P) ∪ {G}.

  ii) If θ is a computed answer for comp(P') ∪ {G} then θ is a correct answer for comp(P) ∪ {G}.

  iii) If P' ∪ {G} has a finitely failed SLDNF-tree, then G is a logical consequence of comp(P).

The proofs of the theorems 2, 3, 4 and 5 can be found in [34].

3 Relative merits of Partial Deduction

Partial deduction in logic programming is a generalization of an unfold transformation method and instantiation. Its additional strength comes from the ability of unification to propagate data structures via derivation trees to resultants and, consequently, to new definitions. This is a form of data flow analysis. It was called *forward* and *backward data structure propagation* in my thesis [25].

The "unfold" and "fold" transformations were originally introduced by Burstall and Darlington [9]. Their goal was to create a system for 'computer-aided programming'. Clark [11] and Hogger [22] showed that such transformations provide an additional advantage when applied to logic programs, since the transformed program is logically implied by the original program. However, such transformations (both in the functional and logic framework) are only *partially correct*. One of the remaining problems was total correctness. Theorem 1 shows that partial deduction is indeed totally correct. Informally, the residual program terminates if and only if the original program does. There is an important difference, however. Partial deduction specializes programs, while program transformation is concerned with obtaining equivalent programs. The relationship between partial deduction and program transformation is discussed in detail by Proietti and Pettorossi in [38].

Another reason explaining the power of partial deduction seems to be related to the properties of the high-level language of logic programs. It encourages writing complex, but lucid and relatively easy to simplify programs. Meta-linguistic abstractions are well-supported and the overhead of meta-interpretation can be removed by partial deduction.

The most important advantage of partial deduction is that more efficient programs can be generated. Efficiency gains for a partially deduced program can be discussed in terms of the size of derivation trees. One efficiency factor is the length of derivation. The length of derivation is usually shortened since input clauses in a residual program are resultants. At the same time, pruning, or elimination of the failing branches reduces the number of alternatives or the range of the derivation tree.

We also note that residual clauses are more specific, in the sense that variables in the original clauses may become instantiated to non-variable terms. One advantage is that unification can be performed at earlier stages of derivations when using more specific clauses. This implies that failed derivations can be shorter. Another advantage is that compilers and tools which analyze Prolog programs often work better with procedures with instantiated arguments than with procedures with variables which are instantiated in the bodies. (cf $fact(0,1)$. which is preferred to $fact(X,Y) \leftarrow X = 0, Y = 1$.) For instance, if heads are more specific then compilers can generate better control information.

Despite its relative merits, partial deduction needs to be extended if it is to cope satisfactorily with the combinatorial explosion of the generated code and to handle recursion in some cases. Since this issue is related to the termination problem, no general criterion can be provided. The best we can hope for are (correct) heuristics.

4 Partial Deduction in Computing

The principle of partial deduction is strikingly simple. Interpreted as a principle of logic it computes approximations of a consequence operator starting with the identity transformation and ending in a model of the theory (if this is possible). In the latter case, if applied to Prolog, it computes facts that follow from the program. In spite of its simplicity PD has found several applications in computational logic and in several complex problems of computing. A possible explanation is that logic has recently made strong inroads in computing, and thus applying partial deduction has become possible.

4.1 Towards automatic synthesis of software

Notwithstanding relative merits of partial deduction, a few problems have to be solved before it can be effectively used. For example, partial deduction as defined above does not include folding-like transformations, and some important results known in the case of transformation of functional programs can only be obtained if folding is included (e.g. a transformation from an exponential to a linear Fibonacci function).

Another problem is that the knowledge that is required to answer the question when to unfold or fold is very complex and rather hard to establish. Hence, it is difficult to build a sophisticated autonomous system that would understand partial deduction and the program being specialized, that could trade space for time, and that would know when to prune the derivation tree. In fact, the problem of deriving new recursive definitions for folding is generally unsolvable, although algorithms for some folding cases have been designed [37].

Rather than attempting to design an autonomous partial deduction system, we defined two tactics, opening and abbreviating, that let the user indicate where and how partial deductions can be performed [27]. They were proven correct in [29]. Opening corresponds to unfolding, and abbreviating to folding. In this paper abbreviating is slightly modified to account for non-recursive abbreviations and to exclude an introduction of new local variables.

4.2 Opening and Abbreviating

The first tactic uses the information encoded in the structure of specifications, which consist of the abstract data types part and of the logic part of the specification. In informal terms, the atoms which can be unfolded are the ones which define an implementation of the abstract data types used by the specification. The first mention of opening can be found in [30].

DEFINITION 11 (OPENING OF PREDICATES) *Let* A *be a finite, independent set of atoms,* G *a definite goal, and* P *a definite program where* $P = P_1 \cup S$ *for some disjoint* P_1 *and* S*, where* S *is a set of definitions of some predicate symbols. We call* P_1 *a* source*, and* S *an* axiomatization*. Let* P' *be a partial deduction of* P *wrt* A *such that* $P' \cup \{G\}$ *is* A*-closed, where the predicates of the atoms selected by the computation rule are the predicates of the heads of the clauses in* S*. Such a partial deduction is called* opening of predicates S in P wrt A.

Remark Opening is trivially correct by the partial deduction theorem.

Informally, the purpose of opening is to identify where partial deduction steps should take place. This is in contrast to other approaches to partial deduction that advocate processing of entire programs rather than their fragments. Although efficiency gains do not necessarily have to be very big, at times they can be substantial. Consider a very simple example of a parser for the language defined by $S \rightarrow aSb|c$. Let the source and axioms defining the parser be:

EXAMPLE 4 *A parser*

```
nont(T, R) ← ta(T, V), nont(V, R1), tb(R1, R).      % Source
nont(T, R) ← tc(T, R).

ta([a | Es], Es).      % Axioms
tb([b | Es], Es).
tc([c | Es], Es).
```

Opening the axioms in the source wrt $\{nont(T, R)\}$ leads to the following program:

EXAMPLE 5 *Residual parser*

```
nont([a | Es0], R0) ← nont(Es0, [b | R0]).
nont([c | Es0], Es0).
```

Comment The initial program would make $2n+1$ calls. The residue will need only $n + 1$ calls and one unification of a list of length n. In most cases this is more efficient.

Another example is the so called McCarthy's verification problem of finding the fringe of a binary tree, where the fringe of a tree is defined to be a list of the leaf nodes of the tree in the left to right order. Partial deduction helps obtain automatically an inductive schema defining a program linear in time.

EXAMPLE 6 *Fringe*

```
fringe(leaf(L), LeafList) ← make_list(L, LeafList).      % Source
fringe(tree(L, R), Leaves) ←
   fringe(L, LeftFringe), fringe(R, RightFringe),
   conc(LeftFringe, RightFringe, Leaves).

make_list(Elt, [Elt | Es] - Es).      % Axioms
conc(X - Y, Y - Z, X - Z).
```

The result of opening the axioms wrt $\{fringe(T, L)\}$ is:

EXAMPLE 7 *Fringe: residue*

```
fringe(leaf(L), [L | Rest] - Rest).
fringe(tree(L, R), LeftFringe - RightFringe) ←
   fringe(L, LeftFringe - Rest), fringe(R, Rest - RightFringe).
```

This synthesis requires a degree of "eureka" since the axioms have to be suggested. (Another possible selection could be standard lists. This would lead to a quadratic algorithm.) Note that the use of difference lists does not create any problem here.

More interesting than the efficiency improvements is the prospect of an emerging support for representing design choices. Namely, assume that there is a library of implementations of abstract data-types, control structures and other constraints (all called axiomatizations). The design process consists in explicit selection of a particular axiomatization such as a data-type. Partial deduction, using the opening heuristics, automatically constructs a program that (efficiently) uses the data-type (see [28]). Such choices of an axiomatization of a data-type are similar to the issues of automatic implementation of abstract data-types.

The second tactic is a generalization of instantiation and folding that was introduced by [9]. Folding in the case of logic programming was first defined in [45], and investigated later in [41]. Roughly speaking, the tactic is concerned with the logic of a specification and is called abbreviating.

We introduce first some notational conventions. Let $\widehat{C_k}$ denote a conjunction of literals $C_1, \ldots, C_k$, that is, let $\widehat{C_k} = C_1, \ldots, C_k$, for $k \geq 0$. In this notation clause $H \leftarrow B_1, \ldots, B_n$ is written $H \leftarrow \widehat{B_n}$.

DEFINITION 12 (ABBREVIATING) *Let P be a program, $\mathbf{A}$ a finite, independent set of atoms, $H \leftarrow \widehat{L_k}$ a (standardized apart) clause of P, and $A\sigma \leftarrow \widehat{L'_k}, \widehat{B_n}$ a resultant in an A-closed SLDNF-derivation of $P \cup \{\leftarrow A\}$. Finally, let θ be a substitution such that $\widehat{L'_k} = \widehat{L_k}\theta$, with $H\theta$ not unifiable with any other head of a clause of P. If opening $H\theta \leftarrow \widehat{L_k}\theta$ in $A\sigma \leftarrow H\theta, \widehat{B_n}$ produces a variant of $A\sigma \leftarrow \widehat{L'_k}, \widehat{B_n}$ then $A\sigma \leftarrow H\theta, \widehat{B_n}$ is the result of abbreviating $\widehat{L'_k}$ with H and is called* abbreviated resultant.

Partial deduction can be correctly extended with abbreviating steps in derivations due to the following corollary.

Corollary Abbreviated resultant is entailed by P if and only if the corresponding non-abbreviated resultant does, that is, resultant $A\sigma \leftarrow \widehat{L'_k}, \widehat{B_n}$ is entailed by P if and only if resultant $A\sigma \leftarrow H\theta, \widehat{B_n}$ does.

The use of abbreviating is illustrated with the following example.

EXAMPLE 8 *Specification of sort*
```
srt(X, Y) ← perm(X, Y), ord(Y).     % Source
perm([], []).        % Axioms
perm([A | X], Y) ← perm(X, Z), ins(A, Z, Y).
```

Opening the axioms wrt $\{srt(X, Y)\}$ leads to:

EXAMPLE 9 *Sort: 1st residue*

```
srt(□, □) ← ord(□).
srt([A0 | X0], Y0) ← perm(X0, Y1), ins(A0, Y1, Y0), ord(Y0).
```

An invariant $ins(A0, Y1, Y0), ord(Y0) \models ord(Y1)$ is added to the body of the second resultant:

EXAMPLE 10 *Sort: add invariant*

```
srt(□, □) ← ord(□).
srt([A0 | X0], Y0) ←
   perm(X0, Y1), ord(Y1), ins(A0, Y1, Y0), ord(Y0).
```

We notice that $perm(X0, Y1), ord(Y1)$ is an instance of the source (cf Example 8). This enables abbreviating which produces the final resultants:

EXAMPLE 11 *Sort: abbreviated resultants*

```
srt(□, □) ← ord(□).
srt([A00 | X00], Sorted0) ←
   srt(X00, Y0), ins(A00, Y0, Sorted0), ord(Sorted0).
```

This example was originally presented (with a minor mistake) in [45]. The abbreviated resultants give a $O(n^3)$ algorithm as opposed to $O(n!)$ of the specification.

4.3   Generation of compiled code from interpreters

One of the particularly fruitful applications of PD (and partial evaluation in functional programming) is generation of compiled code from meta-interpreters. It is known as the *1st Futamura projection* [17]. In this application, a meta-interpreter is specialized with respect to an object program. The resulting residual program has the functionality of the interpreter embedded in it, but the extra level of interpretation is removed. The following example is chosen to illustrate this and some other ideas.

EXAMPLE 12 *Compilation of a meta-interpreter*

```
solve(□).        % Source: meta-interpreter
solve([G] ) ← reduce([G | Body]), solve(Body).
solve([G1, G2 | Gs]) ← solve([G1]), solve([G2 | Gs]).

reduce([tc(X, Y), r(X, Y)]).    % Axioms: object program
reduce([tc(X, Z), r(X, Y), tc(Y, Z)]).
reduce([r(a, b)]).
reduce([r(b, c)]).
```

The object program defines a recursive program for transitive closure. Let $A = \{solve([tc(X, c)]), solve([r(X, Y)])\}$. By opening the object program in the meta-interpreter wrt to A the following residue is produced.

EXAMPLE 13 *Residual meta-interpreter*

```
solve([tc(X0, c)]) ← solve([r(X0, c)]).
solve([tc(X0, c)]) ← solve([r(X0, X1), tc(X1, c)]).
solve([r(a, b)]).
solve([r(b, c)]).
```

A syntactic transformation gives an object program.

EXAMPLE 14 *Projecting the residue to the object level*

```
stc(X0, c) ← sr(X0, c).
stc(X0, c) ← sr(X0, X1), stc(X1, c).
sr(a, b).
sr(b, c).
```

In general, such a renaming is not a trivial transformation. For a discussion of this issue see, for instance, [18]. Takeuchi and Furukawa provided the first example of partially deducing a meta-interpreter [44].

4.4 Other applications

Other applications of partial deduction include refinement calculi for specifications, optimization of deductive database queries, machine learning, knowledge and inheritance, planning, etc.

A form of machine learning, as defined by the Explanation-Based Generalization, can be expressed in terms of partial deduction [21]. There is still some discussion whether the equation EBG=PD is correct or not. Let us remark that it is rather difficult to compare the two notions, since EBG is described only through informal definitions and examples. In any case, we have performed an experiment in which all published examples of this form of learning were translated to logic programming and partially deduced. Furthermore, we used PAL, our PArtiaL deduction environment to run mechanical versions of these experiments. Opening (and abbreviating) were sufficient to regenerate all the solutions [35].

Example 12 shows that partial deduction is applicable to deductive databases. A particularly elegant application is by F. Bry, who showed that a Magic Set can be obtained by partial deduction of a Backward Fix-point Procedure and the intensional and extensional database [8].

5 Partial Deduction and Prolog

The theoretical results on partial deduction can be applied to Prolog and to the Prolog family of languages. The theorems 2 and 3 are proven for all computation rules used in synthesizing a residual program P'_G from P. However, it may be so that $P' \cup \{G\theta\}$ has a refutation using a different computation rule than $P \cup \{G\theta\}$.

Unsafe computation rules, such as Prolog's first literal, must be excluded if applied to negative literals. Since all computation rules are equivalent for

SLD-resolution, the results for definite programs and the success part of the theorems are better. One computation rule can be used to obtain a residual program, another one for derivations from the residual program, and a third for derivations with the initial program. In some cases the results hold for fixed computation rules. If Prolog's computation rule is used in all the three cases, the results hold also for the failure part of the theorems. The results will hold as well, if any fair rules, not necessarily the same ones, are used.

For Prolog's depth-first search it is necessary to preserve the order of resultants. They must be ordered according to the lexicographic order of the clauses used in their derivations.

The results on partial deduction can be potentially applied to the Prolog family of languages such as Parlog, Concurrent Prolog, GHC, etc.

5.1 Implemented Systems

Problems with partial deduction of full Prolog programs, i.e. programs that contain side-effects, are numerous. They are due to the non-logical constructs and to the built-in procedures of Prolog. Since the non-logical and built-in constructs are usually imperative, the name partial *evaluation* is kept for the case of full Prolog, while partial deduction is maintained for logic programming and pure Prolog.

There are several implemented systems for Prolog. Lack of space in this article does not allow their review here. The reader is referred to [32] for an evaluation of some of the earlier systems. A more recent is Dahlin's Mixtus [40]. The PRINCE project lead by J. Gallagher is concerned with practical applications of partial deduction. His systems follows Lloyd and Shepherdson's formalization.

Our system, PAL, also follows this formalization. It adds opening, abbreviating and external predicates. A particular aim of PAL is to provide a support for sequences of synthesizing steps [31]. PAL is an LPA-MacProlog implementation of an environment for partial deduction. It runs under MacProlog 3.5 and is available for distribution. We note that all of the examples used in this paper were generated in PAL.

6 Research issues

As it was indicated earlier (Section 4.1), there are several difficult issues in partial deduction. Many of them are connected to control, others to the failure-oriented semantics.

There are two termination problems. Both of them are concerned with unfolding: one is terminating recursive unfolding, and the other − deciding when to terminate any unfolding. The issues are generally undecidable and can be only solved by heuristics (that must be proven correct). For the second issue, and even without recursive definitions, unleashed partial deduction will produce combinatorially many resultants. A related problem is where to start partial deduction.

It is likely that the first termination problem of partial deduction will profit from the research on the general termination of logic programs such as [1], [39] and [43]. In fact, R. Bol has recently applied a loop checking mechanism to termination of partial deduction, [5]. See also a tutorial on logic program termination by D. De Schreye [42].

In my thesis I took a conservative approach to the second problem and suggested processing the so called *deterministic procedures*. (A deterministic procedure is a procedure that has only a one-clause definition.) The opening tactic that was defined earlier in this paper allows the programmer to indicate which definitions are to be processed thereby providing some additional control. It is an open problem to define heuristics for finding such definitions. In [28] I describe an approach in which the first synthesis is done by the programmer and the following ones reuse the synthesis by replacing axioms which define data structures by equivalent ones.

A related problem is that of prohibiting a failure and a subsequent pruning. If in Example 1 the selection rule had been defined so that *mother/2* could be chosen then most of the tree would have been pruned. *grandfather(george, jan)* would be the only resultant. This is clearly undesirable in incremental program synthesis because definitions may be supplied later. We solved this problem in PAL by letting the user to indicate which definitions are external to the current program.

Several similar ideas were investigated independently by Benkerimi and Lloyd who contributed to solving the problem of termination, combinatorial explosion and failure-avoidance in their work on a procedure for partial deduction [3].

For completeness of partial deduction it is required that derivations be closed. Consider the problem of finding a residue for the transitive-closure meta-interpreter (Example 12). It is natural to begin with a wrt set $A = \{solve([tc(X, c)])\}$. In that case the residue will contain only the first two resultants of (Example 13), and the condition of closedness will not be satisfied. One has to add $solve([r(X, Y)])\}$ to A. Natural questions are an efficient choice of the new wrt set and incrementality of partial deduction.

Partial deduction is related to partial evaluation in general (e.g. [16, 14, 10]), and to partial evaluation of Lisp (e.g. [2, 20, 13]) and functional languages in particular (e.g. [23, 19]). For a collection of articles on partial evaluation see [4].

Although some issues in partial deduction (e.g. extensions to the wrt set) are of a different nature than in partial evaluation, there is a need for a better communication of results between the two communities. Self-applicability has been solved in the case of partial evaluation [6], but solutions for partial deduction [15, 7] seem to need further work. It would be interesting to reformulate Futamura projections [17] for partial deduction.

Abstract interpretation of logic programs [12] is an important tool that should enhance partial deduction.

Partial deduction has been defined so far for logic programming and, in particular, using an Abstract Prolog Machine (i.e. operational semantics) [25, 26]. It was reformulated later in terms of another operational (i.e. proof-theoretic) semantics [34]. These semantics are based on SLD- and SLDNF-resolution. It is interesting to investigate other semantics, such as well-founded semantics, and other logical formalisms for partial deduction.

I believe that another fruitful area will be an investigation of partial deduction as an approximating operator for computing consequences (a closure) of a theory. More applications are likely to be found in machine learning and deductive databases. The primary examples are: certain forms of machine learning that are concerned with making explicit what already belongs to the closure of a theory, recursive query answering mechanisms in deductive databases that find specializations for faster processing, and intensional query answering (cf a resultant is an intensional answer). Related issues are belief revision, negation and non-monotonic partial deduction.

Although partial deduction is applicable to other languages in the Prolog family, there are specific issues that need a thorough investigation, for instance, the commit operator in concurrent Prologs, or constraint solving and partial deduction in the CLP languages.

I have been particularly interested in the application of partial deduction to software synthesis. One topic for discussion is the relationship between synthesis by partial deduction and the proofs-as-programs methodology. Partial deduction is correct and complete, and constructive in some sense, but it does not guarantee termination in the same way as the other method. Are the differences only due to non-determinism that is inherent to logic programming? There seem to be fine points here which are worth investigation.

Acknowledgments

Thanks are due to the organizers of the META'92 Workshop on Meta-programming for inviting me to present this material. John Gallagher, Maurizio Proietti and an anonymous reviewer were helpful with their suggestions for improvements. Alberto Pettorossi is kindly acknowledged for his insightful comments, encouragement, and support while writing the paper. Agata and Jacek Wrzos-Kamińscy read several drafts of the paper and indicated several betterments. And, finally, thanks to Anna.

References

[1] K. Apt and D. Pedreschi. Proving termination of genreal prolog programs. In *Proceedings of the International Conference on Theoretical Apsects of Computer Science*, Sendai, 1991.

[2] L. Beckman et al. A partial evaluator, and its use as a programming tool. *Artificial Intelligence*, 7(4):319–357, 1976.

[3] K. Benkerimi and J. Lloyd. A Procedure for the Partial Evaluation of Logic Programs. TR-89- 04, Department of Computer Science, University of Bristol, 1989.

[4] D. Bjørner, A. Ershov, and N. Jones, editors. *Partial Evaluation and Mixed Computation. Proceedings of the IFIP TC2 Workshop, Gammel Avernæs, Denmark, October 1987.* North-Holland, 1988. 625 pages.

[5] R. Bol. Loop checking in partial deduction. CS-R 9134, Center for Mathematics and Computer Science, Amsterdam, 1991. To appear in the Journal of Logic Programming.

[6] A. Bondorf. *Self-Applicable Partial Evaluation.* PhD thesis, DIKU, University of Copenhagen, Denmark, 1990. Revised version: DIKU Report 90/17.

[7] A. Bondorf, F. Frauendorf, and M. Richter. An experiment in automatic self-applicable partial evaluation of Prolog. Technical Report 335, Lehrstuhl Informatik V, University of Dortmund, West Germany, 1990. 20 pages.

[8] F. Bry. Query Evaluation in Recursive Databases: Bottom-up and Top-down Reconciled. In *Proc. 1st Int. Conf on Deductive and Object-Oriented Databases,* December 1989. Kyoto, Japan.

[9] R. M. Burstall and J. Darlington. Some transformations for developing recursive programs. *Journal of ACM,* 24(1):44–67, 1977.

[10] T. E. Cheatham, G. H. Holloway, and J. A. Townley. Symbolic evaluation and analysis of programs. *IEEE Transactions on Software Engineering,* 5(4), 1979.

[11] K. Clark. Negation as failure. In H. Gallaire and J. Minker, editors, *Logic and Data Bases,* pages 293–322. Plenum Press, 1978.

[12] P. Cousot and R. Cousot. Abstract interpretation and application to logic programming. Rapport de Recherche 8, Ecole Polytechnique, Laboratoire d'Informatique, 91128 Palaiseau Cedex France, 1992.

[13] P. Emanuelson and A. Haraldsson. On compiling embedded languages in Lisp. In *1980 Lisp Conference, Stanford, California,* pages 208–215, 1980.

[14] A. Ershov. On the essence of compilation. In E. Neuhold, editor, *Formal Description of Programming Concepts,* pages 391–420. North-Holland, 1978.

[15] H. Fujita and K. Furukawa. A self-applicable partial evaluator and its use in incremental compilation. *New Generation Computing,* 6(2,3):91–118, 1988.

[16] Y. Futamura. Partial evaluation of computation process – an approach to a compiler-compiler. *Systems, Computers, Controls,* 2(5):45–50, 1971.

[17] Y. Futamura. Program evaluation and generalized partial computation. In *International Conference on Fifth Generation Computer Systems, Tokyo, Japan*, pages 1–8, 1988.

[18] J. Gallagher and M. Bruynooghe. Some low-level source transformations for logic programs. In M. Bruynooghe, editor, *Proceedings of the Second Workshop on Meta-Programming in Logic, April 1990, Leuven, Belgium*, pages 229–246. Department of Computer Science, KU Leuven, Belgium, 1990.

[19] C. Gomard and N. Jones. A partial evaluator for the untyped lambda-calculus. *Journal of Functional Programming*, 1(1):21–69, January 1991.

[20] A. Haraldsson. *A Program Manipulation System Based on Partial Evaluation*. PhD thesis, Linköping University, Sweden, 1977. Linköping Studies in Science and Technology Dissertations 14.

[21] F. van Harmelen and A. Bundy. Explanation-Based Generalisation=Partial Evaluation. *Journal of Artificial Intelligence*, 36:401–412, 1988.

[22] C. J. Hogger. Derivation of logic programs. *Journal of ACM*, 28(2):372–392, 1982.

[23] N. Jones, P. Sestoft, and H. Søndergaard. Mix: A self-applicable partial evaluator for experiments in compiler generation. *Lisp and Symbolic Computation*, 2(1):9–50, 1989.

[24] S. Kleene. *Introduction to Metamathematics*. D. van Nostrand, Princeton, New Jersey, 1952.

[25] J. Komorowski. *A Specification of An Abstract Prolog Machine and Its Application to Partial Evaluation*. PhD thesis, Department of Computer and Information Science, Linköping University, Linköping, 1981.

[26] J. Komorowski. Partial evaluation as a means for inferencing data structures in an applicative language: a theory and implementation in the case of Prolog. In *Proc. of the ACM Symp. Principles of Programming Languages*, pages 255–267. ACM, 1982.

[27] J. Komorowski. Towards synthesis of programs in the framework of partial deduction. In *Proc. of the Workshop on Automating Software Design, XIth International Joint Conference on Artificial Intelligence*. Kestrel Institute, August 1989.

[28] J. Komorowski. Elements of a programming methodology founded on partial deduction - part 1. In Z. Ras, editor, *Proc. of the Int. Symp. on Methodologies for Intelligent Systems*. North Holland, October 1990. Knoxville, Tennessee.

[29] J. Komorowski. Towards a programming methodology founded on partial deduction. In *Proc. of the European Conference on Artificial Intelligence*. Pitman Publ. Co., August 1990.

[30] R. Kowalski. *Logic for Problem Solving*, volume 7 of *Artificial Intelligence Series*. North Holland, 1979.

[31] J. Lahtivuori. An Environment for Partial Deduction and Its Use for Structuring Logic Programs. Master's thesis, Åbo Akademi University, 1990. In Swedish.

[32] J. Lam and A. Kusalik. A partial evaluation of partial evaluators for pure prolog. TR-90 1, Department of Computational Science, University of Saskatchewan, 1990.

[33] J. W. Lloyd. *Foundations of Logic Programming*. Springer Verlag, second edition, 1987.

[34] J. W. Lloyd and J. C. Shepherdson. Partial evaluation in logic programming. *Journal of Logic Programming*, 1991. (also, TR-87-09, Dept. of Comp. Sc., U. of Bristol).

[35] Å. Mæhle. Explanation-based learning and partial deduction. In *Proc. of Norsk Informatikk Konferanse*, pages 175–188, Trondheim, Norway, 1991.

[36] U. Nilsson and J. Małuszyński. *Logic, Programming and Prolog*. John Wiley and Sons, 1990.

[37] A. Pettorossi and M. Proietti. Decidability results and characterization of strategies for the development of logic programs. In *Proc. of the International Conf. on Logic Programming*, Lisabon, Portugal, 1989. MIT Press.

[38] A. Pettorossi and M. Proietti. The loop absorption and the generalization strategies for the development of logic programs and partial deduction. In J. Komorowski, editor, *Special Issue of the Journal of Logic Programming on Partial Deduction*. North-Holland, 1992. to appear.

[39] L. Plümer. *Termination proofs of logic programs*, volume LNCS 446 of *lncs*. Springer Verlag, C, 1990.

[40] D. Sahlin. *An Automatic Partial Evaluator for Full Prolog*. PhD thesis, Kungliga Tekniska Högskolan, Stockholm, Sweden, 1991. Report TRITA-TCS-9101, 170 pages.

[41] T. Sato. An equivalence preserving first order unfold/fold transformation system. In H. Kirchner and W. Wechler, editors, *Algebraic and Logic Programming, Second International Conference, Nancy, France, October 1990. (Lecture Notes in Computer Science, vol. 463)*, pages 173–188. Springer-Verlag, 1990.

[42] D. De Schreye. Termination of logic programs: Tutorial notes. In A. Pettorossi, editor, *Proceedings of the Workshop on Metaprogramming*, Uppsala, 1992. Springer Verlag. to appear.

[43] D. De Schreye, K. Verschaetse, and M. Bruynooghe. A framework for analysing the termination of definite logic programs. In *Proceedings of FGCS92*, 1992. to appear.

[44] A. Takeuchi and K. Furukawa. Partial evaluation of Prolog programs and its application to meta programming. In H.-J. Kugler, editor, *Information Processing 86, Dublin, Ireland*, pages 415–420. North-Holland, 1986.

[45] H. Tamaki and T. Sato. Unfold/fold transformation of logic programs. In S.-Å. Tärnlund, editor, *iclp*, pages 127–138, 1984.

[46] D. S. Warren. Memoing for logic programs. In *Special Issue of the CACM on Logic Programming*. ACM, March 1992.

Tutorial on Termination of Logic Programs

Danny De Schreye, Kristof Verschaetse

Department of Computer Science, Katholieke Universiteit Leuven
Celestijnenlaan 200A, 3001 Heverlee, Belgium
email: dannyd@cs.kuleuven.ac.be

Abstract. We present a general introduction to termination analysis for logic programs, with focus on universal termination of SLD-derivations and on definite programs. We start by providing a generic definition of the termination problem. It is parametrised by the sets of goals and the sets of computation rules under consideration. We point out a distinction between two streams of work, each taking a different approach with respect to the undecidability of the halting problem. We then recall the notions of recurrency and acceptability from the works of Apt, Bezem and Pedreschi. We illustrate how these notions provide an elegant framework for reasoning about termination. We then identify four basic components that are present in any approach to termination analysis. We point out the interdependencies between these components and their relevance for the termination analysis as a whole. We also use these components to illustrate some differences between automatic approaches to termination analysis and the more theoretically oriented frameworks for termination.

1. Introduction

An immediate observation that one can make when going through the literature on termination of logic programs is that there exists a variety of different definitions of termination. This can be related to three causes:

1. Logic programming represents not one single language, but a family of programming languages, each of which has a different operational semantics. More specifically, there are potentially infinitely many different *computation rules* (subgoal selection) and *selection rules* (clause selection). Each choice gives rise to a different operational behaviour and can therefore have different termination properties.

2. The various degrees of freedom for the use of a given program. We assume that a *program* is a theory of Horn clauses which does *not* include the goal. The degrees of freedom relate to the various *modes* in which a program can be called and to the different *types* of data it uses. The typical example is append/3, which can be used to concatenate two lists, or alternatively, to split up a given list into two sublists. Moreover, the members of these lists may take various types and the lists could be either of fixed length or non-fixed length (open ended). In general, termination of a program may depend on all these choices.

3. Logic programs can be nondeterministic. This creates a choice between either considering *universal termination*, where we are interested in the termination of the entire computation (after *all* solutions have been generated), or *existential termination*, where the computation should either terminate with failure or should return *at least one* solution (after which it may, or may not enter an infinite computation).

In this paper we restrict our attention to universal termination (see [24] for comments on existential termination). This immediately implies that the choice of the selection rule (point 1.) becomes irrelevant as well: Given any program P, goal G and computation rule R, the condition that all derivations for (P, G) under R should be finite is completely independent from the applied selection rule.

On the level of the computation rule, let **CR** denote the set of all possible computation rules. In principle, we could be interested in the termination properties of a pair (P,G) under any element of **CR**. More generally, given any subset CR_0 of **CR**, we could be interested in finding the set of pairs (P,G) which terminate under every computation rule R in CR_0. In the literature, two instances of this general formulation have attracted specific attention. The first is the case where CR_0 = **CR**. Here, we are interested in universal termination *under any computation rule*, which could truly be regarded as termination of a *logic program*. The second is the case where CR_0 is the singleton {LD}, where LD denotes the left-to-right computation rule of Prolog. The termination properties studied under this second case are usually referred to as *left-termination*.

Regarding the choice of the set of goals for which one aims to prove termination (point 2.), there is much more divergence in the literature. A simple approach is to investigate termination of a program for a single goal. More frequently, however, we are interested in the termination properties of an entire set of goals, all of which have some common characteristics (e.g. all atomic goals with a same predicate symbol and for which a fixed argument position has a ground value). So, denoting the set of all possible goals as G and given a program P, we aim to investigate whether all pairs (P,G), $G \in S$, $S \subset G$, are (left-)terminating.

In this paper, we further restrict our attention to atomic top-level goals. We denote the set of all such goals by **AG**.

More formally, termination can now be characterised as follows:

Definition 1.1 Given a definite program P, a subset S of **AG** and a subset CR_0 of **CR**, denote

$$SLD(P, S, CR_0) = \{\tau \mid \tau \text{ is the SLD-tree for (P,G) under R,}$$
$$\text{for some } G \in S \text{ and some } R \in CR_0\}$$

The program P *terminates with respect to S and* CR_0, denoted Terminates(P, S, CR_0), iff all SLD-trees in SLD(P, S, CR_0) are finite.

The parameter S in this definition gives rise to many different notions of termination. Some choices for S taken in the literature are:

1. S is the set of all ground atomic goals.

   — In combination with $CR_0 = CR$, this corresponds to the notion of termination presented in [5] and - in the context of normal programs - [1].

   — For left-termination, $CR_0 = \{LD\}$, it corresponds to the notion dealt with in [2] and - for normal programs - [3].

2. Given a predicate symbol p/n and a set of argument positions for p/n, a_1, a_2,..., a_m, $1 \leq a_i \leq n$, let S be the set of all atomic goals with predicate symbol p/n and such that their terms on the argument positions a_1, a_2,..., a_m are ground. This corresponds to the notions for termination dealt with in [22], [17] and [18]. In the case of Plümer, $CR_0 = \{LD\}$. The work of Ullman and Van Gelder also assumes that CR_0 is a singleton, but in this case, the single computation rule is nonstandard and automatically produced by the NAIL-system.

3. Generalisations of the above cases 1. and 2., where groundness is replaced by either *boundedness* or *rigidity* with respect to some measure function (see Sections 2 and 3 for definitions of these notions). Such generalisations are considered in the works of Apt, Bezem and Pedreschi mentioned above (in fact, here termination for ground goals and for bounded goals are shown to be equivalent). Generalisations involving rigidity are considered in [7], [8] and [19].

4. S is any set of atomic goals ([13] and [25]).

Of course, *termination analysis* is not concerned with the termination of one specific triplet (P, S, CR_0). If we denote the set of all possible definite programs for a given (countable) first order language by **P** and the power set of any set X by D(X), termination analysis addresses the problem:

Given CR_0 and a function $F: \mathbf{P} \rightarrow D(D(AG))$, where F(P), for any $P \in \mathbf{P}$, is the set of all sets $S \in D(AG)$ the particular analysis technique is designed for, define the relation Terminates/2 on $\{(P,S) \mid P \in \mathbf{P}, S \in F(P)\}$, by Terminates(P, S) = Terminates(P, S, CR_0).

To clarify the role of the F-function, in the context of the second type of sets S presented above (Ullman, Van Gelder, Plümer), we could define F(P) = $\{\{\leftarrow p(t_1,...,t_n) \mid t_{i_1},..., t_{i_m}$ are ground terms$\} \mid$ p/n is a predicate symbol in P and $1 \leq i_j \leq n\}$.

It is well-known that for all interesting, nontrivial functions F the predicate Terminates/2 is undecidable, so that no complete (returning the value "true" for all terminating pairs) and safe (always terminating) procedure computing its truth-values can be provided. For certain instances of the general formulation, it is of course semi-decidable. In particular, if each F(P) consists of a single goal and $CR_0 = \{R\}$, merely executing each pair (P, F(P)) under R is a complete, but in general not safe, semi-deciding procedure. Often, in termination analysis, one is more interested in providing procedures which are safe but not complete. In view of these observations, work in termination analysis can take one of the following two approaches:

- Provide an incomplete definition for the Terminates/2 relation, which only expresses a sufficient condition for termination, but which can be automatically verified. In what follows, this approach will be referred to as *automatic termination analysis*.

- Provide a complete definition, expressing a necessary and sufficient condition for termination. The point of this approach is that, in addition to being undecidable, the definition of the Terminates/2 relation given above is even hard for humans to reason about. This is due to the fact that we need to analyse potentially infinite sets of SLD-trees. The idea is to provide equivalent, but more practical conditions, that can more easily be verified manually.

  Another potential use for approaches of this second type is that they could serve as theoretical frameworks on which an automatic termination analysis can be based. In particular, certain parts of such frameworks could involve decidable properties and they could directly be included in an automatic approach, while other parts could be replaced by decidable, but weaker conditions. Approaches of the second type will therefore be referred to in what follows as *frameworks* for termination analysis. Unfortunately, there are very few frameworks available which actually provide support of this kind.

Yet another approach to termination is the following one: given some function F and a set of computation rules CR, find an (interesting) subset P_0 of P, such that Terminates/2 becomes decidable if it is restricted to the domain $\{(P,S) \mid P \in P_0, S \in F(P)\}$.

In [14] such a decision procedure is given for the case where P_0 consists of all definite programs with one single *binary* (of the type $p(s_1,..., s_n) \leftarrow p(t_1,..., t_n)$) clause and one single *linear* (no variable occurs twice) fact and for $F(P) = \{\{\leftarrow A\} \mid A$ is a linear atom with predicate symbol in $P\}$.

In the next section, we briefly describe one particular approach to termination analysis as an example and an introduction to how the problem can be handled. In the third section we present an overview of the considerations that are needed when dealing with the termination analysis. On the basis of this overview, we point out various choices that can be made for the basic components of the analysis and we relate some existing techniques on the basis of the actual choices made within them.

As already remarked, we restrict our attention to definite programs. Some comments on the treatment of normal programs and on the relation between termination analysis and e.g. stratification, are included in the discussion.

2. Recurrency and acceptability

In this section, we briefly present the approaches proposed in [5] and [2] and we illustrate them by an example. These methods are instances of the *framework*-type approach.

Our simple example is the well-known permutation program.

Example 2.1

perm([], []).
perm([X1|X], [Y1|Y]) ← delete(Y1, [X1|X], Z), perm(Z,Y).

delete(X1, [X1|X], X).
delete(Y1, [X1|X], [X1|Z]) ← delete(Y1, X, Z).   ∎

As mentioned in Section 1, both [5] and [2] address termination analysis with respect to the set of all ground goals. Thus, for each $P \in P$, $F(P)$ is the singleton with as its single element the set of all ground atomic goals with predicate symbols in P. [5] focuses on termination ($CR_0 = CR$), [2] on left-termination. Both rely on the following notion of a *level mapping* (see also [10]). Let us denote the Herbrand Base of the first order language underlying a given program P as B_P.

Definition 2.2 A *level mapping* is a function $|.|: B_P \to IN$.

The key concept in [5] is *recurrency*. It is the specialisation of the concept of *acyclicity* to definite programs.

Definition 2.3 A definite program P is *recurrent* if there exists a level mapping, $|.|$, such that for each ground instance $A \leftarrow B_1,..., B_n$ of a clause in P, $|A| > |B_i|$, for each i=1,...,n.

Interpreting B_P as a set of atomic goals instead of atoms, one of the main results in [5] can now be reformulated as follows:

Theorem 2.4 Given a definite program P, P is recurrent iff Terminates(P, B_P, CR).

Notice that the notion of recurrency assigns a measure function to the program. This measure is not directly related to the SLD-trees in SLD(P, B_P, CR), but to the ground instances of clauses in P. As a result, reasoning about the recurrency condition is easier than dealing with the termination problem itself, because the syntactic structure of the clauses in P is fixed and finite. Still, it trivially follows from Theorem 2.4 that recurrency remains an undecidable property. This is reflected in 1) the actual level mapping that must be provided, 2) the potentially infinite number of ground instances of clauses in P that need to be checked.

Let us now define a measure function which is useful for generating level mappings in many cases. Let $Term_P$ denote the set of all terms in the first order language underlying to P.

Definition 2.5 The *list-length norm* is a function listl: $Term_P \to IN$, defined as:

listl($[t_1 | t_2]$) = 1 + listl(t_2), with t_1 and t_2 any terms,
listl(t) = 0   otherwise.

Example 2.6 First, consider the delete/3 procedure of the permutation program. Let $|.|$ be the level mapping defined as:

$|$delete(t_1, t_2, t_3)$|$ = listl(t_2), for any $t_1, t_2, t_3 \in Term_{perm}$.

To prove recurrency, take any ground instance of the second clause for delete/3, say:

$$\text{delete}(s_1, [t_1|t], [t_1|r]) \leftarrow \text{delete}(s_1, t, r).$$

We have that:

$$|\text{delete}(s_1, [t_1|t], [t_1|r])| = \text{listl}([t_1|t]) > \text{listl}(t) = |\text{delete}(s_1, t, r)|$$

so that delete/3 is recurrent and therefore terminating with respect to B_{perm} and CR.

The permutation program as a whole is not recurrent. Proving non-termination through non-recurrency is often difficult, since one then needs to prove that no level mapping can satisfy the recurrency condition. So, in this case, we directly show that the program is non-terminating instead.

Consider the ground goal $\leftarrow$ perm([1], [1,2]). After one derivation step, we obtain the new goal $\leftarrow$ delete(1, [1], Z), perm(Z, [2]). Assume that the computation rule selects perm(Z, [2]). A next derivation step, using the second clause for perm/2, gives rise to the goal

$$\leftarrow \text{delete}(1, [1], [Z1 \mid Z2]), \text{delete}(2, [Z1| Z2], Z3), \text{perm}(Z3, [])$$

with binding Z = [Z1 | Z2]. Finally, assume that the computation rule now selects the atom delete(2, [Z1| Z2], Z3) and, in following derivation steps, its direct descendents. It should be clear that this includes an infinite derivation.

∎

As a final remark on the recurrency condition, observe that, although the set of all ground goals is very important in the context of declarative semantics, it is not particularly useful in programming practice. [5] shows that Theorem 2.4 can be extended to more practical sets as follows.

Definition 2.7 Let P be a definite program and |.| a level mapping. An atom A is *bounded with respect to* |.| if the set $\{|A\theta| \mid \theta$ a grounding substitution for A$\}$ is bounded in $\mathbb{N}$. Given P and |.|, we denote the set of all bounded atoms in the language underlying to P and with respect to |.| as $Bound_{P,|.|}$.

Theorem 2.4 can be generalised as follows.

Theorem 2.8 Given a definite program P, if P is recurrent with respect to a level mapping |.| then Terminates(P, $Bound_{P,|.|}$, CR). Conversely, if Terminates(P, $Bound_{P,|.|}$, CR) for some level mapping |.|, then P is recurrent.

Notice that this is not formulated as an iff statement. This is because in the second implication, P is not necessarily recurrent *with respect to* |.|.

Example 2.9 For the permutation example, we may now conclude that delete/3 terminates for any goal $\leftarrow$ delete(t_1, t_2, t_3), such that t_2 is a list of fixed length (but not necessarily ground).

Observe that in Example 2.6 we could also have defined the level mapping as:

$$|\text{delete}(t_1, t_2, t_3)| = \text{listl}(t_3), \text{ for any } t_1, t_2, t_3 \in Term_{perm}.$$

With this level mapping, delete/3 is also recurrent. However, now we find that delete/3 terminates for all goals $\leftarrow$ delete(t_1, t_2, t_3), such that t_3 is a list of fixed length. In fact, by defining the level mapping as the sum of the two level mappings above, we get termination for the union of the two corresponding sets.

∎

In [2] the results of [5] are reformulated in the context of left-termination. Here, the basic concept replacing recurrency is *acceptability*.

Definition 2.10 A definite program P is *acceptable* if there exists a level mapping |.| and a model I for P, such that for each ground instance $A \leftarrow B_1,..., B_n$ of a clause in P, $|A| > |B_i|$, for each $i = 1,..., n$, such that $I \models B_j$, for all $j = 1,..., i-1$.

Intuitively, the definition expresses that, for ensuring termination, the level mapping only needs to decrease between the head of the ground instance of the rule and a corresponding body atom, B_i, if all atoms to the left of B_i already follow from the model. If one of these atoms to the left of B_i would not follow from the model then resolution under the LD computation rule would never reach the point in which B_i is selected. The refutation would fail before that. So, there is no reason to impose that the level mapping should also decrease for such atoms.

The following theorem rephrases Theorem 2.4 in the context of left-termination. For the case of bounded goals, we refer to [2].

Theorem 2.11 A definite program P is acceptable iff Terminates(P, B_P, {LD}).

Example 2.12 We show acceptability of perm/2. The level mapping is defined as follows. Let perm(t_1, t_2) and delete(t_1, t_2, t_3) be any atoms in B_{perm}:

$$|perm(t_1, t_2)| = listl(t_1)+1,$$
$$|delete(t_1, t_2, t_3)| = listl(t_2)$$

For the model I, let the domain of the interpretation be $\mathbb{N}$. The pre-interpretation, is determined by the function J: $Term_{perm} \rightarrow \mathbb{N}$, $J(t) = listl(t)$. This means that J maps every constant in $Term_{perm}$ to 0, the list-functor to the function $(x,y) \rightarrow y+1$ and every other functor in the language to the constant 0-function. Finally, I(perm/2) is the binary predicate on $\mathbb{N}$ which is true everywhere and I(delete/3) is the relation

$$\{(X, Y, Z) \in \mathbb{N}^3 \mid Y = Z+1 \}.$$

Observe that I is a model for the permutation program. For the perm/2-predicate this is obvious, since I(perm/2) is true everywhere. In the case of delete/3, for all ground terms t_1, t_2, t_3 such that t_3 is the list obtained from the list t_2 by deleting its member t_1, we have that $listl(t_2) = listl(t_3)+1$.

Since the only nontrivial clause for delete/3 has only one atom in its body, the definition of acceptability of delete/3 with respect to the level mapping |.| and model I above coincides with the definition of recurrency with respect to |.|. This, we already proved in Example 2.6.

Next, we consider the recursive clause for the perm/2 predicate. There are two inequalities that we need to prove for it. The first is that for any ground terms t_1, t, s_1, s and r:

$$|perm([t_1|t], [s_1|s])| > |delete(s_1, [t_1|t], r)|$$

This reduces to $listl([t_1|t])+1 > listl([t_1|t])$, which is true. The second inequality is that for any ground terms t_1, t, s_1, s and r

$$|perm([t_1|t], [s_1|s])| > |perm(r,s)|$$

should hold, given that I $|=$ delete(s_1, $[t_1|t]$, r). This reduces to $listl([t_1|t])+1 > listl(r)+1$, given that $listl([t_1|t]) = listl(r) + 1$. Again, this is clearly the case, so that perm/2 is acceptable and left-terminating.   ■

3. Basic components of termination analysis

In this section we describe four components which are (sometimes implicitly) present in almost any termination analysis. We use them to point out the major differences between different existing approaches.

3.1 Well-founded partially ordered sets

An SLD-tree can be regarded as a *strictly partially ordered set*, denoted s-poset in what follows. The nodes of the tree form a set. The tree structure defines a strict partial order on this set. Under this ordering, for any two nodes N_1 and N_2, respectively labeled by goals G_1 and G_2, $N_1 < N_2$ iff G_1 descends from G_2 in the SLD-tree.

An s-poset, [A,>], is *well-founded* iff it contains no infinite descending sequence $a_1 > a_2 > a_3 > ...$

Obviously, all SLD-trees in SLD(P, S, CR_0) are finite iff they are well-founded under their natural tree-ordering (by König's lemma, a finitely branching tree is finite iff all its branches are finite).

So, the starting point of any termination analysis is to try to provide evidence for well-foundedness of each SLD-tree in the set. Although formulated in terms of SLD-trees here, this approach is not typical for logic programs. Well-foundedness is commonly used in computer science as a basic concept for proving termination of programs (see e.g. [16], [11]).

Note that the reason why we use the notion of well-founded sets, instead of just *finite sets* to model termination, is that we are usually not interested in proving finiteness of a single SLD-tree, but, given some SLD(P, S, CR_0) we are interested in finiteness of every SLD-tree in it. Now, the union of all the nodes in these trees still forms an s-poset under the strict ordering inherited from the individual SLD-trees. Even if every SLD-tree in SLD(P, S, CR_0) is finite, this s-poset is not necessarily finite. However, all SLD-trees are finite iff the s-poset associated to SLD(P, S, CR_0) is well-founded.

A technique to prove the well-foundedness of a given s-poset [A,>], is to provide a *monotonic* function, f: [A,>] → [W,>$_W$], to some s-poset [W,>$_W$], which is known to be well-founded. An obvious candidate for $[W, >_W]$ are the natural numbers, $[\mathbb{N}, >]$, with the usual ordering. Due to monotonicity of the mapping f: $[A, >] \to [\mathbb{N}, >]$, the s-poset $[A, >]$ inherits well-foundedness from $[\mathbb{N}, >]$.

As a result, termination techniques make use of some type of measure function defined on some syntactical structures (goals, atoms or terms) related to the nodes of the SLD-trees and taking values in the natural numbers. In [5] and [2], these measure functions are the *level mappings*.

A first main distinction between automatic approaches and formal frameworks is situated on the type of measure functions that are considered. In the automatic approaches, only functions whose values can be computed on the basis of syntactical properties of their input goals are used. In formal frameworks, no such restriction exists and the definition of the measure function (e.g. level mapping) may rely on other sources of information, such as for instance the semantical properties of the program under consideration (see e.g. the GAME example in [3]).

In automatic approaches, measure functions are usually defined on the basis of *norms*.

Definition 3.1 A norm is any function $\|.\|$: $\text{Term}_P \to \mathbb{N}$.

Usually, norms are assumed to be invariant under variable renaming. This can be formalised by defining them on the extended Herbrand universe underlying to P, $U^E{}_P$, defined as $\text{Term}_P/_\sim$, where $\sim$ is equivalence under variable renaming, instead of on Term_P itself.

Typical examples of norms that are often used in practice are *list-length*, listl (see the previous section), and *term-size*, defined as:

Definition 3.2

$\quad\quad$ termsz$(f(t_1, \cdots, t_n)) = 1 + \Sigma_{i=1,n}$ termsz(t_i) with f any function symbol and n>0,

$\quad\quad$ termsz$(x) = 0$ otherwise.

Norms give rise to measure functions defined on goals by considering, for each given goal, a predefined set of argument positions of atoms in the goal and by computing the sum (or some linear combination) of the norms of the terms occurring in these argument positions.

Of particular interest are *semi-linear* norms, defined in [8].

Definition 3.3 A norm $\|.\|$ is *semi-linear* if it is recursively defined by means of the following schema:

$\quad\quad \|V\| = 0$ if V is a variable, and
$\quad\quad \|f(t_1,..., t_n)\| = c + \|t_{i_1}\| + ... + \|t_{i_m}\|$ with $c \in \mathbb{N}$, $i_j \in \{1,...,n\}$
$\quad\quad\quad\quad\quad\quad$ and c, i_1, ..., i_m only dependent of f/n.

Both list-length and term-size are semi-linear norms. An exception is *term-depth*, which computes the length of the longest branch in the tree-representation of a term. The relevance of semi-linear norms relates to the notion of *rigidity* (see [7]).

Definition 3.4 Let $\|.\|$ be a norm and $t \in \text{Term}_P$. We say that t *is rigid with respect to* $\|.\|$ if for any substitution θ, $\|t\theta\| = \|t\|$.

The fact that the norm of a term is invariant under substitution is important. Deciding on the *size* of a nonground term occurring in some derivation would otherwise become difficult, since its size could change in subsequent derivation steps. Rigid terms do not suffer from this problem. Bounded atoms (see the previous section) are more general and serve the same purpose. Although these are not invariant under substitution, subsequent instantiation of a bounded atom can only cause its size to grow up to a finite upper bound. Therefore, for bounded atoms, we can still compare the sizes of different nonground atoms e.g. by comparing the lub's of the sizes for all their ground instances. In fact, it is precisely by taking the lub over all its ground instances that the value of a level mapping on a bounded atom is defined.

Of course, in general it may be impossible to decide which terms are rigid (the definition above is particularly impractical). This is why semi-linear norms are introduced. Terms that are rigid with respect to a semi-linear norm can be syntactically characterised. Essentially, a term is rigid with respect to a semi-linear norm $\|.\|$ iff recursively decomposing the term over the relevant argument positions $i_1,...,i_m$ for its principle functor (see Definition 3.3), does not produce any variables. In other words, no variable can occur at any (nested) relevant argument position of the given norm (see [8] for a more formal treatment).

Example 3.5 Let $f/3$ and $g/2$ be the functors in the language underlying to P. Let $\|.\|$ be the semi-linear norm

$\|V\| = 0$ if V is a variable, and
$\|f(t_1, t_2, t_3)\| = 1 + \|t_1\| + \|t_3\|$
$\|g(t_1, t_2)\| = \|t_1\|$
$\|a\| = 0$
$\|b\| = 0$

Then, $g(f(a, g(X,Y), g(b, X)), Z)$ is rigid (the norm of any of its instances is 1), while $g(f(X, a, b), c)$ is not (taking $X = f(a, a, a)$ its norm is 2; taking $X = a$ its norm is 1). ∎

To conclude the section, we briefly comment on the measure functions used in the other techniques mentioned in Section 1. The method of Ullman and Van Gelder ([22] is inherently restricted to the use of the list-length norm. [17] and [18] use *linear* norms. These are a special case of semi-linear norms, with term-size as their most important example. List-length however, is not linear. In [19], Plümer extends his techniques to semi-linear norms.

In [12] and [25] so called *natural* level mappings are induced from semi-linear norms. There is a special focus on *type*-norms, which are a refinement of the term-size norm, taking information obtained by type inference (through abstract interpretation) of the given program into account.

3.2 Propagation of the set of goals

Recall that the S parameter in Definition 1.1 presents the set of all (atomic) goals for which termination has to be proved. Since nontermination of pure logic programs can only be caused by recursive procedures, proving termination for S can only be achieved through a termination proof for all goals recursively descending from elements of S, by applying clauses of P. As an example, given $S = \{\leftarrow p(t_1, t_2) \mid t_1$ is ground$\}$, $CR_0 = CR$ and a program P containing the clause

$$p(X, Y) \leftarrow q(X), r(X, Y).$$

as the definition for p/2, we will need to prove termination for (P, S_1, CR) and for (P, S_2, CR), where $S_1 = \{\leftarrow q(t) \mid t$ is ground$\}$ and $S_2 = \{\leftarrow r(t_1, t_2) \mid t_1$ is ground$\}$.

This implies that we need some method for predicting the way in which the initial set of goals S gives rise to new sets of goals, say S_i, at lower levels in the SLD-trees. This is referred to as the *propagation* of S.

To relate this (on an intuitive basis) to the previous subsection, the measure functions are essentially used to monitor that at each derivation step (or more generally, at each predefined finite sequence of derivation steps) some amount of input data is consumed, and that, as a result, the measure function decreases. In this context, prediction of the sets S_i is needed to ensure that at lower levels in the SLD-trees, such input data will still be available.

Some approaches have been proposed in the literature to compute the sets S_i. It should be clear however, that if the set S can be expressed through the abstract properties of its elements (e.g. through modes, types or combinations of both), rather than through enumeration of its elements, then the adequate tool for computing all descending S_i is abstract interpretation over an abstract domain that captures the properties of interest. In particular, [22], [17] and [18] rely on a simple mode analysis. [12] uses abstract interpretations over various abstract domains: modes, rigid types and integrated types (see [15]), depending on the required precision.

[8] and [19] use the very general notion of pre- and post-conditions for calls, which can in principle capture any syntactical property of the terms occurring in the call. In these approaches, the pre-and post-conditions are assumed to be given in advance.

At first sight, the recurrency and acceptability approaches of the previous section seem to form an exception with respect to this part of the analysis: No explicit mention of any propagation is made. Here however, the result of propagation implicitly follows from the approach as a whole. Given a program P which is recurrent (or acceptable) with respect to some level mapping |.| and a (|.|)-bounded atomic) goal ←A for P, it is straightforward to prove that any other atom occurring in any SLD-tree in SLD(P, {←A}, CR_0) is also bounded. So, in these approaches all sets S_i are identical to S and no explicit propagation analysis is needed.

3.3 Syntactic structures for proving well-foundedness

No termination analysis applies the measure function and the propagation of S directly to the actual SLD-trees to obtain the termination proof. In fact, this is one of the main differences between (compile-time) termination analysis and loop-avoidance, as it is for instance applied in partial deduction (see [6] and [9]).

Instead, a simpler syntactic structure is introduced and it is proved that if (and -for frameworks- only if) the measure function, with respect to the propagations of S, satisfies certain properties on these simple syntactic structures, then all SLD-trees in SLD(P, S, CR_0) are finite.

In the case of recurrency, these syntactic structures are the clauses of the given program. Here, taking the propagation of S into account corresponds to checking a condition for all ground instances of these clauses (although this is not very precise, since the propagation for ground atomic goals results in bounded descending goals, but -due to technicalities beyond the scope of this tutorial- verification of the recurrency condition for all ground instances of clauses entails the "recurrency" for all bounded instances). Finally, it needs to be proven that the actual condition on these structures -in the case of recurrency, $|A| > |B_i|$ for every body-atom B_i of the ground instances $A \leftarrow B_1,..., B_n$ of the clauses- implies finiteness of all SLD-trees.

There is one main difference between the way recurrency and acceptability deal with this part of the analysis and the way other techniques approach it. The other techniques do not impose a decrease of the measure function for every atom in the body of a clause, but instead, impose some form of decrease between each call to a recursive predicate and each of its descending recursive calls. To simplify the discussion, we restrict our attention to programs which only involve direct recursion. With directly recursive programs, we mean programs for which the transitive closure of the *predicate dependency* relation is anti-symmetric. Furthermore, for the moment we restrict our attention to termination (CR_0 = CR), leaving left-termination for the next subsection.

The syntactical structures are again the clauses of the program. Let |.| be the measure function, defined on atoms. The termination condition can then be formulated as:

- for all $p(r_1, ..., r_n)$ in the propagation of S (in other words, all atoms which occur in a goal of some SLD-tree for an atom of S),

- and, for each recursive clause, say $p(t_1, ..., t_n) \leftarrow B_1, ..., B_n$,
  such that $\theta = mgu(p(r_1, ..., r_n), p(t_1, ..., t_n))$ exists

- and, for every body-atom with the same predicate symbol, say $B_i = p(s_1, ..., s_n)$:

$$|p(r_1, ..., r_n)\theta| > |p(s_1, ..., s_n)\theta| .$$

Now, assume that the measure function |.| only takes into account those argument positions of predicate symbols p/n such that the terms occurring on these argument positions in the propagation of S are always *rigid* with respect to the measure function. As a result, for any atom A in the propagation and any substitution θ,

$|A\theta|=|A|$. Then every SLD-tree in SLD(P, S, CR_0) is finite, because the value of the measure function decreases at every recursive call in every derivation.

Example 3.6 Consider again the delete/3 predicate in the permutation program. Assume that $S = \{delete(t_1, t_2, t_3)|\ t_2$ is a list of fixed length$\}$. Observe that in this case the propagation of S is identical to S (which can easily be inferred by type-analysis). As in Example 2.12, take $|delete(t_1, t_2, t_3)| = listl(t_2)$, for any delete($t_1, t_2, t_3$) in S. Notice that this level mapping is rigid with respect to the elements of S.

The only recursive clause is:

delete(Y1, [X1|X], [X1|Z]) ← delete(Y1, X, Z).

Taking, $\theta = mgu(delete(t_1, t_2, t_3), delete(Y1, [X1|X], [X1|Z]))$, we have:

$$|delete(t_1, t_2, t_3)| = listl(t_2) = listl(t_2\theta) = listl([X1|X]\theta)$$
$$= listl(X)\theta+1 > listl(X)\theta) = |delete(Y1, X, Z)\theta|\ .$$

So, delete/3 terminates with respect to the set S. ∎

Note that the above example is too simple to illustrate all differences with the recurrency approach. In particular, the body of the recursive clause does not contain any atoms with a predicate symbol which differs from the predicate symbol of the head. For such atoms, as opposed to recurrency, the latter approach does not impose a decrease of the measure function.

On the other hand, the latter approach is restricted to directly recursive programs, which is not the case for recurrency. For indirectly recursive programs, the condition formulated above is not sufficient for proving termination.

A detailed discussion of the many issues involved in dealing with indirect recursion and of the approaches proposed for solving them is outside the scope of this tutorial. We only provide some basic intuitions.

It is well known that there are several transformations that allow to transform indirectly recursive programs into equivalent directly recursive ones. In particular, the transformed program has the same termination properties as the original one. Observe that mere unfolding is insufficient to achieve this. As an example, consider the following program scheme:

p(...).
p(...) ← p(...), q(...).

q(...).
q(...) ← q(...), p(...).

Using unfolding as the only basic transformation step, the indirect recursion cannot be removed. Further transformations, introducing a new predicate are needed.

So, a simple approach to deal with indirect recursion is to eliminate the indirect recursion through transformation and then to apply the techniques of the directly recursive case (e.g. [17]). A problem with this approach is that, although the termination properties are preserved under the transformation, in practice, the measure functions needed to prove the termination of the transformed programs tend to be more complex. The reason is that the recursive structure of the transformed program does not necessarily keep pace with the data-consumption formulated in the original program. As a result, automatic techniques for termination analysis, which are based on simple measure functions defined in terms of list-length and term-size, are often unable to decide on the termination of such programs.

An alternative approach is taken in [8] and [12]. The idea here is to replace the simple syntactic construct of a program clause, used in the directly recursive case, by more complex syntactic constructs related to the indirect recursion. The constructs are finite sets of clauses that can be computed on the basis of the cycles in the predicate dependency graph or in the strongly related specific graph (see [8]). The disadvantage of these approaches is that the termination condition becomes much more complex. The advantage is that, even with very simple measure functions, termination proofs can automatically be generated for more programs.

To conclude this subsection, we return to the recurrency and acceptability conditions of Apt, Bezem and Pedreschi. Notice that in these approaches no special considerations for indirect recursion are needed. The reason is that a condition such as recurrency, where for each ground instance

$$A \leftarrow B_1, ..., B_n.$$

of a clause, $|A| > |B_i|$, *for all* i=1,...,n holds, fulfills three different functions at the same time:

1. It does impose that the measure function must decrease at each (directly) recursive call.

2. Indirectly recursive calls also obtain a decreased measure, since *all* atoms in the body have a decreased measure with respect to the head.

3. Due to that same decrease for each atom, descending calls are bounded (and no explicit propagation is needed).

The price for obtaining all that power from one simple condition is that straightforward measure functions, such as list-length or term-size, can seldom be used. Since the level mapping has to decrease on all the atoms in the body, these functions will in general not fulfill the recurrency condition. So, for each new example, some specially designed level mapping has to be provided. As an example, notice the summand +1 in the level mapping assigned to the perm/2 predicate in Example 2.12. In general these functions may become very complex. This is why we claimed earlier that frameworks provide insufficient direct support for automated techniques.

3.4 Dealing with existentially quantified variables

Recall that the discussion in the previous subsection is restricted to termination (CR$_0$=CR). This subsection deals with left-termination. Consider the clause:

$$\text{perm}([X1|X], [Y1|Y]) \leftarrow \text{delete}(Y1, [X1|X], Z), \text{perm}(Z, Y).$$

Assume that we aim to prove left-termination with respect to the set $S = \{\text{perm}(t_1, t_2)|\ t_1 \text{ is ground}\}$.

For left-termination, the propagation of a set S consists of all atoms that can occur as the left-most subgoal in any goal of an LD-tree for a goal in S. In the example, mode analysis is sufficient to derive that the perm/2 atoms in the propagation of S are precisely those of S itself. Thus, a good candidate to measure these atoms is the function $|\text{perm}(t_1, t_2)| = \|t_1\|$, where $\|.\|$ is some norm, i.e. list-length.

Now, assume that we would follow the same reasoning as explained in the previous subsection for directly recursive programs and for termination with respect to CR. For the above clause, one can notice that proving the inequality

$$\|[X1|X]\theta\| > \|Z\theta\|,$$

where θ is any mgu of an atomic perm/2 goal in S with the head of the clause, is impossible. The reason was already given in Example 2.6. The problem is caused by the variable Z, which does not occur in the head of the clause, and hence, does not get instantiated by applying the mgu.

However, since we consider left-termination, we can take advantage of the fact that Z gets instantiated after solving the intermediate delete/3 call. The recursive call will be $\text{perm}(Z,Y)\theta\sigma$, where σ is an LD-computed answer substitution for $\text{delete}(Y1,[X1|X],Z)\theta$. So, in order to prove left-termination, we need to prove the inequality $\|[X1|X]\theta\| > \|Z\theta\sigma\|$, for every computed answer substitution σ for $\text{delete}(Y1,[X1|X],Z)\theta$.

Now, note that due to soundness and completeness of SLD-resolution, σ is a computed answer substitution for some goal $\leftarrow \text{delete}(Y1, [X1|X], Z)\theta$ if and only if for any model I for delete/3 (or, more generally for the permutation program) we have that $I \models \text{delete}(Y1, [X1|X], Z)\theta\sigma$. In addition, if we consider only ground atoms and clauses, σ is a computed answer substitution if the atom $\text{delete}(Y1, [X1|X], Z)\theta\sigma$ is true in any model I for delete/3. This -again- explains why the desired condition in the case of acceptability is the inequality

$$|\text{perm}([t_1|t], [s_1|s])| > |\text{perm}(r,s)|$$

for any ground t_1, t, s_1, s and r, such that there exists a model I for the permutation program, with $I \models \text{delete}(s_1, [t_1|t], r)$.

In order to relate all this to automatic techniques, observe that we do not really need the information contained in the computed answer substitutions, or equivalently, in the substitutions for which the intermediate body-atoms follow from some model for the program. The only purpose we have for these substitutions, is to check some inequalities between natural numbers (in our example: $\|[X1|X]\theta\| > \|Z\theta\|$).

So, what we are actually interested in is a relation over the natural numbers, say delete_nat/3, such that $\text{delete\_nat}(n_1, n_2, n_3)$ holds for all triples of numbers obtained as the norms for the arguments of $\text{delete}(t_1, t_2, t_3)$, such that $I \models \text{delete}(t_1, t_2, t_3)$ for some model I.

Example 3.7 For the delete/3 example and using the list-length norm, one can easily verify that the relation is $\{(n_1, n_2, n_3) \in \mathbb{N}^3 \mid n_2 = n_3 + 1\}$.

∎

Such relations are usually referred to as *interargument relations* or *size relations* in the literature.

Existing automatic techniques differ essentially in the type of interargument relations they are capable of inferring. [22] considers relations of the form $n_i + c \geq n_j$, for some i,j and some $c \in \mathbb{Z}$, expressing that for each $p(t_1,..., t_m)$ in B_P, $listl(t_i)+c \geq listl(t_j)$ should hold. Hence, they essentially restrict the discussion to inequality relations between the sizes of exactly two arguments in a predicate.

[17] extends their technique, by considering relations of the form $\Sigma_{i\in I}\, n_i + c \geq \Sigma_{j\in J}\, n_j$, where I is a set of input argument positions and J a set of output argument positions.

Example 3.8 For the delete/3 example and the list-length norm, both Ullman & Van Gelder and Plümer would produce either the relation $n_2 -1 \geq n_3$, or the relation $n_3+1 \geq n_2$, depending on the mode in which the program is used.

∎

[26] uses relations over the natural numbers that can be obtained as the solution of a system of linear equations, with coefficients in $\mathbb{Z}$. This extends the approaches of Ullman & Van Gelder and Plümer, in the sense that linear combinations are used instead of sums and that the relation is defined as the solution of a system of linear expressions, instead of one such expression. Again, abstract interpretation is used to compute these systems of linear equations. In [23], a similar approach is taken, extending our approach to inequalities. Finally, in [8], the pre- and post-conditions we mentioned before are also used to solve the problem of existentially quantified variables. As before, this provides a very rich expressivity, but is hard to automate.

4. Discussion

This paper is intended as a tutorial, not as a complete survey of the work on termination analysis. Several approaches to termination have not been mentioned, some of which even fall outside the scope of the four-components treatment that we presented in the previous section. The most notable case is the work of Rao, Kapur and Shyamasundar ([20], [21]). This work is based on the observation that termination issues have received much more attention in the context of term rewrite systems than in logic programming. The authors take two approaches. In [20] they provide a transformation from logic programs to term rewrite systems and prove that for well-moded queries and programs, termination properties are preserved under the transformation. Then, they use termination conditions from term rewriting to ensure the termination of the logic programs. In [21], a more interesting approach is taken. Here, a start is made of directly translating some of the more powerful termination criteria for rewrite systems into a logic programming context. Further work in this direction seems promising.

Another approach which is not covered by our four components is that in [4]. This technique associates a termination theorem to a given program and then attempts to prove this theorem using theorem proving techniques.

Normal (also referred to as general) programs have only received limited attention in the literature so far. Most notable are the works of Apt, Bezem and Pedreschi, [1] and [3], generalising the notions of recurrency and acceptability to the context of normal programs.

A point which has not been addressed in this paper is the efficiency of the described automatic techniques. The techniques in [22] and [14] have polynomial complexity (the ones in [14] are even linear). This is not the case for those of Plümer and for our own techniques. In fact, polynomial complexity has been one of the main concerns in the work of Ullman & Van Gelder. As one might expect, efficiency decreases with the precision of the analysis. In particular, incorporating abstract analysis over refined abstract domains (e.g. integrated types) currently seems very expensive in terms of efficiency.

Finally, there is the relation between termination analysis and conditions imposed in the context of normal programs to guarantee the existence of a *perfect* model. The concept of recurrency (or more generally acyclicity) is very similar to the concept of local stratification. Essentially, the only syntactic difference is that in local stratification, for each ground instance $A \leftarrow B_1,..., B_n$ of a clause, $|A| > |B_i|$ should only hold if B_i is a negative literal, while $|A| \geq |B_i|$ is sufficient for the positive ones. As an immediate consequence, acyclicity (the generalisation of recurrency to normal programs) implies local stratification. The fact that local stratification imposes little or no restrictions on the positive literals in the body of a clause should not be surprising. Notions like stratification, local stratification and weak stratification have been introduced with the purpose of ensuring that a perfect model can be characterised as the fixpoint of a bottom-up immediate consequence-like operator. Termination of the bottom-up computation is not an issue here. It is sufficient that the fixpoint is reached at the ω-ordinal power. The only thing that is required is that atoms can be ordered in such a way that no two atoms rely on each other through negation. As a result, we obtain a condition which looks very similar to a termination condition, but it acts as if negation is the only "recursive predicate". This ensures that for any atom in the model, only a finite number of passes through negation are needed.

Acknowledgements

Danny De Schreye is supported by the Belgian National Fund for Scientific Research. Kristof Verschaetse is supported by the Belgian Diensten voor Programmatie van het Wetenschapsbeleid, under contract RFO-AI-02. We thank Alan Bundy, Bern Martens, Alberto Pettorossi and Maurizio Proietti for useful comments on draft versions of this paper and Krzysztof Apt for the interesting lectures on this subject he gave in Leuven.

References

1. K.R. Apt and M. Bezem, Acyclic programs, New Generation Computing, 9, 1991, pp. 335-363.

2. K.R. Apt and D. Pedreschi, Studies in pure Prolog: termination, in Proceedings of the Esprit symposium on computational logic, ed. J.W. Lloyd, 1990, pp. 150-176.

3. K.R. Apt and D. Pedreschi, Proving termination of general Prolog programs, in Proceedings International Conference on Theoretical Aspects of Computer Science, Sendai, Japan, 1991.

4. M. Baudinet, Proving termination properties of Prolog programs: a semantic approach, in Proceedings of the 3rd IEEE symposium on logic in computer science, Edinburgh, 1988, pp. 336-347. Revised version to appear in Journal of Logic Programming.

5. M. Bezem, Characterising termination of logic programs with level mappings, Proceedings NACLP89, eds. E.L. Lusk and R.A. Overbeek, 1989, pp. 69-80. Revised version will appear in Journal of Logic Programming.

6. R. N. Bol, Loop checking in partial deduction, Technical Report CS-R9134, Centre for mathematics and computer science, Amsterdam, 1991.

7. A. Bossi, N. Cocco and M. Fabris, Proving termination of logic programs by exploiting term properties, in Proceedings CCPSD-TAPSOFT '91, Springer-Verlag, LNCS 494, 1991, pp. 153-180.

8. A. Bossi, N. Cocco and M. Fabris, Norms on terms and their use in proving universal termination of a logic program, Technical Report 4/29, CNR, Department of Mathematics, University of Padova, March 1991.

9. M. Bruynooghe, D. De Schreye and B. Martens, A general criterion for avoiding infinite unfolding during partial deduction, in Proceedings ILPS'91, San Diego, 1991, MIT Press, pp. 117-131.

10. L. Cavedon, Continuity, consistency, and completeness properties for logic programs, in Proceedings ICLP89, eds. G. Levi and M. Martelli, 1989, pp. 571-584.

11. N. Dershowitz, Termination of rewriting, Journal of Symbolic Computation, 3, 1987, pp. 69-116.

12. D. De Schreye and K. Verschaetse, Termination analysis of definite logic programs with respect to call patterns, Technical Report CW 138, Department Computer Science, K.U.Leuven, January 1992.

13. D. De Schreye, K. Verschaetse and M. Bruynooghe, A framework for analysing the termination of definite logic programs, in Proceedings FGCS92, ICOT, 1992, pp.:481-488.

14. P. Devienne, Weighted graphs: a tool for studying the halting problem and time complexity in term rewriting systems and logic programming, Theoretical Computer Science, 75 (1&2), 1990, pp. 157-215.

15. G. Janssens and M. Bruynooghe, Deriving descriptions of possible values of program variables by means of abstract interpretation, Technical Report CW 107, Department of Computer Science, K.U.Leuven, March 1990. To appear in Journal of Logic Programming.

16. Z. Manna and S. Ness, On the termination of Markov algorithms, pp. 784-792 in Proc. 3rd Hawaii Int. Conf. on Syst. Sci., Honolulu, Hawaii (1970).

17. L. Plümer, Termination proofs of logic programs, LNCS 446, Springer-Verlag, 1990.

18. L. Plümer, Termination proofs for logic programs based on predicate inequalities, in Proceedings ICLP'90, Jerusalem, 1990, MIT Press, pp. 634-648.

19. L. Plümer, Automatic termination proofs for Prolog programs operating on nonground terms, in Proceedings ILPS'91, San Diego, 1991, MIT Press, pp. 503-517.

20. M. R. K. Krishna Rao, D. Kapur and R. K. Shyamasundar, A transformational methodology for proving termination of logic programs, in Proceedings Computer Science Logic, CSL91, 1991.

21. R. K. Shyamasundar, M. R. K. Krishna Rao and D. Kapur, Rewriting concepts in the study of termination of logic programs, in Proceedings ALPUK92, 1992.

22. J.D. Ullman and A. Van Gelder, Efficient tests for top-down termination of logical rules, J. ACM, 35(2), 1988, pp. 345-373.

23. A. Van Gelder, Deriving constraints among argument sizes in logic programs, in Proceedings 9th symposium on principles of database systems, ACM Press, 1990, pp. 47-60.

24. T. Vasak and J. Potter, Characterisation of terminating logic programs, in Proceedings 1986 symposium on logic programming, Salt Lake City, 1986, pp. 140-147.

25. K. Verschaetse, Static termination analysis for definite Horn clause programs, Ph.D.-thesis, K.U. Leuven, 1992.

26. K. Verschaetse and D. De Schreye, Derivation of linear size relations by abstract interpretation, in Proc. PLILP92, 1992, to appear.

Definable Naming Relations in Meta-level Systems

Frank van Harmelen

S.W.I., University of Amsterdam
e-mail: frankh@swi.psy.uva.nl

"Don't stand chattering to yourself like that," Humpty Dumpty said, looking at Alice for the first time, "but tell me your name and your business."
"My *name* is Alice, but —"
"It's a stupid name enough!" Humpty Dumpty interrupted impatiently. "What does it mean?"
"*Must* a name mean something?" Alice asked doubtfully.
"Of course it must," Humpty Dumpty said with a short laugh: "*my* name means the shape I am — and a good handsome shape it is, too. With a name like yours, you might be any shape, almost."

"Through the Looking Glass and What Alice Found There",
Lewis Caroll, 1871.

Abstract. Meta-level architectures are always, implicitly or explicitly, equipped with a component that establishes a relation between their object- and meta-level layers. This so-called *naming relation* has been a neglected part of the architecture of meta-level systems. This paper argues that the naming relation can be employed to increase the expressiveness and efficiency of meta-level architectures, while preserving known logical properties. We argue that the naming relation should not be a fixed part of a meta-level architecture, but that it should be *definable* to allow suitable encoding of syntactic information. Once the naming relation is definable, we can also make it *meaningful*. That is, it can also be used to encode pragmatic and semantic information, allowing for more compact and efficient meta-theories. We explore the *formal constraints* that such a definable naming relation must satisfy, and we describe a *definition mechanism* for naming relations which is based on term rewriting systems.

1 Introduction

An important property of any meta-level system is the connection between its object- and meta-level layers. A crucial aspect of this connection is what is known as the *naming relation*. A naming relation associates elements of the object-layer with their names in the meta-layer, and thereby allows the meta-layer to refer to, and express properties of, the elements of the object-layer.

Below, we will first discuss the naming relations that have been used (either implicitly or explicitly) in the literature (Sect. 2). We will argue that these naming relations have been used in a very limited way, and that they can be used more effectively if made *definable* (Sect. 3). Section 4 discusses the advantages of a definable naming relation, and Sect. 5 gives some examples of definable naming relations from different areas of AI. Naming relations cannot be defined arbitrarily, but must satisfy certain formal constraints which will be explored in Sect. 6. we describe a particular formalism based on term rewriting for constructing definable naming relations (Sect. 7). Comparison with related work (Sect. 8) concludes this paper.

Before we proceed a remark is in order on the scope of this paper. The discussion in this paper will be in the context of systems that use logic as their representation language: they consist of theories built out of logical expressions and are equipped with inference rules. However, many of the notions, arguments and conclusions of this paper can be easily generalised to meta-systems based on other types of languages, such as object-oriented languages [9], functional languages [12] or mixed representation languages [4].

2 Existing Naming Relations

As many important concepts in knowledge representation, the notion of a naming relation originates in work by logicians [13]. A naming relation is a mapping from syntactic constructs in the object-language $\mathcal{L}_O$ to variable free terms in the meta-language. $\mathcal{L}_M$ Through this mapping, object-level constructs become available for discourse in the meta-level theory. This makes it possible in the meta-level theory to quantify over object-level constructs while staying within a first order framework, without having to resort to second order logic to express properties of object-level predicates.

So far, we have been deliberately vague about the domain of the naming relation, and have described it with the term "object-level constructs". There is a considerable amount of freedom in choosing the domain of a naming relation. This choice will depend on what constructions we wish to predicate on in the meta-theory, but in general the domain of the naming relations will always consist of constructions over the elements of $\mathcal{L}_O$. In the simplest case, the domain of the naming relation is $\mathcal{L}_O$ itself, thus providing names for object-level sentences and terms. However, the domain can also be restricted to subsets of $\mathcal{L}_O$ (say only terms), or extended to other constructions over $\mathcal{L}_O$ such as object-level proofs (trees or sequences of object-sentences), object-theories (sets of object-sentences), etc.

Even though the domain of a naming relation can include any of these possibilities, for the purpose of examples in the remainder of this paper, we will often assume that the naming relation ranges over the elements of $\mathcal{L}_O$, although most of our arguments and conclusions will not crucially depend on this.

The co-domain of the naming relation must always be a set of variable free terms from $\mathcal{L}_M$. The fact that names are terms enables the meta-theory to express properties of object-level expressions while staying within a first order language. Names must be variable free in order to allow for the correct semantic interpretation for them in the model of a meta-theory (see Sect. 6 below).

Note that nothing in either the definition of the domain or co-domain of the naming relations excludes the possibility that we choose $\mathcal{L}_O \subset \mathcal{L}_M$, or even $\mathcal{L}_O = \mathcal{L}_M$. This possibility (discussed in [1]) allows for the famous self-referential constructions known from [5], but nothing in this paper depends on the equality or otherwise of $\mathcal{L}_O$ and $\mathcal{L}_M$.

Traditionally, two types of naming relations have been employed by logicians, starting again from [13], called *quotation-mark names* and *structural description names*. The simplest possible names are the quotation-mark names, which associate object-level expressions from $\mathcal{L}_O$ with atomic constants from $\mathcal{L}_M$. For mnemonic reasons, the name of an object-expressions $\phi \in \mathcal{L}_O$ is often written as the constant $\lceil \phi \rceil \in \mathcal{L}_M$, but the name can of course be any arbitrary constant in $\mathcal{L}_M$.

Structural descriptive names turn object-expressions into complex terms in the meta-level theory, where these terms reflect the syntactic structure of the object-level expressions. An example would be the name

$$\text{all(var1, all(var2, imply(2pred(p,var1,var2),2pred(p,var2,var1))))}$$

in $\mathcal{M}$ for the following sentence in $\mathcal{L}_O$, which expresses the commutativity of the predicate p:

$$\forall x \forall y [p(x, y) \rightarrow p(y, x)]$$

(Notice that the object-variables x and y are named by meta-constants $var1$ and $var2$ to obtain a ground term in $\mathcal{L}_M$, and that all predicates, logical constants and quantifiers of $\mathcal{L}_O$ are named by function symbols or constants of $\mathcal{L}_M$).

A large number of AI systems with a meta-level architecture have implemented a naming relation in the sense described above. For example, the Socrates knowledge representation system [8], directly implements a quotation-mark naming relation, whereas the logic programming systems Gödel [6] and Reflective Prolog [3] implement a structural descriptive naming relation

3 Naming Relations Should Be Definable

In all systems described in the literature (with the possible exception of FOL, but see Sect. 8 for more discussion on this system) the naming relation has been attributed an "external" status: In presentations by logicians, the naming relation is assumed to be fixed, and systems are studied with respect to different sets of axioms and rules of inference in $\mathcal{O}$ and $\mathcal{M}$, but no variation in the naming relation is ever considered. Similarly, in presentations of AI systems, the user or programmer is allowed to define the contents of $\mathcal{O}$ and $\mathcal{M}$ (and sometimes even the languages $\mathcal{L}_O$ and $\mathcal{L}_M$ and their rules of inference), but the naming relation is assumed to fixed, and not available for redefinition.

This state of affairs seems to be based on a lack of appreciation of the importance of the choice of the naming relation. It appears to be a neglected but nevertheless crucial fact that the nature of the naming relation determines for a large part the expressiveness of the meta-theory. As argued in [11], a meta-theory is *model relative*, which means that the expressiveness of a meta-theory depends on the model it has of the object-theory. A meta-theory can only refer to properties of object-expressions that are made visible on the meta-level through the naming relation, because it is the naming relation that determines for a large part the model that the meta-level has of the object-level theory. If, for example, we look at the following two naming relations:

$$object\text{-}sentence : p(f(x)) \wedge q(g(x,y)) \tag{1}$$

$$quotation\ name : \lceil p(f(x)) \wedge q(g(x,y)) \rceil \tag{2}$$

$$structural\ name : p'(f'(var1)) \wedge' q'(g'(var1, var2)) \tag{3}$$

(where $\wedge'$ is a binary function symbol of $\mathcal{L_M}$, written in infix notation), then it is clear that (3) is more expressive than (2), thereby allowing meta-theories to express properties which would otherwise not be expressible. An example is the statement which expresses that the commutativity of a given object-predicate p is provable in the meta-theory. Using the structural naming relation from (3), this statement can be expressed as:

$$\forall x \forall y [prove(p'(x,y)) \rightarrow prove(p'(y,x))] \tag{4}$$

but a similar statement would not be expressible using the quotation naming relation (2). The closest approximation that would be possible with this poorer naming relation is:

$$prove(\lceil \forall x \forall y [p(x,y) \rightarrow p(y,x)] \rceil)$$

which is different from (4) since it expresses that the commutativity of p is provable in the object-theory, whereas (4) asserts that this property of p is provable in the meta-theory. However, even though the structural names from (3) are more powerful than the atomic names from (2), they still restrict the expressibility of the meta-theory. For instance, they do not allow for quantification over object-level function symbols, since these are represented by meta-level function symbols, and this would require second order quantification in the meta-language. An alternative name of (1) is illustrated in the following example

$$alternative\ name : p'(func(f, [var1])) \wedge' q'(func(g, [var1, var2])) \tag{5}$$

which does allow quantification over object-level function symbols. For instance the sentence

$$\forall f \forall arg[\ldots func(f, [arg]) \ldots]$$

is quantified over all unary function-symbols. However, this naming relation does not allow for quantification over object-level predicate symbols. Ever more complicated naming relations can be invented, for instance allowing quantification over predicate symbols, which would make the name of (1) look like:

$$pred(p, [func(f, [var1])]) \wedge' pred(q, [func(g, [var1, var2])])$$

or even allowing quantification over logical constants, yielding:

$$logical\text{-}const(and, pred(p, [func(f, [var1])]), pred(q, [func(g, [var1, var2])]))$$

as the name of (1).

However, there is of course no end to the amount of properties that we may want to have available in the meta-theory: we may want to quantify over the arity of object-level function- or predicate-symbols, in a sorted logic we may want to include the sorts of object-level terms in their names, etc.

On the other hand, if none of these properties of object-level expressions is needed for a given meta-theory, it is advantageous to employ as simple a naming relation as possible, in order to reduce the complexity of the expressions in the meta-theory.

The conclusion of this progression of ever more general naming relations and the trade-off with complexity is of course that no single naming relation can be found that is optimal for all meta-theories, but that the richness (and there-fore the complexity) of the naming relation *should be adapted to the particular requirements of a given meta-theory*. Since in many systems the definition of the contents of the meta-theory is up to the user or programmer, so should the nature of the naming relation between object- and meta-theories be left to the user or programmer. In other words, the naming relation between object- and meta-theories should be made *internal* to the system, *definable* by the user, in-stead of being external to the system and fixed once and for all by the system's designers, as is current practice.

4 Advantages of Definable Naming Relations

The previous section introduced the notion of a definable naming relation. In this section, we shall argue why this notion is indeed a useful one.

The first advantage of definable names has already been described above: they allow the complexity of names in $\mathcal{L_M}$ to be tailored to the requirements of the specific meta-theory $\mathcal{M}$ in which they are used. This means that names will only be as complex as they are required to be, and will not encode any syntactic details of expressions from $\mathcal{L_O}$ which are not used by the axioms in $\mathcal{M}$. This is beneficial both from a human standpoint (since meta-level expressions will be easier to comprehend) and from a computational standpoint (since meta-level names will be easier to compute).

Readers will of course have recognised this argument as a special case of the general lesson in Computer Science that it is good to allow flexible use of datastructures, allowing users to formulate their own definitions separately from the rest of the program. We have simply applied this argument to the special case of naming relations in meta-level systems.

A second, and more fundamental advantage of a definable naming relation is that they can be used to reduce the computational complexity of $\mathcal{M}$. This will be the subject for the remainder of this section.

A property shared by all naming relations exploited in existing systems and in the literature, is that the names are defined uniformly across the syntax of $\mathcal{L_O}$: syntactically similar expressions will have similar names. However, if the naming relation is made into a definable component of the system, then there is no longer a need to only define the naming relation uniformly across the syntax of $\mathcal{L}_O$, but we may employ the naming relation to introduce other distinctions between expressions of $\mathcal{L}_O$. In particular, it becomes possible to encode semantic and pragmatic aspects of expressions from $\mathcal{L}_O$ into their names in $\mathcal{L}_M$. We have coined the phrase *"meaningful naming"* for this idea (as opposed to the more traditional "syntactic naming"), because it becomes possible to encode more than only syntactic structure into names, but we use the names also to encode both semantics (meaning) and pragmatics (use).

As we will show, this mechanism of meaningful names can be used to increase the efficiency and the intelligibility of the meta-theory, by reducing both the number of meta-level axioms and the size of the remaining axioms.

In first order logic, properties of individuals are represented by unary predicates applied to constants. Thus, properties of expressions from $\mathcal{L}_O$ are expressed by unary predicates of from $\mathcal{L}_M$. However, the notion of definable names allows us to encode such properties in a different way, namely no longer as unary predicates, which will occur in the axioms of $\mathcal{M}$, but instead in the names themselves. After all, it is no longer necessary to define names uniformly across the syntax of elements in $\mathcal{L}_O$ (which would require that similar expressions have similar names); we can now take into account non-syntactic properties of expressions from $\mathcal{L}_O$, and encode these properties in their names. Thus, if two expressions in $\mathcal{L}_O$ are syntactically similar, but one has a certain property that we want to represent in $\mathcal{M}$, and the other does not, then we can assign different names to these expressions, which will encode these differences, even though the expressions are syntactically similar. In this way, we can exploit the naming relation to eliminate as many of the unary predicates from $\mathcal{M}$ as we would like. We know from previous results that a transformation which removes unary predicates from a theory brings substantial gains. The transformation from an unsorted to a sorted first order theory, as described for instance in [14] also results in the removal of unary predicates from a theory (in that case, they are encoded in the sort structure), and problems which are known to be combinatorially explosive in unsorted theories can be efficiently dealt with in sorted theories (e.g. Schubert's Steamroller). This close parallel between the introduction of sorts and the introduction of meaningful names shows how meaningful names will lead to more compact and efficient meta-theories in meta-level systems.

5 Examples of Naming Relations

In this section, we will illustrate the idea of exploiting the naming relation to encode properties of object-expressions in their names in the meta-language through a number of small examples.

5.1 Control Information in PRESS

A well-known meta-level system, and in fact one of the early systems to propagate the use of a meta-level architecture, is the equation solving system PRESS

[2]. A with many meta-level systems, the purpose of the meta-theory $\mathcal{M}$ in PRESS is to express constraints on the search space of the object-theory $\mathcal{O}$. In PRESS, $\mathcal{O}$ consists of algebraic rewrite rules, such as

$$log_U V = W \Longrightarrow V = U^W \qquad (6)$$
$$T \times W + V \times W = U \Longrightarrow (T + V) \times W = U \qquad (7)$$

PRESS's meta-theory $\mathcal{M}$ classifies all these rewrite rules into a number of categories, depending on the way the rule should be used in the process of solving an algebraic equation. For instance, PRESS distinguished between isolation rules and collection rules. Isolation rules isolated the single occurrence of the unknown on one side, thereby solving the equation, whereas collection rules combined multiple occurrences of the unknown into one, thereby enabling the application of isolation rules. Some axioms of PRESS's meta-theory are given in Fig. 1 (all variables are universally quantified if not stated otherwise). PRESS used structural names rather than the atomic names in Fig. 1, which have been introduced for simplified presentation. The crucial point here of course is that PRESS used *syntactic* names, be they atomic or structural.

```
• isolation-rule(V, rule-6)
• collection-rule(W,rule-7)
• one-occurrence(var,lhs=rhs) ∧
  isolation-rule(var, rule) ∧
  apply(rule, lhs=rhs, var=rhs') → solve(var, lhs=rhs, rhs')
• multiple-occurrence(var, lhs=rhs) ∧
  collection-rule(var, rule) ∧
  apply(rule, lhs=rhs, lhs'=rhs') ∧
  solve(var, lhs'=rhs', rhs'') → solve(var, lhs=rhs, rhs'')
```

Fig. 1. A meta-theory of PRESS

If we would employ a definable naming relation, we could exploit the names of rules (6) and (7) to encode that they are of different categories. Instead of the undescriptive names *rule-6* and *rule-7* used in Fig. 1, we could use the more descriptive names *isolation-rule*$(V, 6)$ and *collection-rule*$(W, 7)$. This would enable a new formulation of the meta-theory as in Fig. 2.

Clearly, the axiom set from Fig. 2 is smaller than the one from Fig. 1: there are fewer axioms (2 instead of 4), and the axioms that remain are simpler (both implications have lost one conjunct from their antecedent).

This transformation of the theory from Fig. 1 to the theory from 2 may look insignificant, but this is mainly due to the simplified nature of our example. As argued above, we know from parallels with sorted logic that such transformation do in fact have substantial benefits.

- one–occurrence(var, lhs=rhs) ∧
  apply(isolation-rule(var,rule), lhs=rhs, var=rhs') →
  solve(var, lhs=rhs, rhs')
- multiple-occurrence(var, lhs=rhs) ∧
  apply(collection-rule(var, rule), lhs=rhs, lhs'=rhs') ∧
  solve(var, lhs'=rhs', rhs'') → solve(var, lhs=rhs, rhs'')

Fig. 2. A PRESS meta-theory using meaningful names

5.2 Control Information in Logic-programming

We can use definable names to construct a naming relation that encodes information about the degree of instantiation of formulae from $\mathcal{L}_\mathcal{O}$. If, for example, we have an object-theory encoding graphs as $edge(node_1, node_2)$ predicates, plus an axiom defining the transitive closure as

$$connected(N_1, N_2) \leftarrow edge(N_1, N_3) \wedge connected(N_3, N_2)$$

then we can construct a meta-theory containing the control assertion that the conjunction defining $connected(N_1, N_2)$ should be executed from left to right if N_1 is given, but from right to left if N_2 is given. We can exploit the naming relation for this purpose by assigning different names to formulae from $\mathcal{L}_\mathcal{O}$ depending on their degree of instantiation:

formula from $\mathcal{L}_\mathcal{O}$	name in $\mathcal{L}_\mathcal{M}$
$edge(node15, X))$	$groundvar(\lceil edge \rceil, \lceil node15 \rceil, \lceil X \rceil)$
$connected(X, node15)$	$groundvar(\lceil connected \rceil, \lceil X \rceil, \lceil node15 \rceil)$
$edge(X, node15)$	$varground(\lceil edge \rceil, \lceil X \rceil, \lceil node15 \rceil)$
$connected(X, node15)$	$varground(\lceil connected \rceil, \lceil X \rceil, \lceil node15 \rceil)$

This enables us to write down the meta-axiom:

$$before(groundvar(\_, \_, \_), varground(\_, \_, \_)),$$

which can be taken into account by a meta-interpreter in $\mathcal{M}$ for $\mathcal{O}$.

We deliberately give this well-known example from meta-programming in logic-programming to show how a definable naming relation can be used to achieve more compact meta-theories.

5.3 Knowledge Roles in KBS

Knowledge-roles are a concept from the development of knowledge-based systems used to indicate the role that particular expressions play in the inference process. An example of knowledge roles (a term introduced by [10]) is for instance the difference between a causation-relation and a specialisation-relation. These will be used in very different ways in the inference process, even though they may be logically represented in similar ways. We can capture such different roles by using a definable naming relation, as follows:

formula from $\mathcal{L_O}$	name in $\mathcal{L_M}$
acute-meningitis→bact-meningitis	type-of($\lceil$acute-meningitis$\rceil$,$\lceil$bact-meningitis$\rceil$)
meningococcus→bact-meningitis	causes($\lceil$meningococcus$\rceil$,$\lceil$bact-meningitis$\rceil$)

and a meta-theory specifying the inference steps can then use these names to distinguish between the use of causation-relations and specialisation-relations.

6 Properties of Definable Names

Once a naming relation is a definable instead of a predefined component of a meta-level architecture, it becomes important to know what the formal properties are that a naming relation must satisfy, since these properties can no longer be enforced once and for all by the system implementer, but must be checked on the definition supplied by the user.

For the purpose of this section, we will use the 2-place symbol N to denote the naming relation, and we will use the notation $\overline{\phi}$ to denote any name of the object-expression ϕ. Thus, we have $\mathcal{N}(\phi, \overline{\phi})$ for any name $\overline{\phi}$ of ϕ. (Notice that the relation $\mathcal{N}$ is not definable in either object- or meta-theory, but is outside the scope of both. We will continue to use $\lceil \phi \rceil$ for the quotation name of ϕ.

6.1 Definable Names Are a Conservative Extension

It is easy to see that the introduction of non-syntactically determined names are "only" a notational device, and do not extend the logical power of our formalism. For any expression $\phi \in \mathcal{L_O}$ with a meaningful (= non-syntactically determined) name $\overline{\phi} \in \mathcal{L_M}$, we can use simply the syntactic name (say the atomic name $\lceil \phi \rceil$), and assert the additional equality $\overline{\phi} = \lceil \phi \rceil$ in $\mathcal{M}$. This shows that definable names are only a conservative extensions (everything that is expressible and provable in both object- and meta-theory with definable names is also expressible and provable without them). However, as we have argued in the previous section, the *combinatorial* properties of $\mathcal{M}$ change for the better with the introduction of definable names.

6.2 Relational Properties of Naming

In Sect. 2, we have already discussed the domain and range of naming relations. If we write, $\mathcal{N} : \mathcal{D} \times \mathcal{R}$, then the range $\mathcal{R}$ of the naming relation must be a set of ground terms from $\mathcal{L_M}$. The domain $\mathcal{D}$ can vary from definition to definition, and can be such things as sets of object-level terms, formulae, proofs, theories, etc.

Four general properties that characterise any relation are the following:

total: $\forall x \in \mathcal{D} \, \exists y \in \mathcal{R} : \mathcal{N}(x, y)$;
surjective: $\forall y \in \mathcal{R} \, \exists x \in \mathcal{D} : \mathcal{N}(x, y)$;
functional: $\forall x \in \mathcal{D} \, \forall y_1, y_2 \in \mathcal{R} : \mathcal{N}(x, y_1) \wedge \mathcal{N}(x, y_2) \rightarrow y_1 = y_2$;
injective: $\forall x_1, x_2 \in \mathcal{D} \, \forall y \in \mathcal{R} : \mathcal{N}(x_1, y) \wedge \mathcal{N}(x_2, y) \rightarrow x_1 = x_2$;

Of these, none is required (and even desired) with the exception of the last (injectivity). This is in contrast with the naming relations as defined by logicians and as implemented in the systems mentioned above, which are all total, functional and injective.

- A *non-total* naming relation implies that not all elements of $\mathcal{D}$ have names in $\mathcal{L_M}$, and thus some elements of $\mathcal{R}$ cannot be described in $\mathcal{M}$. This can be used in systems where only some parts of $\mathcal{L_O}$ need to be visible outside the theory. This corresponds closely to the "export" operation that is well known from programming and specification languages, and will lead to a reduction in the search space of proofs in $\mathcal{M}$.
- Surjectivity would imply that $\mathcal{L_M}$ contains only terms that are names of elements of $\mathcal{D}$, and no terms denoting anything else. It would exclude such obvious meta-theories as those that reason about the length of proofs, which require natural numbers as terms besides terms that refer to $\mathcal{L_O}$. This would be an absurd requirement, which is not enforced in any naming relation in the literature, either from logic or from AI. There is one case where some form of surjectivity would seem reasonable: if the meta-language $\mathcal{L_M}$ is a sorted language, and contains sorts whose terms are meant to be only terms from $\mathcal{L_M}$ that are names for $\mathcal{D}$, then we would expect $\mathcal{N}$ to be surjective onto such sorts
- A non-functional $\mathcal{N}$ can assign more than one name to a single element of $\mathcal{D}$. This is unusual in syntactic naming relations, but is useful in meaningful definition of $\mathcal{N}$, since it is well possible that we may want to reason about different properties of an element of $\mathcal{D}$. In such a case, it is more convenient to assign different names to an element from $\mathcal{D}$, each name encoding a different property, rather than constructing a single complicated name to encode all properties at once.
- The only property that is formally required of $\mathcal{N}$ is its injectivity. This becomes clear when we realise that the proper semantics of a naming relation, as pointed out by [13], is that the object-theory should be regarded as a partial model of the meta-theory, in the sense that the denotations of names in $\mathcal{M}$ (the elements of $\mathcal{R}$) are the corresponding expressions in $\mathcal{O}$ (the elements of $\mathcal{D}$), in other words:

$$\forall x \in \mathcal{D} \forall y \in \mathcal{R} : [\![y]\!] = x \text{ iff } \mathcal{N}(x, y),$$

where $[\![y]\!]$ is the semantic denotation of y. This makes N the inverse relation of semantic denotation. Thus, if $\overline{\phi}$ is the name of ϕ, then ϕ is the denotation of $\overline{\phi}$. Since semantic denotation is required to be functional (no term is allowed to have more than one denotation), N must therefore be injective.

6.3 The Complexity of the Naming Relation

In the literature, naming is often tacitly assumed to be an operation of trivial complexity (e.g. quoting, or some direct form of syntactic isomorphism). The following example (due to Luciano Serafini at IRST) illustrates that we must indeed put an upper bound on the complexity of N. Let $\mathbb{N}$ be the standard model of arithmetic, and define N as follows:

$$\mathcal{N}(\phi, \overline{\phi}) \leftrightarrow \begin{cases} \overline{\phi} = true(\lceil \phi \rceil) & \text{if } \mathbb{N} \vdash \phi \\ \overline{\phi} = false(\lceil \phi \rceil) & \text{otherwise} \end{cases}$$

Suppose we have the following natural deduction rules:

$$\frac{\phi}{prove(true(\lceil \phi \rceil))} \text{ if } \mathcal{N}(\phi, true(\lceil \phi \rceil)) \qquad\qquad \frac{\phi}{prove(false(\lceil \phi \rceil))} \text{ otherwise}$$

and

$$\frac{prove(true(\lceil \phi \rceil))}{\phi} \qquad\qquad \frac{prove(false(\lceil \phi \rceil))}{\neg\phi}$$

In this way, the object-theory would be a consistent and complete axiomatization of the standard model of arithmetic, something which we know to be impossible. The construction of this example relied of course on the definition of $\mathcal{N}$, and in particular on the fact that the above definition was not a recursive function. Thus, we will have to require that $\mathcal{N}$ is at least computable, but in practical systems we would expect it not only to be computable, but tractable as well. After all, a major reason for organising a system as a meta-architecture is to control the combinatorial explosion in the inference process, and we would thus expect N not to contribute to this explosion (for instance by being of exponential cost in the length of the object-formulae).

6.4   Equality of Names

If we allow the naming relation N to be non-functional (that is: some elements of $\mathcal{D}$ have more than one name in $\mathcal{R}$), what is the relation among the different names $\overline{\phi}_i$ of an element $\phi \in \mathcal{D}$? The choice here is whether we want all names $\overline{\phi}_i$ of an element $\phi \in \mathcal{D}$ to be provably equal in the meta-theory or not. In other words, the question is whether we want the following:

$$\text{if } \mathcal{N}(\phi, \overline{\phi}_1) \text{ and } \mathcal{N}(\phi, \overline{\phi}_2) \text{ then } \mathcal{M} \vdash \overline{\phi}_1 = \overline{\phi}_2. \tag{8}$$

From a logical point of view, this is perfectly natural, since it amounts to having equality in the model as equality in the language Taking the object-theory as a partial model for the meta-theory, as suggested above, this simply amounts to keeping $\mathcal{M}$ complete with respect to its model $\mathcal{O}$.

It is less clear that (8) is also a desirable property from a computational point of view. After all, we would like the meta-level search space to be smaller than the object-level space. However, if we keep $\mathcal{M}$ complete, this implies that everything provable in $\mathcal{O}$ will also be provable in $\mathcal{M}$ (and possibly more), thus resulting in an even larger search space. Thus, from a computational point of view, it may be advantageous to make $\mathcal{M}$ incomplete, and in particular to have certain equalities such as those resulting from (8) *not* provable in $\mathcal{M}$.

Summarising: equality of names in $\mathcal{M}$ is possible (and possibly even desirable) on logical grounds, but problematic on computational grounds.

7 Naming as a Term Rewriting System

7.1 Definitions

If we accept the arguments of the previous sections concerning the need for definable naming relations, we obviously require some mechanism that will enable us to write down definitions of naming relations. We need a formal language that we can use to express naming relations, and we would like to be able to use such a language to show that a particular definition fulfills the formal requirements that have been investigated above.

One possibility is to define a naming relation as a term-rewriting system. The motivation for this choice is the problem that the domain $\mathcal{D}$ of a naming relation can vary between applications, and could consist of formulae, formulae-trees (proofs), formulae-sets (theories), etc. In order to minimise the assumptions we make about $\mathcal{D}$, we will define the naming relation as ranging over arbitrary term-trees, and assume that every possible choice for $\mathcal{D}$ can be presented in such terms.

The choice of using rewrite systems to specify the naming relation is only meant as an example to illustrate the general point that it is both possible and advantageous to have user defined naming relation. Many other formalisms are possible to define the transformation between object-expressions and their meta-level names.

Neither the two place relation $\mathcal{N}$ nor the rewrite rules specifying it are themselves expressible in first-order meta-level terms, since they require second order quantifications. The definition of the naming relation by the user is done *outside* the meta-object theories.

When using a set of rewrite rules to define a naming relation, a naming relation $\mathcal{N} : \mathcal{D} \rightarrow \mathcal{R}$ is realised by a finite set $\mathcal{E}$ of rewrite rules of the form $name(\lambda) \mapsto \rho$, with λ and ρ taken from $\mathcal{D}$ and $\mathcal{R}$ respectively, but allowing for rule-variables ν_i to occur in both λ and ρ, where these rule-variables are taken from a set $\mathcal{V}$ assumed to be disjoint from form $\mathcal{D}$ and $\mathcal{R}$. We also allow the occurrence of subterms $name(\nu_i)$ in ρ, and assume the symbol $name$ does not occur in $\mathcal{R}$.

As usual, a rule $name(\lambda) \mapsto \rho$ is applicable to $\phi \in \mathcal{D}$ *iff* there exists a substitution for rule-variables σ such that $\sigma(\lambda) = \phi$. The result of such a rule application is then $\sigma(\rho)$. We then define the naming relation as the normal form under $\mathcal{E}$:

$$\mathcal{N}(\phi, \overline{\phi}) \ \textit{iff} \ name(\phi){\downarrow} = \overline{\phi}, \ and \ \overline{\phi} \in \mathcal{R} \tag{9}$$

The intuition behind this definition is as follows: we use the rewrite system $\mathcal{E}$ to gradually translate elements of $\mathcal{D}$ into elements of $\mathcal{R}$. Each rule translates part of an element of $\mathcal{D}$, leaving the translation of other parts unspecified (the $name(\nu_i)$ subterms) which have to be computed by further applications of rewrite rules. Whenever the rewriting process terminates (when we have computed a normal form), we will either have completely translated our original element of $\mathcal{D}$ into its name in $\mathcal{R}$, or we will have failed to do so because we have not arrived at a correct element of $\mathcal{R}$. This latter gives rise to a non-total naming relation.

It is important to realise that the set of rule variables $\mathcal{V}$ is disjoint from the variables in either $\mathcal{D}$ or $\mathcal{R}$. During the application of rewrite rules, only the rule variables are bound by matching λ with ϕ. During the rewrite process, the elements from $\mathcal{D}$ are regarded as ground terms, and variables from $\mathcal{O}$ occurring in elements of $\mathcal{D}$ play no distinguished role in the rewrite process.

Notice that in testing whether a rule $name(\lambda) \mapsto \rho$ is applicable to an expression ϕ, only λ can possibly contain rule variables, and not ϕ, and we therefore do not require unification, but only matching in order to test for rule applicability.

As we will see in the examples below, it will turn out to be useful to impose a sort structure on $\mathcal{V}$ and $\mathcal{D}$, and to require that the substitution σ respects this sort-structure.

7.2 Examples

In this section we shall illustrate naming relations as rewrite rules by defining some of the naming relations used earlier in this paper in this way.

Example 1 Binding information.

For defining the naming relation from Sect. 5.2 which encodes instantiation information, we can usefully exploit the obvious sort-structure on syntactic elements of $\mathcal{L}_O$:

$$\ldots$$

$$name(\nu_1^{\mathrm{ps}}(\nu_2^{\mathrm{c}}, \nu_3^{\mathrm{v}})) \mapsto groundvar(\lceil \nu_1^{\mathrm{ps}} \rceil, \lceil \nu_2^{\mathrm{c}}, \rceil \lceil \nu_3^{\mathrm{v}} \rceil)$$
$$name(\nu_1^{\mathrm{ps}}(\nu_3^{\mathrm{c}}, \nu_2^{\mathrm{v}})) \mapsto varground(\lceil \nu_1^{\mathrm{ps}} \rceil, \lceil \nu_3^{\mathrm{c}} \rceil, \lceil \nu_2^{\mathrm{v}} \rceil)$$

$$\ldots$$

(We write ν_i^s for a rule-variable ν_i of sort s, and write ps and fs for predicate and function symbols, c for constants, v for variables, t for terms and s for sentences, arranged in the obvious hierarchy).

Example 2 Knowledge roles.

The definition of the naming relation that captures the different roles that expressions from $\mathcal{L}_O$ play in the inference process (Sect. 5.3) differs from the previous example because it explicitly refers to particular non-logical symbols from $\mathcal{L}_O$:

$$name(\text{acute-meningitis} \rightarrow \text{bact-meningitis}) \qquad \mapsto$$
$$\text{type-of}(\lceil \text{acute-bact-meningitis} \rceil, \lceil \text{bact-meningitis} \rceil)$$
$$name(\text{meningococcus} \rightarrow \text{bact-meningitis}) \qquad \mapsto$$
$$\text{causes}(\lceil \text{meningococcus} \rceil, \lceil \text{bact-meningitis} \rceil)$$

The naming relation used in the example concerning PRESS (Sect. 5.1) has a similar form, showing that the effectiveness of PRESS's meta-theory can be explained as the assignment of different knowledge roles to the rewrite rules from its object-theory.

Example 3 Purely syntactic names.

A segment of a structural naming relation, illustrated above in 3 can be formulated as follows:

$$name(\nu_1^c) \mapsto \lceil\nu_1^c\rceil'$$
$$name(\nu_2^{fs}(\nu_3^t)) \mapsto \lceil\nu_2^{fs}\rceil(name(\nu_3^t))$$
$$name(\nu_4^{ps}(\nu_3^t)) \mapsto \lceil\nu_4^{ps}\rceil(name(\nu_3^t))$$
$$name(\nu_5^s \wedge \nu_6^s) \mapsto name(\nu_5^s) \wedge' name(\nu_6^s)$$
$$\cdots$$

This naming relation is in fact (very close to) the one used in the Gödel logic programming language (see Sect. 8 below).

By contrast, an quotation naming relation can be captured by the following trivial rewrite system:

$$name(\nu_1^{any}) \mapsto \lceil\nu_1^{any}\rceil$$

7.3 Properties

Section 6 argued that naming relations should be allowed to be non-functional and non-total, but are required to be injective. How is all this reflected in our realisation of the naming relation as a finite set of rewrite rules?

It is clear that defining a naming relation as rewrite rules allows for non-total and non-functional naming relations. After all, we can have rule sets with no rule applicable to a particular element of $\mathcal{D}$ (non-total).

Similarly, $\mathcal{E}$ is allowed to be non-canonical, allowing some elements of $\mathcal{D}$ to have more than one normal form, thereby making the naming relation non-functional.

The only requirement that we had to enforce on a naming relation was its injectivity. A well-known result from term-rewriting theory (e.g. [7]) tells us that this property is decidable, since the injectivity of $\mathcal{E}$ corresponds to the set of reverse rules $\mathcal{E}^{-1}$(mapping from $\mathcal{R}$ to $\mathcal{D}$) being canonical, and being canonical is decidable for terminating finite rewrite systems. Clearly, $\mathcal{E}^{-1}$ is finite (since $\mathcal{E}$ is finite), and it is also terminating (proof by induction on the size of terms to be rewritten).

Finally, naming as rewriting as defined in this section is clearly efficient: it is linear in the size of the element of $\mathcal{D}$ to be rewritten.

8 Comparison With Existing Systems

Not very many of the meta-systems in the literature consider the naming relation as a definable part of the system. In this section, we discuss the only exception we know of, namely the theorem-proving system FOL [15].

The FOL system consists of a set of theories, called contexts, some of which may form object-meta pairs, in the sense that terms from one theory refer to formulae of the other. These object-meta pairs are connected by means of the standard reflection rules for upwards and downwards reflection, and by a naming relation as required by these rules.

The FOL system is one of the few in the literature where the notion of a definable name already occurs. To be more precise, in FOL the user does not actually define naming, but instead defines the inverse operation (denotation). It is possible, by means of the attach command, to associate terms from the meta-theory with (the data-structures underlying) the formulae from the object-theory. This denotation relation is required to be functional, and as a result the naming relation satisfies the injectivity requirement from Sect. 6. It is also possible for a single (datastructure underlying an) object-formula to serve as the denotation of multiple meta-level terms, and as a result, naming in FOL is not required to be functional, again in accordance with Sect. 6. Among such multiple names of a single object-formula, FOL distinguishes a *preferred name*. This user-defined naming relation is used in FOL during the application of the reflection rules for the purposes of substituting object-level computation with meta-level computation or vice versa.

The major difference between FOL and the approach taken in this paper is the mechanism by which the naming relation is defined: FOL is restricted to a "point-wise" definition of the naming relation (where names are assigned to individual formulae of $\mathcal{L_O}$, much as in example 2), whereas our rewrite -rule formulation can be used to define the names of classes of expressions. For example, the rule

$$\nu_1^{\mathrm{pr}}(\nu_2^t) \mapsto unary(\lceil \nu_1^{\mathrm{pr}} \rceil, \lceil \nu_2^t \rceil)$$

assigns a name to all unary predicates from $\mathcal{L_O}$, whereas an FOL user would be forced to write down the name for each separate unary predicate from $\mathcal{L_O}$ in turn.

Furthermore, none of the papers on FOL ever use the naming relation for anything else than a purely syntactic (typically structurally descriptive) naming relation. Although FOL's mechanisms seem to allow meaningful naming relations, this idea has not been investigated in the FOL community.

Acknowledgements

Both the contents and the presentation of this paper have greatly benefitted from extensive discussions with Fausto Giunchiglia, Alex Simpson and Luciano Serafini (all at IRST in Trento, Italy), Alan Smaill (University of Edinburgh, Scotland), and Pat Hill (University of Leeds, England). The research reported here was carried out in the course of the REFLECT project. This project is partially funded by the Esprit Basic Research Programme of the Commission of the European Communities as project number 3178. The partners in this project are The University of Amsterdam (NL), the German National Research Institute for Computer-Science GMD (D), the Netherlands Energy Research Foundation ECN (NL), and BSR Consulting (D).

References

1. K.A. Bowen and R.A. Kowalski. Amalgamating language and metalanguage in logic programming. In K. Clark and S. Tarnlund, editors, *Logic Programming*, pages 152–172. Academic Press, 1982.

2. A. Bundy and B. Welham. Using meta-level inference for selective application of multiple rewrite rule sets in algebraic manipulation. *Artificial Intelligence*, 16:189–212, 1981.

3. S. Costantini and G.A. Lanzarone. A metalogic programming language. In G. Levi and M. Martelli, editors, *Proceedings of the Sixth International Conference on Logic Programming*, pages 218–233, Lisbon, 1989. The MIT Press.

4. L. Erman, A. Scott, and P. London. Separating and integrating control in a rule-based tool. In *Proceedings of the IEEE Workshop on Principles of Knowledge Based Systems*, pages 37–43, Denver, Colorado, December 1984.

5. K. Gödel. Über formal unentscheidbare Sätze der Principia Mathematica und verwandter Systeme I. *Monatsh. Math. Phys.*, 38:173–98, 1931. English translation in *From Frege to Gödel: a source book in Mathematical Logic, 1879-1931*, J. van Heijenoort (ed.), Harvard University Press, 1967, Cambridge, Mass.

6. P.M. Hill and J.W. Lloyd. The Gödel report (preliminary version). Technical Report TR-91-02, Computer Science Department, University of Bristol, March (Revised September '91) 1991.

7. G. Huet and D.C. Oppen. Equations and rewrite rules: a survey. In R. Book, editor, *Formal languages: perspectives and open problems*. Academic Press, 1978. Presented at the conference on formal language theory, Santa Barbara, 1979. Also: technical report CSL-111, SRI International, Menlo Park, California.

8. P. Jackson, H. Reichgelt, and F. van Harmelen. *Logic-Based Knowledge Representation*. The MIT Press, Cambridge, MA, 1989.

9. P. Maes. Reflection in an object-oriented language. In P. Maes and D. Nardi, editors, *Meta-Level Architectures and Reflection*, Amsterdam, 1988. North-Holland. Also: AI-Laboratory Memon 86-8, Vrije Universiteit Brussel.

10. J. McDermott. Preliminary steps towards a taxonomy of problem-solving methods. In S. Marcus, editor, *Automating Knowledge Acquisition for Expert Systems*, pages 225–255. Kluwer Academic Publishers, The Netherlands, 1988.

11. B. Smith. Reflection and semantics in a procedural language. Technical Report TR-272, MIT, Computer Science Lab., Cambridge, Massachussetts, 1982.

12. B. Smith. Reflection and semantics in Lisp. In *Proc. 11th ACM Symposium on Principles of Programming Languages*, pages 23–35, Salt Lake City, Utah, 1984. also: Xerox PARC Intelligent Systems Laboratory Technical Report ISL-5.

13. A. Tarski. Der Wahrheitsbegriff in den formalisierten Sprachen. *Studia Philosophica*, 1:261–405, 1936. English translation in *Logic, Semantics, Metamathematics*, A. Tarski, Oxford University Press, 1956.

14. C. Walther. A mechanical solution of Schuberts steamroller by many-sorted resolution. *Artificial Intelligence*, 26(2):217–224, 1985.

15. R. Weyhrauch. Prolegomena to a theory of mechanized formal reasoning. *Artificial Intelligence*, 13, 1980. Also in: *Readings in Artificial Intelligence*, Webber, B.L. and Nilsson, N.J. (eds.), Tioga publishing, Palo Alto, CA, 1981, pp. 173-191. Also in: *Readings in Knowledge Representation*, Brachman, R.J. and Levesque, H.J. (eds.), Morgan Kaufman, California, 1985, pp. 309-328.

Meta for Modularising Logic Programming

A. Brogi, P. Mancarella, D. Pedreschi, F. Turini

Dipartimento di Informatica, Università di Pisa
Corso Italia 40, 56125 Pisa, Italy
e-mail:{brogi,paolo,pedre,turini}@di.unipi.it

Abstract

We use metalogic to define a suitable notion of module in logic programming.
A module is viewed as a pair of logic programs, corresponding to the
visible and the hidden part of the module, respectively. The construction
of a module is implemented through a metalevel composition operator of
object-level logic programs. We introduce metalevel operators for combining
separate modules together. All the operators are straightforwardly defined
by extending the standard vanilla metainterpreter with new clauses. More
importantly, the semantics of modules and module composition operators
is defined in terms of the standard semantics of logic programming in a
compositional way.

1 Introduction

Two observations are at the basis of this work. On the one hand, it is
widely recognised that the logic programming paradigm combines clean and
simple semantics, ease of use and expressive power. On the other hand,
logic programming is not well suited to tackle large applications, for it lacks
linguistic features for mastering complexity. ADA's packages and classes in
object oriented languages are typical examples of such constructs. Indeed
large applications, when developed by big teams, call for a modular and
incremental design process, which is actually feasible only if the language
offers features in support of it. One of our objectives is to equip logic
programming with extensions for programming-in-the-large under a very
tight constraint, that is without blemishing its semantic kernel.

We propose to build on top of logic programming a notion of module
and a set of operators for combining modules. We view a module as a
pair of logic programs, representing the visible and the hidden part of the
module, respectively. An interesting aspect, due to the very nature of logic
programming, is the possibility of managing the notion of exporting in two
different ways. On the one hand, a relation can be made visible to other
modules by exporting its clauses. On the other hand, a relation can be
made usable but only at the provability level. The difference between the

two notions is evident when two modules are composed together via the composition operators.

More precisely, a module is viewed as a pair of logic programs $\langle P, Q \rangle$. The syntax of P is completely visible from the external environment and can be exploited when combining it with other modules. On the other hand, the syntax of the hidden part Q cannot be accessed by the other modules. The set of formulae which are provable in the hidden part of the module can be only referred by the visible part of the same module. Such mechanism offers the possibility of making the provability of part of a module visible without exporting its syntactic structure, and realises a model-theoretic account for implementation hiding.

This work represents a further step in the development of a general methodology, based on metalogic, for extending logic programming. In previous work [3], we have defined a set of metalevel operators for composing logic programs. We have shown that the application of the methodology supports a number of extensions of logic programming for knowledge representation and reasoning [3, 5]. In this paper, we build on top of program composition operators and define a further layer for handling modules.

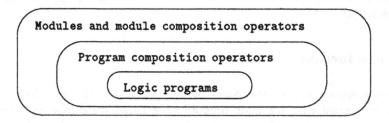

The use of metalogic and the preservation of the standard semantics of logic programming are the core of the methodology. As far as the use of metalogic is concerned, both program and module composition operators are defined by adding new clauses to the standard vanilla metainterpreter in an incremental way. A first advantage of the approach is simplicity. The operational meaning of the various operators is defined by means of straightforward (metalevel) axioms. More importantly, the use of metalogic allows us to express different extensions of logic programming from inside logic programming itself.

As far as the semantics is concerned, the standard semantics of logic programming is retained. It is well known that logic programming can be formalised according to three different, albeit equivalent, approaches: model theoretic, operational and fixpoint [17]. We take the fixpoint approach as the reference semantics since it offers a bridge between the model-theoretic and the operational approach. In a sense, the fixpoint semantics is the best compromise between an abstract, declarative understanding of programs and an operational, computation oriented one. We follow a conservative approach and define extensions of logic programming which can be characterised in terms of the standard fixpoint semantics. More precisely, as in the case of program composition operators, the semantics of a composition of two modules is expressed through a suitable composition of the semantics of the separate modules.

The plan of the paper follows. In section 2, we briefly introduce some background which will be used in the rest of the paper. Sections 3 and 4 contain respectively the intended semantics and the metalogical definition of modules and module composition operators. Other operators for realising import/export relations and finer grained encapsulation operators are presented in section 5. Finally, in section 6 we draw some conclusions and envisage future research.

2 Background

2.1 Semantics

We use standard notations of logic programming [1, 12]. A logic program is a finite set of definite Horn clauses. Programs are identified by constant names and denoted by capital letters such as P, Q (possibly indexed).

We take the *immediate consequence operator* T_P [17] as the semantics of a logic program P. For any program P, T_P maps Herbrand interpretations into Herbrand interpretations, and is defined as follows [1]:

$$T_P(I) = \{A \mid A \leftarrow B_1, \ldots, B_n \in ground(P) \land \{B_1, \ldots, B_n\} \subseteq I\}$$

The powers of T_P are defined as usual:

$$
\begin{aligned}
T_P{\uparrow}0\,(I) &= I \\
T_P{\uparrow}(n+1)\,(I) &= T_P(T_P{\uparrow}n\,(I)) \\
T_P{\uparrow}\omega\,(I) &= \bigcup_{n<\omega} T_P{\uparrow}n\,(I)
\end{aligned}
$$

A well known result by van Emden and Kowalski [17] states that the T_P operator is monotonic and continuous, and that the minimal Herbrand model of a definite logic program coincides with the least fixpoint of T_P (denoted by $T_P{\uparrow}\omega$).

Several studies have demonstrated that the T_P semantics is adequate to deal with the problems of compositionality in the semantics of logic programs. Lassez and Maher [11], and O'Keefe [15] studied the properties of the T_P semantics when considering composition operators over logic programs. Mancarella and Pedreschi [13] have extended those results by defining an algebra of logic programs in terms of the T_P operator. We will recall the main results reported in [13] in subsection 2.3.

2.2 Metalogic

We use metalogic [2] for the definition of program composition operators. We adopt the simple naming relation used in [10] where object-level expressions are denoted by themselves at the metalevel, and object-level provability is defined through metalevel axioms. The metalevel predicate *demo* denotes the provability relation. Namely, $demo(P, G)$ means that the formula G is provable in the object-level program P. For the sake of uniformity, clauses of object-level programs are also denoted by the metapredicate *demo*. Let us formally define the metalevel representation of object-level programs.

Definition 2.1 *The metalevel representation P_m of any object-level program P is defined as follows:*

$$A \leftarrow G \in P \Longleftrightarrow demo(P, A \leftarrow G) \leftarrow \in P_m$$

An example of the metalevel representation of an object-level program follows.

Object-level:
$n(0) \leftarrow$
$n(s(X)) \leftarrow n(X)$

Metalevel:
$demo(nat, n(0) \leftarrow) \leftarrow$
$demo(nat, n(s(X)) \leftarrow n(X)) \leftarrow$

The simplest application of metalogic is the so called *vanilla* metainterpreter, which defines the SLD-resolution procedure of logic programming.

Definition 2.2 *The vanilla metainterpreter is defined by the following axioms:*

(d1) $demo(P, true) \qquad \leftarrow$

(d2) $demo(P, (G1, G2)) \quad \leftarrow \quad demo(P, G1),$
$\qquad\qquad\qquad\qquad\qquad\qquad demo(P, G2)$

(d3) $demo(P, A) \qquad\quad \leftarrow \quad demo(P, A \leftarrow G),$
$\qquad\qquad\qquad\qquad\qquad\qquad demo(P, G)$

Let us now state the soundness of the vanilla metaprogram, which has been formally proved in [8].

Proposition 2.3 *Let P be an object-level program, and M the metaprogram composed by P_m and the vanilla metaprogram. For any object-level ground atomic formula A:*

$$M \vdash demo(P, A) \Longleftrightarrow A \in T_P \!\uparrow\! \omega$$

As far as a single program is concerned, the first argument of *demo* (i.e. P) can be omitted altogether. However, in a multi-program framework, where several programs can be dynamically composed, the first argument of *demo* becomes relevant.

2.3 Union and intersection of logic programs

Following [3, 13], we now introduce two basic program composition operators of logic programs: *union* ($\cup$) and *intersection* ($\cap$). Informally, *union* yields a combined program where the original programs cooperate with each other during deduction, in the sense that partial conclusions of either program may possibly serve as premises for the other one. On the other hand, *intersection* yields a combined program where the original programs are forced to agree during deduction, in the sense that they must agree on every single partial conclusion (i.e. at each step of the computation). The logical justification for the former definition derives from the completion semantics of logic programs [6]. In fact, $\cup$ and $\cap$ correspond to disjunction and conjunction of completed definitions [13].

Before showing how $\cup$ and $\cap$ can be actually realised, we introduce the intended semantics of the operators. In [13], a compositional semantics of logic programs has been defined by taking the immediate consequence transformation T_P as the semantics of a program P. In order to denote the $\cup$ and $\cap$ operators, the immediate consequence operators $T_{P \cup Q}$ and $T_{P \cap Q}$ are defined in terms of T_P and T_Q [13].

Definition 2.4 *Given two programs P and Q:*

   i) $T_{P \cup Q} = \lambda I.T_P(I) \cup T_Q(I)$
   ii) $T_{P \cap Q} = \lambda I.T_P(I) \cap T_Q(I)$

Program composition operators can be defined using metalogic in a quite elegant way, by extending the definition of the *demo* metapredicate.

Definition 2.5 *The metainterpretive definition of $\cup$ and $\cap$ is given by the following three axioms to be added to the vanilla metainterpreter:*

   (d4) $demo(P \cup Q, A \leftarrow G)$ $\leftarrow demo(P, A \leftarrow G)$

   (d5) $demo(P \cup Q, A \leftarrow G)$ $\leftarrow demo(Q, A \leftarrow G)$

   (d6) $demo(P \cap Q, A \leftarrow G1, G2)$ $\leftarrow demo(P, A \leftarrow G1),$
                                         $demo(Q, A \leftarrow G2)$

Note that the first argument of *demo* is a term constructed with program composition operators, suitably represented by infix function symbols. The meaning of the metainterpreter is straightforward. For example, in the case of $\cup$, the metainterpretive definition says that a clause $A \leftarrow G$ belongs to $P \cup Q$ if it belongs either to P or to Q. The following proposition [3] states that the above metainterpreter actually corresponds to the semantics of the $\cup$ and $\cap$ operators.

Proposition 2.6 *Let P and Q be object-level programs, let V be the vanilla metaprogram consisting of axioms $(d1)-(d6)$ and M be the metaprogram composed by P_m, Q_m and V. For any object-level ground atomic formula A:*

$$i) \quad M \vdash demo(P \cup Q, A) \iff A \in T_{P \cup Q} \!\uparrow\! \omega$$
$$ii) \quad M \vdash demo(P \cap Q, A) \iff A \in T_{P \cap Q} \!\uparrow\! \omega$$

3 Modules

In the previous section, we have introduced two basic metalevel operators for composing logic programs. As shown in [3, 5], the use of such operators in a multi-program framework supports a number of programming techniques, including forms of non-monotonic, hypothetical and hierarchical reasoning. From a software engineering perspective, however, the granularity of programs and program composition operators is too coarse.

Most of the languages which have been designed for programming-in-the-large contain some notion of module and of information hiding. These languages also allow a programmer to define import-export relations among separate modules, and to combine modules together. For example, the specification of inheritance relations among modules can be used to organise a system in a hierarchical way. As far as logic programming is concerned, several solutions have been proposed for modularising logic programs. Some of these proposals augment logic programming with extra-logical constructs for declaring and using modules, which affect the computational model of logic programming. Typical examples are existing commercial Prolog systems, as well as the declarative language Gödel [9] and the calculus described in [16]. Other proposals introduce notions of modules in logic programming inspired by other logics, such as intuitionistic [14] or modal logic [7]. Our approach aims at providing mechanisms to build modules as combination of programs. The main advantage is that the semantics of modules is a conservative extension of the standard semantics of logic programming.

Moreover, logic programming provides two different forms of exporting knowledge from a module: at the intensional level (e.g. clauses, predicate or constant names) and at the extensional level (i.e. provability). The former is due to the possibility of spreading the clauses for the same predicate over different modules, the latter is due to the possibility of exporting only the logical consequences of a (sub)module, thus providing a model- theoretic account of implementation hiding. The proposed definition of module supports both forms of interaction and provides a suitable integration of them. After introducing the notion of module in a formal way, two operators for composing separate modules are presented.

3.1 Definition of modules

We view a module as a pair of programs $\langle P, Q \rangle$ where P is the visible part and Q is the hidden one. Roughly, the visible part is visible by other modules and supports the exportation of syntactic parts of the module. On the other hand, the hidden part cannot be accessed directly by the other modules. The set of formulae which are provable in the hidden part of the module can be referred to only by the visible part of the same module.

Definition 3.1 *Any pair of programs $\langle P, Q \rangle$ forms a* module.

Let us formally state what is the intended semantics of a module.

Definition 3.2 *Given a module $\langle P, Q \rangle$:* $T_{\langle P,Q \rangle} = \lambda I \ . \ T_P(I \cup T_Q \uparrow \omega).$

The semantics of a module is its immediate consequence operator, which is defined in terms of the programs composing the module. It is worth observing that the standard semantics of logic programming is retained. Indeed, if the hidden part Q of a module $\langle P, Q \rangle$ is empty then the semantics of the module coincides with the semantics of the program P, that is of the visible part of the module.

Definition 3.2 conveys the intuition that

1. The intension (viz. the code) of Q is hidden to P, as the minimal model $T_Q \uparrow \omega$ of Q is adopted instead of T_Q, and
2. Both the intension and the extension of Q are hidden to other modules, as $T_Q \uparrow \omega$ occurs as an argument of T_P only.

We now give two examples of modules.

Example 3.3 Consider the following programs P and Q containing the definition of a (naive) reverse and of an append operation on lists, respectively.

```
P:                                  Q:
reverse([ ],[ ])  ←                 append([ ],Ys,Ys)  ←
reverse([X|Xs], Ys)  ←              append([ ], Ys, Ys)  ←
   reverse(Xs, Zs),                    append(Xs,Ys,Zs)
   append(Zs, [X], Ys)
```

Starting from the above programs, one can build the module $\langle P, Q \rangle$ where the definition of **reverse** is visible, while the definition of **append** is hidden. This means that other modules may call the **reverse** operation in $\langle P, Q \rangle$, but they do not have visibility of the **append** operation.

Definition 3.1 does not impose any condition on the programs which form a module. This allows one to spread the definition of certain relations over the module, and to hide only part of the definition of a relation, as illustrated by the following example.

Example 3.4 Suppose that a bank is supported by an expert system for granting credits to customers. One of the critical procedures of the expert system is about judging the reliability of customers. It might be useful to hide part of the definition of the procedure as it is done in the following.

```
P:                                Q:
gives_credit(X)  ←                good_customer(X)  ←
   good_customer(X)                  in_politics(X)
good_customer(X)  ←               has_account(smith,..)  ←
   customer(X),                   in_politics(andreotti)  ←
   has_account(X,T),              threshold(..)  ←
   threshold(Y),
   greater_than(T,Y)
customer(smith)  ←
```

3.2 Composition of modules

We now introduce two basic operators for module composition: union ($\sqcup$) and intersection ($\sqcap$) of modules. These operators actually correspond to $\cup$ and $\cap$ on programs (introduced in section 2). The following definition of the intended semantics of the operators formalises this intuition.

Definition 3.5 *Given two modules M_1 and M_2:*

$$i) \quad T_{M_1 \sqcup M_2} = \lambda I.T_{M_1}(I) \cup T_{M_2}(I)$$
$$ii) \quad T_{M_1 \sqcap M_2} = \lambda I.T_{M_1}(I) \cap T_{M_2}(I)$$

The expected behaviour of the compositions is that the visible parts of the two modules cooperate with each other while exploiting their hidden parts separately. To illustrate this behaviour better, we unfold definition 3.5. For example, let $M_1 = \langle P_1, Q_1 \rangle$ and $M_2 = \langle P_2, Q_2 \rangle$:

$$T_{M_1 \sqcup M_2} = \lambda I.T_{\langle P_1, Q_1 \rangle \sqcup \langle P_2, Q_2 \rangle}(I) = \lambda I.T_{P_1}(I \cup T_{Q_1}\!\uparrow\!\omega) \cup T_{P_2}(I \cup T_{Q_2}\!\uparrow\!\omega)$$

Example 3.6 Let us consider the following modules $M_1 = \langle P_1, Q_1 \rangle$ and $M_2 = \langle P_2, Q_2 \rangle$.

```
P₁:                               P₂:
p(X)  ← f(X)                      p(X)  ← g(X)


Q₁:                               Q₂:
f(0)  ←                           g(0)  ←
f(s(s(X)))  ← f(X)                g(s(s(s(X))))  ← g(X)
```

Each module defines in its hidden part a relation over natural numbers. Relation f in Q_1 defines the multiples of 2, while g in Q_2 defines the multiples of 3. Both modules contain in their visible part a relation p which is defined in terms of f and g, respectively. The fixpoint of T_{M_1} and of T_{M_2} follow.

$$T_{M_1}\!\uparrow\!\omega = \{p(0)\} \cup \{p(s^{2n}(0)) \mid n \in N^+\}$$

$$T_{M_2}\!\uparrow\!\omega = \{p(0)\} \cup \{p(s^{3n}(0)) \mid n \in N^+\}$$

We now consider the union of the two modules $M_1 \sqcup M_2$. The visible parts of the two modules cooperate with each other, and retain the capability of exploiting their hidden parts separately. Such a behaviour is reflected by the semantics of $\sqcup$:

$$T_{M_1 \sqcup M_2} \!\uparrow\! \omega = \{p(0)\} \cup \{p(s^k(0)) \mid \exists n \in N^+ : (k = 2n \vee k = 3n)\}$$

We observe that the relation p computed by the union of the two modules defines all the natural numbers which are either multiples of 2 or multiples of 3. On the other hand, in the intersection of the modules, p holds for all the natural numbers which are multiples of 6, that is which are multiples both of 2 and of 3. Formally:

$$T_{M_1 \cap M_2} \!\uparrow\! \omega = \{p(0)\} \cup \{p(s^k(0)) \mid \exists n, m \in N^+ : (k = 2n \wedge k = 3m)\}$$

4 Modules: metainterpretive view

In the previous section, an abstract notion of module has been formally introduced, along with suitable operators of module composition. We now present a realisation of modules in terms of metalogic, by mapping back the representation of modules onto programs via the definition of a suitable functor. This also shows how our approach does not betray the spirit of logic programming, in that it can be expressed as an application of metalogic.

4.1 Construction of modules

The idea is to introduce a binary functor *mod* which, when applied to two programs P and Q, gives the metalevel representation of a module $\langle P, Q \rangle$. In the spirit of the metalevel approach discussed in subsection 2.3, we extend the definition of *demo* to properly deal with the *mod* functor.

Definition 4.1 *The metalevel operator mod is defined by the following axiom to be added to the vanilla metaprogram.*

$$(d7) \quad demo(mod(P, Q), A \leftarrow G) \leftarrow demo(P, A \leftarrow G, G'),$$
$$demo(Q, G')$$

Roughly, axiom $(d7)$ states that clauses in $mod(P, Q)$ are obtained from clauses of P by hiding part of their body, provided that it is provable in Q, and that the resulting bindings are applied to the visible part of the body. Notice that the definition of *mod* actually corresponds to the abstract specification of a module given in definition 3.2:

$$T_{\langle P, Q \rangle} = \lambda I \; . \; T_P(I \cup T_Q \!\uparrow\! \omega).$$

The call $demo(Q, G')$ mirrors the fact that P can only exploit the set of formulae provable in Q.

For the sake of simplicity, in definition 4.1, we have assumed that clauses in P are of the form $A \leftarrow G, G'$ where the subgoal G' contains the calls to the predicates of Q. It is worth noting that the former assumption is not restrictive, as it is easy to define a transformation which maps clauses into the required form.

The following proposition states the soundness of the metalogical definition of mod with respect to the basic semantics of modules given in section 3.1.

Proposition 4.2 *Let* P *and* Q *be object-level programs, let* V *be the vanilla metaprogram consisting of axioms* $(d1), (d2), (d3), (d7)$ *and* M *be the metaprogram composed by* P_m, Q_m *and* V. *For any object-level ground atomic formula* A:

$$M \vdash demo(mod(P, Q), A) \iff A \in T_{\langle P,Q \rangle} \uparrow \omega$$

4.2 Composition of modules

We now turn our attention to the module composition operators ⊔ and ⊓ introduced in subsection 3.2. As we have seen, modules are represented at the metalevel as a composition of programs through the mod operator. The main advantage of this approach is that the operators for module composition can be mapped back onto the operators for program composition. Union and intersection of modules (⊔, ⊓) are directly expressed in terms of union and intersection of programs (∪, ∩).

Definition 4.3 *The* ⊔ *and* ⊓ *operators are defined by the following axioms to be added to the vanilla metaprogram.*

(d8) $\quad demo(M1 \sqcup M2, A \leftarrow G) \leftarrow demo(M1 \cup M2, A \leftarrow G)$

(d9) $\quad demo(M1 \sqcap M2, A \leftarrow G) \leftarrow demo(M1 \cap M2, A \leftarrow G)$

It is worth noting that, in the above definition, $M1$ and $M2$ stand for two generic module expressions. For example, to prove a goal G in the union of two modules $\langle P_1, Q_1 \rangle \sqcup \langle P_2, Q_2 \rangle$, the metalevel representation

$$demo(mod(P_1, Q_1) \sqcup mod(P_2, Q_2), G)$$

is adopted. The following proposition states the soundness of the metalevel realisation of module composition operators. All the proofs are omitted here and reported in [4].

Proposition 4.4 *Let* P_1, Q_1, P_2, Q_2 *be object-level programs. Let* V *be the vanilla metaprogram consisting of axioms* $(d1)$—$(d9)$ *and* M *be the metaprogram composed by the metalevel representation of the object-level programs and by* V. *For any object-level ground atomic formula* A:

i) $\quad M \vdash demo(mod(P_1, Q_1) \sqcup mod(P_2, Q_2), A) \iff A \in T_{\langle P_1, Q_1 \rangle \sqcup \langle P_2, Q_2 \rangle} \uparrow \omega$

ii) $\quad M \vdash demo(mod(P_1, Q_1) \sqcap mod(P_2, Q_2), A) \iff A \in T_{\langle P_1, Q_1 \rangle \sqcap \langle P_2, Q_2 \rangle} \uparrow \omega$

5 Other operators

Starting from the two basic operators for composing modules, we define a kit of operators for modularising logic programming. More precisely, we introduce operators for constructing modules out of logic programs, for encapsulating modules and for realising different compositions of modules.

5.1 From programs to modules

Software re-use is one of the main requirements of program construction. Any framework for modularising logic programming must provide means for constructing modules from standard logic programs. There are at least two possible ways for building a module out of a logic program.

One way is to define the module in an *intensional* way, so that the whole program will be visible to other modules. In other words, the program becomes the visible part of a module with empty hidden part. Alternatively, a module can be constructed from a program in an *extensional* way so that only the set of provable sentences of the program will be visible to other modules. We call *int* and *ext* the corresponding operators, with the following expected semantics.

Definition 5.1 *Given a program P:*

$$i) \quad T_{int(P)} = \lambda I.T_P(I)$$
$$ii) \quad T_{ext(P)} = \lambda I.T_P{\uparrow}\omega$$

Intuitively, the above definition states that the semantics of $int(P)$, that is of an intensionally constructed module, coincides with the semantics of the initial program P. On the other hand, the semantics of $ext(P)$ coincides with the least fixpoint of the semantics of P.

The metalogical definition of *int* can be naturally expressed by the following axiom:

$$demo(int(P), A \leftarrow G) \leftarrow demo(P, A \leftarrow G)$$

For the sake of homogeneity, however, we prefer to express *int* in terms of *mod*.

Definition 5.2 *The int operator is defined by the following axiom:*

$$(d10) \quad demo(int(P), A \leftarrow G) \leftarrow demo(mod(P, \{\}), A \leftarrow G)$$

where $\{\}$ denotes the empty program.

A module can be constructed out of a program in an extensional way. This means that only the set of provable sentences of the program are exported to the external world. Again, the metalevel definition is straightforward:

$$demo(ext(P), A \leftarrow true) \leftarrow demo(P, A)$$

The *ext* operator is actually a special case of a more general operator of *encapsulation* of modules which will be introduced in the next subsection. We will show that, to construct a module extensionally, we can first construct it intensionally and then encapsulate it.

5.2 Module encapsulation

The process of encapsulation is twofold. An encapsulated module cannot refer to the contents of other modules, while other modules can exploit the encapsulated module only at the provability level. The expected semantics is the following.

Definition 5.3 *Given a module* $\langle P, Q \rangle$:

$$T_{enc(\langle P,Q \rangle)} = \lambda\ I.\ T_{\langle P,Q \rangle} \uparrow \omega$$

We now provide the encapsulation operation with a metalevel definition.

Definition 5.4 *The enc operator is defined by the following axiom:*

(d11) $demo(enc(M), A \leftarrow true) \leftarrow demo(M, A)$

It is easy to see that the former is a faithful implementation of *enc*.

Proposition 5.5 *Let P and Q be object-level programs, let V be the vanilla metaprogram consisting of axioms* (d1)—(d11) *and M be the metaprogram composed by* P_m, Q_m *and V. For any object-level ground atomic formula A:*

$$M \vdash demo(enc(mod(P, Q)), A) \iff A \in T_{\langle P,Q \rangle} \uparrow \omega$$

It is interesting to point out that, as mentioned previously, *ext* can be defined in terms of *enc* and *int*. Indeed: $\forall G.\ demo(ext(P), G) \equiv demo(enc(int(P)), G)$.

5.3 Disjoint and asymmetric compositions

The *enc* operator allows us to define a number of new operators. In particular disjoint and asymmetric composition of modules can be defined in terms of *enc* and the basic union and intersection operators. All these new definitions are justified by our basic objective of setting up a rich configuration language over modules. For example, disjoint union and intersection (denoted by $\vee$ and $\wedge$, respectively) allow one to build modules which interact only at the extensional level, whereas asymmetric composition allows one to enforce the distinction between an "importing" module and an "exporting" module.

Definition 5.6 *Disjoint union and intersection of modules are defined by the following axioms to be added to the vanilla metainterpreter:*

(*d*12) $demo(M1 \vee M2, A \leftarrow G) \leftarrow demo(enc(M1) \sqcup enc(M2), A \leftarrow G)$

(*d*13) $demo(M1 \wedge M2, A \leftarrow G) \leftarrow demo(enc(M1) \sqcap enc(M2), A \leftarrow G)$

According to the former definition, the set of atomic formulae provable in $M_1 \vee M_2$ ($M_1 \wedge M_2$) coincides with the union (the intersection) of the sets of atomic formulae provable in M_1 and M_2. Notice that the following stronger axioms could be equivalently used instead of axioms (*d*12) and (*d*13):

$$demo(M1 \vee M2, A) \leftarrow demo(M1, A)$$

$$demo(M1 \wedge M2, A) \leftarrow demo(M2, A)$$

We now consider an operation of asymmetric composition of modules $M_1 \ll M_2$. The intuition is that M_1 imports from M_2 or, equivalently, M_2 exports to M_1. In other words, M_1 has access to the definitions in M_2, but not the other way around. Moreover, M_1 has access to the definitions in M_2 only at the extensional level. The expected semantics is that the exporting module contributes to the union only at the provability level.

Definition 5.7 *Given two modules M_1 and M_2:*

$$T_{M_1 \ll M_2} = \lambda I.\ T_{M_1}(I) \cup T_{M_2}\uparrow\omega$$

The $\ll$ operator can be defined in terms of the operators previously introduced.

Definition 5.8 *The $\ll$ operator is defined by adding the following axiom to the vanilla metainterpreter:*

(*d*14) $demo(M1 \ll M2, A \leftarrow G) \leftarrow demo(mod(M1 \sqcup enc(M2), A \leftarrow G)$

It is worth observing how the derived operators provide a kernel configuration language capable of assembling separate modules together to form large systems. For example, consider a hierarchy of four modules $M_1, \ldots, M_4$, such that M_1 imports both from M_2 and M_3, and both M_2 and M_3 import from M_4. The overall system is configured by the expression: $M_1 \ll ((M_2 \vee M_3) \ll M_4)$.

6 Conclusions and future work

We have studied a notion of module for logic programming along with a set of module composition operators. The preservation of the standard semantics of logic programming and the use of metalogic are the distinguishing features of our approach.

Metalogic offers a simple and elegant way for modularising logic programming. The addition of a few clauses to the vanilla metainterpreter actually supports a set of module composition operators. Such an implementation of modules and module composition operators offers the advantage of maintaining object-level programs separated. For example, a certain object-level program Q can be used to define different modules, e.g. $\langle P_1, Q \rangle$ and $\langle P_2, Q \rangle$, without duplicating the code of Q. This is very convenient, for example, when dealing with deductive databases. It is much more efficient to work with virtual composition of files rather than to actually merge databases together to evaluate a query.

Metalevel implementations are however costly in terms of efficiency. Partial evaluation techniques can be applied to reduce the computational overhead due to metainterpretation. As an alternative, we have developed a compilation oriented approach, much in the spirit of the transformational definition of the operators on programs introduced in [3]. In [4], both the interpretive and the compilative implementation are presented and compared in full detail. We also consider the possibility of mixing the two implementation techniques in order to obtain a configuration language for modules with great flexibility. Experimentations with such implementations are currently carried on in the logic programming language Gödel [9].

The complete description of our proposal for modularising logic programming is reported in [4]. We have studied a number of algebraic properties of the operators, capable of providing a formal framework in which module configuration expressions can be compared. Finally, an algebra of modules (and module compositions) is defined by extending the algebra of logic programs presented in [13].

Acknowledgments This work has been partially supported by ESPRIT BRA 3012 Compulog and by Progetto Finalizzato Sistemi Informatici e Calcolo Parallelo under grant 9100880.PF69.

References

1. K. R. Apt. Logic programming. In J. van Leeuwen, editor, *Handbook of Theoretical Computer Science*, pages 493–574. Elsevier, 1990. Vol. B.
2. K.A. Bowen and R.A. Kowalski. Amalgamating Language and Metalanguage in Logic Programming. In K.L. Clark and S.A. Tarnlund, editors, *Logic Programming*, pages 153–173. Academic Press, 1982.
3. A. Brogi, P. Mancarella, D. Pedreschi, and F. Turini. Composition Operators for Logic Theories. In J.W. Lloyd, editor, *Computational Logic, Symposium Proceedings*, pages 117–134. Springer-Verlag, 1990.
4. A. Brogi, P. Mancarella, D. Pedreschi, and F. Turini. Modules as Logic Programs. Technical report, University of Pisa, 1992.

5. A. Brogi and F. Turini. Metalogic for Knowledge Representation. In J.A. Allen, R. Fikes, and E. Sandewall, editors, *Principles of Knowledge Representation and Reasoning: Proceedings of the Second International Conference*, pages 100–106. Morgan Kaufmann, 1990.
6. K. Clark. Negation as failure. In H. Gallaire and J.Minker, editors, *Logic and Data Bases*, pages 293–322. Plenum, 1978.
7. L. Giordano and A. Martelli. A modal reconstruction of blocks and modules in logic programming. In *Proc. International Logic Programming Symposium*, pages 239–253. The MIT Press, 1991.
8. P.M. Hill and J.W. Lloyd. Analysis of metaprograms. In H.D. Abramson and M.H. Rogers, editors, *Metaprogramming in Logic Programming*, pages 23–52. The MIT Press, 1989.
9. P.M. Hill and J.W. Lloyd. The logic programming language Gödel. Technical report, University of Bristol, 1991.
10. R.A. Kowalski and J.S. Kim. A metalogic programming approach to multi-agent knowledge and belief. In V. Lifschitz, editor, *Artificial Intelligence and Mathematical Theory of Computation*. Academic Press, 1991.
11. J.L. Lassez and M. Maher. Closures and fairness in the semantics of logic programming. *Theoretical Computer Science*, 29:167–184, 1984.
12. J.W. Lloyd. *Foundations of logic programming*. Springer-Verlag, second edition, 1987.
13. P. Mancarella and D. Pedreschi. An algebra of logic programs. In R. A. Kowalski and K. A. Bowen, editors, *Proc. Fifth International Conference on Logic Programming*, pages 1006–1023. The MIT Press, 1988.
14. D. Miller. A logical analysis of modules in logic programming. *Journal of Logic Programming*, 6:79–108, 1989.
15. R. O'Keefe. Towards an algebra for constructing logic programs. In J. Cohen and J. Conery, editors, *Proceedings of IEEE Symposium on Logic Programming*, pages 152–160. IEEE Computer Society Press, 1985.
16. D.T. Sannella and L.A. Wallen. A calculus for the construction of modular Prolog programs. *Journal of Logic Programming*, 12:147–177, 1992.
17. M.H. van Emden and R.A. Kowalski. The semantics of predicate logic as a programming language. *JACM*, 23(4):733–742, 1976.

Compiler Optimizations
for Low-level Redundancy Elimination:
An Application of Meta-level Prolog Primitives

Saumya K. Debray

Department of Computer Science

University of Arizona

Tucson, AZ 85721, USA

Abstract

Much of the work on applications of meta-level primitives in logic programs focusses on high-level aspects such as source-level program transformation, interpretation, and partial evaluation. In this paper, we show
how meta-level primitives can be used in a very simple way for low-level
code optimization in compilers. The resulting code optimizer is small,
simple, efficient, and easy to modify and retarget. An optimizer based on
these ideas is currently being used in a compiler that we have developed
for Janus [6].

1 Introduction

Much of the work on applications of meta-level primitives in logic programs
focuses on high-level aspects such as source-level program transformation, interpretation, and partial evaluation. In this paper, we consider instead the use
of meta-level Prolog primitives in low-level code optimization in compilers. We
show how such primitives can be used in a very simple way for a low-level code
optimization called *common subexpression elimination*. The resulting code optimizer is small, simple, efficient, and easy to modify and retarget. Because
it requires only a very simple logical description of the instruction set under
consideration, correctness is simple to guarantee.

The application we describe is not profound—indeed, its primary appeal
to us is its simplicity (it happens also to be efficient, effective, and easy to
implement). An experienced Prolog programmer might very well consider it to
be an "obvious hack," and if the only application we could find for it was in
Prolog compilers written in Prolog, then we would consider it to be too narrow
an application to be much more than a curiosity. One of the contributions of
this paper is to show that these ideas are applicable, not only to the compilation
of logic programming languages, but also to compilers for traditional functional
and imperative languages. A related technique is dicussed by Komorowski in
the context of partial evaluation [8].

Conceptually, the optimization algorithm described here can be thought of as a scheme to transform an intermediate representation of a program from a tree to a DAG via some sort of *value numbering* scheme [1]. This, however, will be true for essentially every algorithm for common subexpression elimination. First, almost any scheme for common subexpression elimination can be thought of in terms of merging distinct computations ("nodes") that are equivalent in some sense, i.e., in terms of a transformation from a tree to a DAG. Moreover, unless the implementation is entirely naive, checking whether two instruction sequences represent the computation of a common subexpression will be carried out, wherever possible, without exhaustively comparing the two sequences: this can typically be formulated in terms of some kind of value numbering scheme. What we feel is interesting and elegant about the approach described in this paper is that the use of unification, together with meta-level primitives, allows much of the low-level clutter associated with these operations to be avoided by the compiler writer. As a result, the code generator produced is transparent, easy to understand, verify, modify, and retarget.

2 Background

2.1 Common Subexpressions

During compilation, a program is typically translated from the source language to a lower level intermediate language. The program resulting from the translation to intermediate code may contain *common subexpressions*. An occurrence of an expression E is called a common subexpression if E was previously evaluated, and the values of variables occurring in E have not changed since the previous computation [1]. Common subexpressions may arise in a program either because there are multiple occurrences of an expression in the source program (which may happen, for example, after macro expansion), or because the translation to intermediate code makes explicit lower-level operations that are not visible at the source level. As an example, the Fortran code fragment

```
A[i] = A[i] + 1
```

may translate, on a machine with 4-byte words, to the intermediate code sequence

```
t0 := 4*i
t1 := A[t0]
t2 := t1+1
t3 := 4*i
A[t3] := t2.
```

Here, the expression `4*i` is a common subexpression: it arises because the low-level details of array subscripting are not visible at the source level.

In general, the performance of a program can be improved by eliminating code to recompute common subexpressions, and using the previously computed value instead. In practice, common subexpression elimination incurs a cost, since it may tie up a machine register to hold the value of the expression for subsequent uses, or result in stores and loads of the saved value from memory. In general, a compiler has to weigh the savings realized from common subexpression elimination against the costs incurred to determine whether a particular common subexpression is worth eliminating (e.g., see [14]). Thus, common subexpression elimination involves

1. identifying common subexpressions;

2. deciding which of these are worth eliminating; and

3. transforming the code sequence to eliminate such common subexpressions.

Since these three points are essentially orthogonal to each other, we will focus on points (1) and (3) in this paper.

2.2 Static Single Assignment Form

As we will see, our compilation model views a variable as a logical entity that can be defined at most once. In an imperative language, it may happen that the source program contains multiple assignments to a variable. It is possible to transform such programs to *static single assignment form* [4, 5] to conform to our compilation model. In static single assignment form, a program is transformed so that the program text contains at most one assignment to any variable along any execution path. The significance of this transformation stems from the fact that (even in imperative languages) single-assignment source variables are desirable for a number of optimizations, including parallelism detection [4, 10, 13]. Dynamically, a program with loops may assign many times to the same variable, even if only one assignment appears in the program text (a scheme to get around this problem, involving the creation of new variables dynamically, is discussed in [4]). In this section, we briefly review the transformation of a program to static single assignment form, as described in [5].

First, consider the transformation for a single basic block with multiple definitions of a variable. The transformation simply renames all definitions of the variable except for the final one, and their corresponding uses, as illustrated below:

Original		**Renamed**	
X = ...	/* define X */	X1 = ...	/* define X1 */
...		...	
= X	/* use X */	= X1	/* use X1 */
...		...	
X = ...	/* define X */	X2 = ...	/* define X2 */
...		...	
= X	/* use X */	= X2	/* use X2 */
...		...	
X = ...	/* define X */	X = ...	/* define X */

Since only the last definition of a variable within a block can be used outside that block, this transformation does not affect any use of that variable in other basic blocks. Within a basic block, each use of a variable must be renamed, if necessary, to match the corresponding definition.

Next, consider a case where multiple definitions reach a use. In this case, we first identify the "join birthpoints" of variables in the control flow graph, i.e., points in the control flow graph where several definitions of a variable meet on different incoming edges for the first time. Consider a join birthpoint for a variable X where k paths, each with a definition for X, meet. We transform the program by renaming X along each path, ensuring that this renaming does not introduce the same name along different paths. Suppose that for each path i, $1 \leq i \leq k$, the last definition of X has been renamed in this process to X_i. We then ensure that the original variable contains the correct value by adding an assignment $X = \phi(X_1, \ldots, X_n)$ for each path i, $1 \leq i \leq k$: here, $\phi(\ldots)$ is a special form of assignment, called a *join-definition*, that assigns the appropriate value depending on which branch of the conditional was taken: in practice, this can be implemented in a fairly straightforward way (see [4]). The approach is illustrated by the following example:

Original	Renamed
X = A+B	X1 = A+B
if (...)	if (...)
then X = X+1	then X2 = X1+1
Z = 0	Z = 0
else X = X+2	else X3 = X1+2
	X = ϕ(X2, X3)
...	...
Y = 2*X	Y = 2*X

This step eliminates multiple assignments to a variable in the absence of loops. A scheme to deal with loops, via dynamic creation of new variables, is described in [4]: since we will be concerned primarily with loop-free programs, this will not be discussed further here.

3 Common Subexpression Elimination: The Traditional Approach

Common subexpression elimination is usually carried out by analyzing the abstract syntax tree for a compilation unit (typically a procedure) to find identical expression subtrees. Suppose that E_1 and E_2 are two such identical subtrees, and E_1 is guaranteed to be evaluated before E_2. Then, if the variables in E_2 can be guaranteed to be unchanged since the evaluation of E_1, then the subtree E_2 can be replaced by a pointer to the subtree E_1. As a result, the syntax tree is transformed into a DAG in which nodes with more than one parent correspond to common subexpressions.

Now suppose we are processing an expression tree E in a procedure. In a naive implementation, determining whether there is another subtree elsewehere in the procedure identical to E might be carried out by actually matching E against the various expression subtrees occurring in the procedure. This, of course, would be hopelessly inefficient in general, even if the search is restricted to "previously computed" expression subtrees. In practice, therefore, a more sophisticated scheme called *value numbering* [3, 9, 10] is used. The essential idea is to assign special symbolic names called value numbers to expressions. Then, if two expressions $E_1 \equiv op_1(t_1, \ldots, t_n)$ and $E_2 \equiv op_2(u_1, \ldots, u_n)$ satisfy (i) $op_1 = op_2$, and (ii) the value number of t_i is the same as that of u_i, $1 \leq i \leq n$, then E_1 and E_2 are guaranteed to compute the same value.

The implementation of value numbering, however, can be considerably more complicated that this description might suggest. For example, [1] describes an implementation scheme that involves using a hash table to keep track of expressions that are potentially common subexpressions. Moreover, if there can be multiple assignments to a variable in the program, then this structure has to be kept consistent with updates.

4 Our Approach

4.1 The Compilation Model

The most significant difference between our compilation model, and that used in traditional compilers, is that we view a (source or temporary) variable as a logical entity whose value can be defined at most once. This turns out to simplify significantly the subsequent reasoning about, and optimization of, the intermediate code program, since there is no need to worry about the value of a variable changing due to multiple assignments to it.

Traditional compilers typically attempt to conserve machine resources by deallocating temporary variables when they are no longer needed, and reusing such deallocated temporaries later if possible. In our model, in contrast, no attempt is made to reuse temporary variables during intermediate code generation. The effects of such reuse are obtained later, during final code generation, when temporary variables are mapped to machine resources such as memory locations or registers: at this time, liveness information can be used to map variables with disjoint lifetimes into the same register or memory location. Multiple assignments to variables in the source program can be handled by transformation to static single assignment form, as discussed earlier.

We also assume that intermediate code instructions are *recyclable*, i.e., can be reused. The idea here is the following: suppose we have two instructions I_1 and I_2, with identical operands, in a basic block. From the assumption that variables and temporaries are single assignment entities, this means that these instructions will have identical operand values at runtime. The assumption of recyclability states that in this case, the result from the first instruction I_1 can be recycled and used in place of the second instruction I_2. While the assumption seems not too unreasonable, it need not be satisfied in practice, e.g., if the instruction under consideration is nondeterministic, or if it involves side effects, e.g., for I/O. In practice, however, our scheme can be used even if not all instructions in the language under consideration are recyclable, as long as we restrict our attention to recyclable instructions.

In the remainder of this paper, we consider common subexpression elimination in loop-free code fragments only (this is not as bad as it may seem, since most compilers restrict themselves to common subexpression elimination within basic blocks or extended basic blocks): in this case, the transformation to static single assignment form suffices to ensure that our assumptions are satisfied.

For the remainder of the paper, intermediate code instructions will be represented as follows unless explicitly mentioned: an instruction with opcode op, operands $In_1, \ldots, In_m$, and results $Out_1, \ldots, Out_n$ will be represented as

$$op([In_1, \ldots, In_m], [Out_1, \ldots, Out_n]).$$

For example, the instruction add([R1, R2], [R3]) indicates that the sum of R1 and R2 is assigned to R3. If an instruction has no operands, then the first argument is the empty list []: for example, an increment instruction might be written 'inc([], [X])'. It is important to note that any entity that may be "read" by an instruction is expected to be listed explicitly as an operand, while any entity that may be "written" by an instruction is expected to be listed as a result: this includes entities, such as stack or heap pointers, that are often treated as implicit operands. Further, the single assignment requirement applies to all operands and results.

4.2 Redundancy Elimination within a Basic Block

Strictly speaking, our approach aims to eliminate redundant instructions rather than common subexpressions. This includes common subexpression elimination as a special case, since a common subexpression manifests itself as a sequence of redundant instructions; however, it also removes certain kinds of redundancies, such as type tests, that might not be considered to be a common subexpression in a traditional compiler. The principle underlying our algorithm is extremely simple and quite obvious: *two (deterministic) instructions that apply the same operator to identical operands will produce the same results*. The determinacy requirement, which says that the instructions compute functions, is important, but not very restrictive for our application: we do not know of any intermediate representation language for compilers that does not satisfy this requirement. It turns out that by using unification and Prolog meta-level primitives, we are able to exploit this obvious fact in a clean and simple way, obtaining simple and efficient code optimizers without having to worry about any of the low-level clutter associated with common subexpression elimination in traditional compilers.

Assume that the instructions in a basic block are represented as a list of Prolog terms (each instruction is a Prolog term of the form described at the end of the previous section). We assume that we have a set of instructions Seen that have already been encountered. The essence of our algorithm is straightforward: given an instruction $I \equiv op(In, Out)$ in the basic block, if there is an instruction $I' \equiv op(In', Out')$ in Seen such that the operands In and In' are identical, then

Input : A basic block B of intermediate code instructions.

Output : A modified basic block B with redundant instructions deleted.

Method :

```
Seen := ∅;
for each instruction I ≡ op(In, Out) in B do
    if ∃ op(In', Out') ∈ Seen such that In == In' then
        unify Out and Out';
        delete I from B;
    else
        add I to Seen;
    fi
od
```

Prolog Realization :

```prolog
cse_elim(B_in, B_out) :- cse_elim(B_in, [], B_out).

cse_elim([], S, []).
cse_elim([I1|Rest], Seen, L) :-
    (find(Seen, I2), eqvt(I1, I2)) ->
        cse_elim(Rest, Seen, L)
    ; (L = [I1|Lrest], add(I1, Seen, Seen0),
       cse_elim(Rest, Seen0, Lrest)
      ).

% eqvt(I1, I2) is true iff the instuctions I1 and I2 have
% identical inputs, i.e., are equivalent.  In this case,
% their outputs are unified.

eqvt(I1, I2) :-
    I1 =.. [Op,In1,Out], I2 =.. [Op,In2,Out], In1 == In2.
```

Figure 1: An Algorithm for Redundancy Removal within a Basic Block

their results Out and Out′ must be equal, so we can simply unify Out and Out′ (so that future references to Out now also reference Out′) and delete I; otherwise, I has not been encountered before, and should be added to Seen. The algorithm, and Prolog code realizing it, is given in Figure 1. The efficiency of the Prolog code may be improved by choosing the data structure for Seen more carefully, e.g., by indexing it by opcode and passing the opcode of the instruction being considered as a third argument to find.

Example 4.1 Consider the following source code statement in a Pascal-like language:

```
a[i, j] := a[i, j] + 1;
```

Assume that the array a is stored in row-major order starting at location 1000, that each array element occupies 4 bytes of memory, and that all of its subscripts range over the interval [1..100]. Then, the address of a[i, j] is given by the following expression (see [1] for details):

$$1000 + 4 * 100 * (i - 1) + 4(j - 1)$$
$$= 4(i * 100 + j) + 596.$$

Code generated directly, without common subexpression elimination, will repeat this address computation:

```
(1)     mult([100, I], [T1])          /* T1 := 100 * I    */
(2)     add([T1, J], [T2])            /* T2 := T1 + J     */
(3)     mult([4, T2], [T3])           /* T3 := 4 * T2     */
(4)     add([T3, 596], [T4])          /* T4 := T3 + 596   */
(5)     indirect_load([T4], [T5])     /* T5 := *T4        */
(6)     add([T5, 1], [T6])            /* T6 := T5 + 1     */
(7)     mult([100, I], [T7])          /* T7 := 100 * I    */
(8)     add([T7, J], [T8])            /* T8 := T7 + J     */
(9)     mult([4, T8], [T9])           /* T9 := 4 * T8     */
(10)    add([T9, 596], [T10])         /* T10 := T8 + 596  */
(11)    indirect_store([T6], [T10])   /* *T10 := T6       */
```

When our algorithm is executed on this code, no instruction will be eliminated until instruction (7) is processed. Since the inputs to instruction (7) are identical to those of (1), this results in the variables T7 and T1 becoming unified and instruction (7) being discarded. The instruction sequence at this point, therefore, is:

```
(1)     mult([100, I], [T1])           /* T1 := 100 * I   */
(2)     add([T1, J], [T2])             /* T2 := T1 + J    */
(3)     mult([4, T2], [T3])            /* T3 := 4 * T2    */
(4)     add([T3, 596], [T4])           /* T4 := T3 + 596  */
(5)     indirect_load([T4], [T5])      /* T5 := *T4       */
(6)     add([T5, 1], [T6])             /* T6 := T5 + 1    */
(8)     add([T1, J], [T8])             /* T8 := T1 + J    */
(9)     mult([4, T8], [T9])            /* T9 := 4 * T8    */
(10)    add([T9, 596], [T10])          /* T10 := T8 + 596 */
(11)    indirect_store([T6], [T10])    /* *T10 := T6      */
```

Notice now that as a result of the unification of T1 and T7 at the previous step, the inputs to instructions (2) and (8) become identical. At the next step, therefore, the variables T2 and T8 will become unified and instruction (8) will be discarded. This process continues, and the code finally generated does not repeat any of the address computation:

```
(1)     mult([100, I], [T1])           /* T1 := 100 * I   */
(2)     add([T1, J], [T2])             /* T2 := T1 + J    */
(3)     mult([4, T2], [T3])            /* T3 := 4 * T2    */
(4)     add([T3, 596], [T4])           /* T4 := T3 + 596  */
(5)     indirect_load([T4], [T5])      /* T5 := *T4       */
(6)     add([T5, 1], [T6])             /* T6 := T5 + 1    */
(11)    indirect_store([T6], [T4])     /* *T4 := T6       */
```

□

The algorithm, as described above, has two minor shortcomings:

1. It may sometimes fail to detect common subexpressions involving copy statements, i.e., assignments of the form

   x := y.

   This is illustrated by the following example: consider the instruction sequence

```
   store([1], [X])                /* X := 1      */
   add([X, Y], [Z])               /* Z := X + Y  */
   add([1, Y], [U])               /* U := 1 + Y  */
```

In this case, the algorithm fails to infer that 1 + Y in the instruction add([1, Y], [U]) is a common subexpression. This problem can be taken care of by carrying out *copy propagation* [1] before common subexpression elimination. *A point to note here is that join definitions, i.e. assignments to a variable introduced at join birthpoints during the transformation to static single assignment form, should not be considered during copy propagation, since otherwise the resulting program may no longer be in static single assignment form.*

2. The algorithm does not know about algebraic properties of operations, e.g. that addition is commutative. As a result, it may sometimes fail to detect some common subexpressions. This is illustrated by the following example:

```
add([1, Y], [Z])              /* Z := 1 + Y */
add([Z, 2], [X])              /* X := Z + 2 */
add([Y, 1], [U])              /* U := Y + 1 */
mult([U, 4], [V])             /* V := U * 4 */
```

In this case, the algorithm fails to infer that `1 + Y` in the instruction `add([Y, 1], [U])` is a common subexpression. This problem can be taken care of by augmenting the Prolog code to express the desired algebraic properties, e.g. by adding clauses of the form

```
eqvt(I1, I2) :-
    1 =.. [Op, [X1,Y1], Out], I2 =.. [Op, [X2,Y2], Out],
    commutative(Op), X1 == Y2, X2 == Y1.

commutative(add).
commutative(mult).
```

Even though this is somewhat more complicated than the original definition given in Figure 1, notice that all that we are doing is elaborating, in a clean and logical way, the notion of "equivalence" between two instructions. The point is that the overall algorithm—and anything that depends on it—is not affected, all we are doing is refining the eqvt/2 relation. Obviously, additional properties could be expressed by suitably elaborating the definition of eqvt/2, without affecting any of the remainder of the algorithm. In our experience, this ability to specify aspects of the instruction set in a clean and declarative way is very helpful for verification and modification of the low-level code optimizer (its simplicity, modularity, declarative reading, and ease of modification contrast very pleasantly with the corresponding code that is typically found in traditional compilers).

4.3 Redundancy Elimination across Basic Block Boundaries

Recall that we are considering only loop-free program fragments. It therefore suffices to consider two cases: (i) a fork point, i.e., where a basic block has more than one successor; and (ii) a join point, i.e., where a basic block has more than one predecessor.

Dealing with fork points is straightforward: if a block B has n successors $B_1, \ldots, B_n$ then the initial Seen set at the entry to each of the blocks $B_1, \ldots, B_n$ is the Seen set at the exit from block B.

For join points, we have to ensure that instructions encountered along one branch leading upto the join point, but not along another branch, are not considered to have been seen when the basic block at the join point is considered. This is easy to handle: consider a basic block B with k predecessors $B_1, \ldots, B_k$, and let the set of instructions seen at the end of a predecessor B_i be $Seen_i$, $1 \leq i \leq k$. Then, the set Seen at the entry to B is given by

$$\text{Seen} = \bigcap_{i=1}^{k} \text{Seen}_i.$$

With this change, the algorithm can be used for common subexpression elimination in any loop-free program.

5 Common Subexpression Elimination in the WAM

The kind of common subexpression most commonly encountered in Prolog programs involves redundant construction of terms [2]. For example, on most Prolog implementations, the clause

```
p([f(X,Y)|L]) :- q(f(X,Y)), p(L).
```

will create two copies of the term $f(X, Y)$ each time around the recursion when executed. Since most high-performance Prolog systems are based on the WAM [12], it would be nice if we could adapt our scheme to the WAM. This cannot be done directly, for the following reasons:

1. WAM instructions use implicit arguments, and as a result are context sensitive. For example, a get_list or get_structure instruction has the registers S (the structure pointer) and H (the heap pointer), as well the mode bit, as implicit outputs. This is not really a significant problem,

since it can be rectified by rewriting the instructions to make all operands explicit.

2. WAM entities are not single assignment. For example, the various registers can be destructively updated. Again, this does not seem to be a significant problem, since in principle we could consider a transformation to static single assignment form, similar to that discussed earlier for imperative programs, before applying our algorithm.

Unfortunately, as far as we can see, our scheme remains inapplicable to reclaiming common substructures even after these problems have been addressed. There are two (related) reasons for this:

1. Consider a clause 'p :− q(f(a), f(a))'. The WAM code for this, after we have transformed the instruction set to make all operands explicit, and further have transformed the resulting code to static single assignment form, would be the following, where S_i and H_i denote different values of the S and H registers respectively:

```
p/0 :   put_structure([f/1, H₀], [A1, H₁, S₁])
        unify_constant([a, H₁, S₁, write_mode], [H₂, S₂])
        put_structure([f/1, H₂], [A2, H₃, S₃])
        unify_constant([a, H₃, S₃, write_mode], [H₄, S₄])
        execute q/2
```

The two put_structure instructions cannot be identified to be equivalent because the register H, which points to the top of the heap and is an input to this instruction, is different for the two instructions. As a result, the two occurrences of the common substructure f(a) cannot be identified and optimized.

2. Suppose, to get around this problem with failing to identify the common substructure f(a), we ignore the value of the H register when determining whether two put_structure instructions are equivalent. In this case, things become even worse: consider the clause

```
p :− q(f(a), f(b)).
```

The code for this is

```
p/0 :   put_structure([f/1, H₀], [A1, H₁, S₁])
        unify_constant([a, H₁, S₁, write_mode], [H₂, S₂])
        put_structure([f/1, H₂], [A2, H₃, S₃])
        unify_constant([b, H₃, S₃, write_mode], [H₄, S₄])
        execute q/2
```

By ignoring the different values of the H register in the two put_structure instructions, we would erroneously infer that these two instructions are equivalent, and optimize away the second. This would yield the following code:

```
p/0 :   put_structure([f/1, H₀], [A1, H₁, S₁])
        unify_constant([a, H₁, S₁, write_mode], [H₂, S₂])
        unify_constant([b, H₁, S₁, write_mode], [H₄, S₄])
        execute q/2
```

This is clearly incorrect: register A2 is not set at all any more, and the constant a written onto the heap by the second instruction is immediately overwritten by the constant b as a result of executing the third instruction.

6   Conclusions

Common subexpression elimination is an important low-level compiler optimization. Traditional approaches to implementing this optimization, typically via value numbering schemes, are complicated by a large amount of low-level clutter involved with the maintenance of information about value numbers. In this paper we propose a much simpler scheme for this transformation, using Prolog meta-language features and unification. The scheme is simple and easy to understand, and easy to implement, modify, and extend, and efficient in practice. It considers loop-free programs and assumes that variables are single-assignment entities: thus, if applied to imperative language programs, it requires that they be transformed to static single assignment form. However, since the static single assignment form is useful for a variety of other optimizations, this need not be a great burden. Our scheme is also very flexible in the sense that, unlike common subexpression elimination in most traditional compilers, it is applicable not only to arithmetic, but also to instructions that test types, bounds, etc.

Acknowledgements: Comments by Mats Carlsson and Jan Komorowski were very helpful in improving the contents of the paper. This work was supported in part by the National Science Foundation under grant number CCR-8901283.

References

[1] A. V. Aho, R. Sethi and J. D. Ullman, *Compilers – Principles, Techniques and Tools*, Addison-Wesley, 1986.

[2] M. Carlsson, personal communication, Jan. 1992.

[3] J. Cocke and J. T. Schwartz, *Programming Languages and their Compilers: Preliminary Notes, Second Revised Version*, Courant Institute of Mathematical Science, New York, 1970.

[4] R. Cytron and J. Ferrante, "What's in a Name? or, The Value of Renaming for Parallelism Detection and Storage Allocation", *Proc. 1987 International Conference on Parallel Processing*, St. Charles, IL, Aug. 1987.

[5] R. Cytron, J. Ferrante, B. Rosen, and M. Wegman, "Efficiently Computing Static Single Assignment Form and the Control Dependence Graph", *ACM Transactions on Programming Languages and Systems* vol. 13 no. 4, pp. 451–490.

[6] S. K. Debray, "QD-Janus: A Prolog Implementation of Janus", research report, Dept. of Computer Science, The University of Arizona, Tucson, May 1991.

[7] A. Houri and E. Shapiro, "A Sequential Abstract Machine for Flat Concurrent Prolog", in *Concurrent Prolog: Collected Papers*, vol. 2, ed. E. Shapiro, pp. 513-574. MIT Press, 1987.

[8] H. J. Komorowski, "Partial Evaluation as a Means for Inferencing Data Structures in an Applicative Language: A Theory and Implementation in the Case of Prolog", *Proc. Ninth ACM Symposium on Principles of Programming Languages*, Albuquerque, NM, Jan. 1982.

[9] J. H. Reif and H. R. Lewis, "Symbolic Evaluation and the Global Value Graph", *Proc. Fourth ACM Symp. on Principles of Programming Languages*, Jan. 1977, pp. 104-118.

[10] B. K. Rosen, M. N. Wegman, and F. K. Zadeck, "Global Value Numbers and Redundant Computations", *Proc. 1988 ACM Symp. on Principles of Programming Languages*, San Diego, CA, Jan. 1988, pp. 12-27.

[11] V. A. Saraswat, K. Kahn, and J. Levy, "Janus: A step towards distributed constraint programming", in *Proc. 1990 North American Conference on Logic Programming*, Austin, TX, Oct. 1990, pp. 431-446. MIT Press.

[12] D. H. D. Warren, "An Abstract Prolog Instruction Set", Technical Note 309, SRI International, Menlo Park, CA, Oct. 1983.

[13] M. N. Wegman and F. K. Zadeck, "Constant Propagation with Conditional Branches", *ACM Transactions on Programming Languages and Systems* vol. 13 no. 2, April 1991, pp. 181-210.

[14] W. Wulf, R. K. Johnsson, C. B. Weinstock, S. O. Hobbs, and C. M. Geschke, *The Design of an Optimizing Compiler*, American Elsevier, New York, 1975.

Reflective Agents in Metalogic Programming

Stefania Costantini(*), Pierangelo Dell'Acqua(*)(**), Gaetano A. Lanzarone(*)

(*) Universita' degli Studi di Milano, Dipartimento di Scienze dell'Informazione
Via Comelico 39/41, I–20135, Milano (+39–2–55006.253/324)
(**) presently at the *Computing Science Department, Uppsala University*
Box 520, S–751 20 Uppsala, Sweden
e–mail: costanti@imiucca.csi.unimi.it, pier@csd.uu.se, lanzarone@hermes.unimi.it

Abstract.. We introduce a representation of agents by means of theories, and a communication among agents based on reflection, within the metalogic programming paradigm. The semantics of these features is shown to be the classical semantics of Horn clauses. The primitives for agents representation and inter–agent communication are very simple, and non–committal w.r.t. any predefined cognitive model or linguistic modality. Yet, it is shown by means of examples that they have enough expressive power for reasoning in non–trivial multi–agent domains, like the three wise men problem, especially when embedded in a powerful language, such as Reflective Prolog that we have previously developed. By integrating agents into Reflective Prolog we get a metalogic language equipped with higher–order–like features, metalevel negation, and theories, all of which rely on logical reflection for a uniform semantics and support each other for greater expressive and problem–solving power.

1. Introduction

The main aim of this paper is to present a new approach to agents and inter–agent communication, based on reflection. This approach has the following basic features. First, it is a general rather than ad hoc approach based on the classical semantics of the Horn–clause language [Ll87] (Least Herbrand Model, characterization as fixpoint of the Van Emden and Kowalski's immediate consequence operator, correctness and completeness of resolution). Second, it is applicable to the Horn–clause language as well as to its extensions and variations, provided that they are based on extensions of the semantics of [Ll87]. Third, it allows great flexibility in inter–agent communication: modes of communication can vary among different agents, or within the same agent in different situations. Fourth, all of the above is obtained not by means of an explicit representation of provability together with metainterpreters/theories composition, but by means of a form of logical reflection in the sense of [Fe62], [Co90]. As one interesting consequence, reflective agents can be usefully combined with other metalevel features having different purposes. Combinations are particularly promising if these features are based on a similar kind of reflection.

In the first part of the paper we introduce the approach, presenting syntax, semantics and some sample applications. In the second part, we show how the approach applies to our language Reflective Prolog, whose declarative and procedural semantics are based on the same form of logical reflection. By means of some examples, we show the expressive power that can be reached by combining intra–agent and inter–agent reflection. Among the examples, we show a solution to the "Three wise men puzzle" based on reasoning about communication rather than about beliefs.

Reflective agents are a first step towards a syntactical rendering of propositional attitudes (such as belief and knowledge), based on logical reflection. We briefly sketch this future work in the last Section.

2. Agents and Communication among Agents

In this Section, we introduce agents and communication among agents in the context of Metalogic Programming.

We assume as basic language and semantics the Horn–clause language and its semantics, as described in [Ll87]. Extensions/variations to the Horn–clause language are also accepted, provided that they are based on limited modifications of the same semantics. In addition, we assume that a naming relation is introduced in the language (and that unification is extended accordingly). This naming relation must provide names at least for terms and atoms, and, possibly, for conjunctions and clauses. Conventionally, we will indicate the name of an object A as 'A (and "A the name of the name of A). Metavariables, to denote terms/atoms containing names, will be indicated with syntax $V.

As a first step, we introduce in the alphabet of a program a finite or denumerable set of constants, called *theory symbols*. Second, every clause in the program and every atom in a query is prefixed with a theory symbol. That is, we associate with an agent a theory, i.e. a set of clauses. The program will be composed of a set of theories.

Definition 1. An annotated atomic formula (*t_atom* for short) is of the form ω:A, where ω is a theory symbol and A is an atomic formula. ◆

Definition 2. An annotated clause is of the form ω:A:–ω:B₁,...,ω:Bn, where ω:A, ω:B₁,...,ω:Bn are t–atoms. The short form ω:A:–B₁,...,Bn is allowed, and ω: is called the *theory prefix* of the clause. ◆

In an annotated clause, every composing atom has the same theory prefix.

Definition 3. A *theory* consists of a finite set of clauses with the same theory prefix. The short form ω:C₁...Cn is allowed for a theory, standing for ω:C₁...ω:Cn, where ω:C₁...ω:Cn are annotated clauses. ◆

Theory symbols can appear within C₁,...,Cn only in referenced form (that is, only the name 'ω of a theory symbol ω is allowed in C₁,...,Cn).

Definition 4. A *program* is a (finite, non–empty) set of theories. ◆

Definition 5. A *goal* is of the form :– ω1:A1,..,ωn:An
(if ω1=...=ωn, the short form :– ω:A1,..,An is allowed). ◆

In a goal, the composing atoms may have different theory prefixes.

In the following, if not explicitly specified differently, by *clauses* we mean annotated clauses, and by program and goal we mean those of Definitions 4 and 5. If the same clause appears in different theories, the abbreviation ω1,...,ωk:C is allowed, standing for ω1:C,...,ωk:C.

An *agent* is composed of a theory (called the *associated theory* of the agent) and, implicitly, by the inference mechanism of the language. That means, declaratively an agent is denoted by the declarative semantics of the associated theory, and procedurally by the closure of the theory itself under resolution. By abuse of notation, we will refer to the "agent ω" meaning the agent whose associated theory is ω:C1...ω:Cn.

We also introduce in the alphabet the distinguished binary predicates tell and told. The syntax is tell('t,'A), told('t,'A), where t must be a theory symbol, and A either an atom or an unprefixed clause.

The intended meaning of t1:tell('t2,'A) is: t1 tells t2 that A, and of t2:told('t1,'A) is: t2 has been told by t1 that A. These two predicates are intended to model the simplest and most neutral form of communication between agents, with no implication about provability (or truth, or whatever) of what is communicated, and no committal about how much of its information an agent communicates and to whom. An agent t1 may communicate to another agent t2 everything it can derive (in its associated theory), or only part of what it can derive, or may even 'lie' (i.e. communicate something it cannot derive).

The intended connection between tell and told is that an agent t may use (by means of told) only the information that another agent has explicitly addressed to it (by means of tell); what use t makes of this information is entirely up to t (depending on what the clauses in theory t are). Thus, the way an agent communicates with the others is not fixed in the language. Rather, it is possible to define in a program different behaviours of different agents, and/or different behaviours of one agent in different situations. The peculiarity of this approach to theories, with respect to others in the literature, is that the primitives introduced in the language are not aimed at structuring programs (e.g. for modularity / reusability), are not intended as schemas of (or means for) composition of theories, and are not concerned with logical/ontological modeling of concepts like belief or knowledge. They are just simple communication means among theories, on top of which more purpose–oriented mechanisms or concepts may possibly be built (some examples of which are given later on).

Example 1. Agent a tells a friend everything it is able to prove, but does not tell the other agents its private matters. In this example we (arbitrarily) assume that the basic language provides a "metavariable" facility similar to that of Prolog, together with some form of referentiation/dereferentiation. We will not be precise about such facilities in this Section, in order to focus on the communication aspects, irrespective of what the basic language is more specifically supposed to be like. In Section 4, all the technical details will be taken care of, in the context of a fully worked–out language.

```
a:
    tell($T,'A):-close_friend($T),A.
    tell($T,'A):-A,not_private('A).
```

Assume that the program includes two other theories, b and t. Assume the following other clauses in the theory a:

```
a:
    close_friend('b).
    not_private('likes("a,'sweets)).
    loves('a,c).
    likes('a,sweets).
```

These clauses allow the conclusions tell('b,'loves("a,'c)), tell('b,'likes("a,'sweets)). About t, the only conclusion is tell('t,'likes("a,'sweets)). The intended way of connecting tell to told should entail:
b:told('a,'loves("a,'c)), b:told('a,'likes("a,'sweets)), t:told('a,'likes("a,'sweets)).
That is, b and t theories should be able to derive that a asserts to like sweets, but only b should be able to derive that a asserts to love c. Assume now that b and t are the following:

```
b,t:
    happy('a):-told('a,'loves("a,$Y)).
```

In this case, we want to be able to derive b:happy('a). ♦

The intended connection between told and tell will be obtained via reflection.

The (Extended) Herbrand base BPE of a program P consists of the set of all ground t_atoms ω:α. An (Extended) Herbrand Interpretation IE of (the language L defined by) P is characterized, as usual, as a subset of BPE where ω:r(a₁,...,aₙ) ∈ IE means that ω:r(a₁,...,aₙ) is true w.r.t. IE.

Definition 6. Let P be a program and L the language of P. A *Reflective Model* for P is an Extended Herbrand interpretation IE for L which is a Model for

$$P' = P \quad \cup \quad \{\omega: \text{told}('\psi,'A) \leftarrow \psi: \text{tell}('\omega,'A)\} \qquad (*)$$

where ω, ψ are theory symbols, $A \in BPE$, $A \neq \text{tell}(...)$. ♦

Definition 7. A t–atom ω:A is a *Reflective Logical Consequence* of a program P if, for every interpretation IE, IE is a Reflective Model for P implies that IE is a Model for ω:A. ♦

Axioms (*) are reflection axioms in the sense of [Fe62], [Co90], and are called *theory axioms*. They state that told('ψ,'A) holds in ω if tell('ω,'A) holds in ψ.

P' can be considered the "extended version" of P, i.e. the program which P –due to the reflection capabilities we are introducing– stands for. Informally, the aim is the possibility of expressing inter–theory communication without either adding new inference rules, or introducing an explicit representation of provability. Instead, we leave the underlying logic unchanged, but implicitly extend the program by adding new clauses (the reflection axioms).

It is easy to see why this approach does not require actual extensions to the Horn–clause language semantics. Annotated clauses are just a syntactic convenience, and can be expressed via simple Horn–clauses. With this assumption, the definition of Extended Herbrand Interpretation reduces to that of Herbrand Interpretation. Then, P' is a Horn–clause program, with an associated Least Herbrand Model. Only, P' is in general infinite (which is allowed in first–order logic). Below we present a variation of the classical definitions of TP (immediate consequence operator) and resolution that, without changing the meaning, allows to keep the program finite. This by means of a form of computational reflection, that takes automatically into account the theory axioms which are implicitly present in the program.

The new operator, defined over the set HE of all the Extended Herbrand Interpretations of P, is obtained as a simple extension to TP.

Definition 8. Let P be a program. The mapping TPE: HE $\rightarrow$ HE is defined as follows (where ω, ψ are theory symbols, T = tell('ω,'B), D = told('ψ,'B), B $\neq$ tell(...)).

$$\text{TPE}(IE) = \{ \omega:A \in BPE :$$
$$\omega: A \leftarrow A_1 ... A_n \text{ is a ground instance of a clause in P,}$$
$$\{\omega:A_1, ... \omega:A_n\} \subseteq IE \} \cup \qquad (1)$$
$$\{ \omega:D \in BPE :$$
$$\psi:T \leftarrow A_1 ... A_n \text{ is a ground instance of a clause in P,} \qquad (2)$$
$$\{\psi:A_1,...,\psi:A_n\} \subseteq IE\} \quad ♦$$

Case (1) applies program clauses, case (2) applies the theory axioms, modeling *inter–theory reflection*.

Proposition 1. Let P be a program. Then the mapping TPE is continuous. ♦

Proposition 2. Let P be a program and I an Extended Herbrand Interpretation of P. Then I is a Reflective Model for P iff TPE(I) $\subseteq$ I. ♦

Theorem 1. (Fixpoint Characterisation of the Least Reflective Herbrand Model) RMP = lfp(TPE) = TPE↑ω. ♦

From Th. 1 and Def. 7 it clearly follows that TPE↑ω is the Least Herbrand Model of P'.

Example 2. Calculate the model of the program of Example 1.

TPE↑0 = ∅

TPE↑1 = TPE↑0 ∪ { a:close_friend('b),a:not_private('likes("a,'sweets)),

```
                    a:loves('a,c),a:likes('a,sweets)}
TPE↑2 = TPE↑1 ∪ {   a:tell('b,'loves("a, 'c)),a:tell('b,'likes("a, 'sweets)),
                    a:tell('b,'not_private("likes('''a,"sweets))),
      a:tell('b,'close_friend("b)),a:tell('t,'likes("a,'sweets)),
      b:told('a,'loves("a,'c)),
      b:told('a,'not_private("likes('''a,"sweets))),
      b:told('a,'close_friend("b)),
      b:told('a,'likes("a,'sweets)),t:told('a,'likes("a,'sweets))}
TPE↑3 = TPE↑2 ∪ {   b:happy('a)}
```

Notice that the whole program has a unique model, including all the agents of the program. Each agent's model can be seen as the corresponding subset of the overall model. I.e., for agent α,
$M(\alpha) = \{ \alpha{:}A \in TPE\uparrow\omega \}$. ♦

We now present the definition of Extended Resolution. Proofs of correctness and completeness w.r.t. the Least Reflective Herbrand Model are not reported here, but they can be obtained quite easily from those in [LI87].

Definition 9. (Extended Resolution) Let G be a goal $\leftarrow \omega_1{:}A_1,\dots,\omega_m{:}A_m,\dots,\omega_k{:}A_k$ and C a clause of program P. If $\omega_m{:}A_m$ is the selected atom in G, then G' is derived from G and C using substitution θ iff one of the following conditions holds:

(i) C is $\omega_m{:}A \leftarrow B_1,\dots,B_q$
θ is an mgu of A_m and A
G' is the goal $\leftarrow(\omega_1{:}A_1,\dots,\omega_{m-1}{:}A_{m-1},\omega_m{:}B_1,\dots,\omega_m{:}B_q,\omega_{m+1}{:}A_{m+1},\dots,\omega_k{:}A_k)\theta$

(ii) A_m is told('ψ,α)
C is ψ:tell('ω_m, β)$\leftarrow B_1,\dots,B_s$
θ is a mgu of α and β
G' is the goal $\leftarrow(\omega_1{:}A_1,\dots,\omega_{m-1}{:}A_{m-1},\psi{:}B_1,\dots,\psi{:}B_s,\omega_{m+1}{:}A_{m+1},\dots,\omega_k{:}A_k)\theta$ ♦

Notice that this is, in substance, just usual resolution on program P': case (i) applies program clauses, case (ii) performs a double resolution step, by first applying the theory axioms, which are implicitly present in the program, and then a program clause from the definition of tell. Notice also that the above definition allows finite application of a denumerable set of axioms, i.e. the theory axioms introduced in Definition 6.

Example 3. With respect to the program of Example 1, prove ?–b:happy('a).

?–b:happy('a)

via case (i) of Extended Resolution and clause
b:happy('a):–told('a,'loves("a,$Y))

?–b:told('a,'loves("a,$Y))

via case (ii) of Extended Resolution,
and theory axiom
b:told('a,'loves("a,'c)):–a:tell('b,'loves("a,'c)).
and clause
tell('b,'loves("a,'c)):– close_friend('b),loves('a,c).
with mgu {$Y/'c}

?–a:close_friend('b),a:loves('a,c)

which obviously succeeds in a.
On the contrary, the goal ?–t:happy('a) fails. ♦

Observation 1. To apply the approach to a language which is an extension/variation of the Horn–clause language, it suffices to adapt the modifications to the particular language semantics. Precisely, the program P' may be possibly obtained from a suitably modified version of P, cases (2) and (ii) may be possibly added to suitably modified versions of TP and resolution respectively. An example of this is developed in Sections 4 and 5 in full technical detail: agents and theories will be added to the metalogic language Reflective Prolog. ◆

3. Applications

The approach to inter–agent communication introduced so far allows a variety of interactions among agents to be represented, depending on the application context, and depending on the capabilities of the logic language to which the approach is applied. We show some examples of possible uses of told and tell.

Example 4. *An agent tells the others a thing it can prove*

ω: tell($X,'p('a)):–p(a). ◆

Example 5. *An agent lies*

ω: tell($X,'not 'p('a)):–p(a).

where not denotes, in the basic language, some kind of negation. ◆

Example 6. *An agent lies to somebody and tells the truth to somebody else*

ω: tell($X,'p('a)):–tell_the_truth_to($X),p(a).
ω: tell($X,'not 'p('a)):–lie_to($X),p(a). ◆

Example 7. *An agent tells another one a thing it can prove within some resource limitations*

ω: tell($X,'p('a)):–limited_prove('p('a)).

The predicate limited_prove must incorporate the desired limitations. ◆

Example 8. *An agent allows nesting of told.*

ω: tell($X,'told("ω,"p("a))):–p(a) .. ◆

Example 9. *An agent tells the others a thing it can prove, provided that some constraints hold*

ω: tell($X,'p($X)):–p(X),constraints(X,$X) . ◆

Example 10. *Different agents have different reasoning powers*

ω1: tell($X,'p('a)):–ω1_prove('p('a)).
ω2: tell($X,'p('a)):–ω2_prove('p('a)) . ◆

The predicates ω1_prove and ω1_prove must incorporate the different reasoning abilities. ◆

Example 11. *An agent trusts another*

ω1: p(a):–told('ω2,'p('a)). ◆

Example 12. *An agent distrusts another*

ω1: p(a):–told('ω2,'not 'p('a)). ◆

Example 13. *An agent is cautious*

ω: p(a):–reliable($A),told($A,'p('a)). ◆

The flexibility and generality of inter–agent communication will depend on the capabilities of the basic language. A naming device alone does not allow much generality. Most of the examples above require at least a referentiation/dereferentiation capability. Others require a full metavariable device, or an explicit representation of provability.

A generalization of the examples above requires a metalevel representation of truth, and some reflection capability: a sentence such as

ω1: $P:–told('ω2,$P).

is otherwise impossible to be expressed and used. In the following Section, we illustrate the application of this approach to our previously–developed language Reflective Prolog.

4. Agents in Reflective Prolog

The metalogic programming and knowledge representation language Reflective Prolog (RP for short, [CL89], [Co90], [CL91], [CL92]) has three basic features. First, a full naming mechanism, which allows the representation of metaknowledge in the form of *metalevel clauses* (that contain, by definition, at least a name term). In the following, the name of a term/atom A in RP is indicated by ↑A. Second, the possibility of specifying *metaevaluation clauses* (which are the metalevel clauses defining the distinguished predicates solve and solve_not), that allow to declaratively extend/restrict the meaning of the other predicates. Last, a form of logic *reflection* which makes this extensions/restrictions effective, both semantically and procedurally [Co90].

An *Extended Herbrand Interpretation* IE of (the language L defined by) a Reflective Prolog definite program P is an Herbrand Interpretation of P which satisfies the condition that solve(↑A) belongs to IE if A belongs to IE. IE can be characterized, as usual, as a subset of the Herbrand Base BPE of P.

The semantics of P is given in terms of *Reflective Model* of P, obtained as follows. First, to characterize metalevel negation (rules for solve_not specify when predicates do not hold) the program P is taken as standing for a program P' obtained by substituting each clause

A*:– B1,...,Bn   (A*=A or A*=solve(↑A), A ≠ solve_not(...))

of P with a clause

A*:– B1,...,Bn,¬solve_not(↑A).

where '¬' is classical negation.

A *Reflective Model* for P' is an Extended Herbrand interpretation IE for L which is a Model for

P* = P' ∪ {A ← solve(↑A) : A ∈ BPE, A ≠ solve_not(...), A ≠ solve(...)}.

The axioms A ← solve(↑A) are called *reflection axioms*. An atom A is a *Reflective Logical Consequence* of an RP program P if, for every interpretation IE, IE is a Reflective Model for P implies that IE is a Model for A. P* can be considered to be the "extended version" of P, i.e. the program which P –due to the reflection capabilities of RP– stands for. This in order to exploit indifferently both object–level knowledge and metaknowledge in proofs: instead of either adding new inference rules, or introducing an explicit representation of provability to perform derivation, the RP semantics leaves the underlying logic unchanged, but on the one hand it implicitly extends the program by adding new clauses (the reflection axioms), and on the other hand it restricts the set of accepted interpretations to those which allow level communication.

Reflective Models are clearly models in the usual sense. For positive programs (programs without solve_not clauses) the *Model Intersection Property* still holds, and therefore there exists a Least Reflective Herbrand Model RMP of a program P. It can be characterized as the least fixpoint of a mapping TPE very similar to the usual TP [Ll87], except that it produces both A and solve(↑A) whenever one of them can be derived. We have [Co90] RMP = lfp(TPE) = TPE↑ω.

For programs with negation, among the minimal models of P* we choose the one that never entails A if it entails solve_not(↑A). It can be obtained [CL91] as the iterated least fixpoint of a mapping T.J(I), where I and J are Extended Herbrand Interpretations. T.J(I) is based on TPE, where an atom A cannot be included in T.J(I) if solve_not(↑A) belongs to J. The definition is the following:

T.J(IE) = {A ∈ TPE(IE) : solve_not(↑A) ∉ J}

The function T.J is monotonic and continuous, independently of J. The *Iterated Reflective Model* of P (IRMP for short) is the least fixpoint of the sequence of *approximate models* defined as follows:

$$M_0 = T_\phi \uparrow \omega$$
$$M_{i+1} = T_{M_i} \uparrow \omega$$

A two–valued IRM is defined a wide class of RP programs, called safe programs [CL91], such that (roughly) solve_not(↑A) never depends (directly or indirectly) on A itself: otherwise, since (by the definition of IRM) A depends implicitly on solve_not(↑A), a circularity occurs. A three–valued IRM is defined on every RP program.

Procedural semantics for positive RP programs is based on RSLD–Resolution (sound and complete w.r.t. RMP). Informally, RSLD–Resolution is able to use clauses with conclusion solve(↑A) to resolve a goal A (*upward reflection*), and vice versa clauses with conclusion A to resolve solve(↑A) (*downward reflection*). That means, RSLD–Resolution is, in principle, just resolution which on the one hand can apply the reflection axioms (implicitly present in the program), and on the other hand can apply the basic property of Extended Herbrand Interpretations, which entail solve(↑A) in consequence of A. NRSLD–Resolution, for programs with metalevel negation, is based on RSLD–Resolution. In particular, if a goal G succeeds with computed answer θ, then solve_not(↑Gθ) is attempted: if it succeeds, then G is forced to failure; otherwise, if solve_not fails the success of G is confirmed. NRSLD–Resolution is [CL92] sound and complete w.r.t. IRMP. It is also independent of the computation rule, and this property extends to Negation as Failure, if explicitly expressed by metaevaluation rules.

As summarized above, declarative and procedural semantics of RP programs with metalevel negation are defined on the basis of declarative and procedural semantics of positive programs. RP semantics is also independent of the naming relation. Therefore, in order to introduce agents in RP, it suffices to introduce the approach to semantics of positive programs presented in Section 2. That means, when introducing agents in RP, both metalevel negation and negation as failure are not influenced at all.

In order to show some examples, we recall the basic features of the RP naming relation NR. The name of a constant or variable c is "c"; the name of a predicate symbol p is <p>; the name of a function symbol f is {f}; the name of either a term f(a1,...,an) or an atom p(a1,...,an) with a1,...,an terms, is {f}(↑a1,...,↑an) or <p>(↑a1,...,↑an) respectively (we remind the reader that ↑a indicates the name of a in RP). The name of a conjunction B1,...Bn is ↑B1^....^↑Bn. The name of a clause A:–T is ↑A<–↑T. The name of a theory symbol ω is "ω". There are four kinds of variables: object variables (syntax V) to denote object–level terms; function metavariables (syntax %V) to denote names of function symbols; predicate metavariables (syntax #V) to denote names of predicate symbols; general metavariables (syntax $V) to denote any metalevel terms. The distinguished predefined unary predicate theory_clause($X) allows the clauses and unit clauses of the program to be inspected.

5. Applications in RP

Due to reflection capabilities of RP, it is possible generalize all the examples of Section 3, like for instance:

ω2:	tell($X,$Y):–solve($Y).	*The sincere*	(1)
ω:	tell($X,<not>$Y):–solve($Y).	*The liar*	(2)
ω1:	solve($P):–told("ω2",$P).	*The credulous*	(3)
ω:	solve($P):–told("ω2",<not>$P).	*The skeptical*	(4)
ω:	solve($X):–reliable($Y),told($Y,$X).	*The cautious*	(5)

The following Examples 14 and 15 are meant to technically illustrate how upward/downward reflection and inter–theory reflection proceed (with the corresponding instantiations) in a derivation through theories.

Example 14. With axiom (3) in ω1, and axiom (1) in ω2, with

$$ω2: p(a). \tag{6}$$

it is possible to prove ?–ω1: p(a) via the following steps:

?–ω1: p(a)	via upward reflection and axiom (3) in ω1 we get
?–ω1: told("ω2",<p>("a")).	via inter–theory reflection and axiom (1) in ω2 we get
?–ω2: solve(<p>("a").	which succeeds by downward reflection and fact (6). ♦

We may have different agents with different reasoning powers without introducing an explicit representation of provability, by equipping the agents with different sets of auxiliary inference rules, represented via solve rules. Reasoning abilities of an agent may be limited via solve_not rules.

Example 15. Agent ω1 knows the meaning of symmetry of relations, while ω2 does not know about symmetry, but knows about transitivity. ω1 tells ω2 all it can prove, and ω2 trusts these conclusions. They have the same knowledge about the world.

ω1:	solve(#R($A,$B)):–symmetric(#R),solve(#P($B,$A)).	(c1)
ω1:	tell("ω2",$Y):–solve($Y).	(c2)
ω2:	solve(#P($X,$Y)):– transitive(#P),theory_clause(#P($X,$Z)),solve(#P($Z,$Y)).	(c3)
ω2:	solve($P):–told("ω1",$P).	(c4)
ω1,ω2:	symmetric(<p>).	(c5)
ω1,ω2:	transitive(<p>).	(c6)
ω1,ω2:	p(a,b).	(c7)
ω1,ω2:	p(c,b).	(c8)

Agent ω2 is not able to prove ?–ω2:p(a,c) without cooperation with ω1, since fact c8 needs to be reversed. This is possible because of the symmetry of p, known to ω1.

?–ω2:p(a,c).	via upward reflection and c3, with {#P/<p>,$X/"a",$Y/"c"}
?–ω2:transitive(<p>),theory_clause(<p>("a",$Z)),solve(<p>($Z,"c")).	via c6 and c7, with {$Z/"b"}
?–ω2:solve(<p>("b","c"))	via c4
?–ω2:told("ω1",<p>("b","c"))	via inter–theory reflection and c2

?—ω1:solve(<p>("b","c"))   via c1,
                                                                 with {#R/<p>,$A/"b",$B/"c"}
?—ω1:symmetric(<p>),solve(<p>("c","b")).   via c5
?—ω1:solve(<p>("c","b")).   succeeds
                                                                 via downward reflection and c8.

♦

Example 16. *Three Reflective Wise men.* As a more elaborate example of the use of agents in RP, we show a metalogical solution to the three wise men puzzle (discussed in [ANS88], [Ko82], [KK90]). The purpose of this example here is to show how the devised inter–theory communication, though very simple, is powerful enough (in the context of a powerful metalogic language) for reasoning in non–trivial multi–agent domains. We follow [ANS88] and [KK90] in using reflection, and [KK90] in using amalgamation of object language and metalanguage. We do not rely on an *explicit* representation of provability. We do not rely on the representation of *belief*, but on *communication* among agents. We do not use *common knowledge*, though our solution might be easily reformulated in this direction. We take the formulation of the puzzle from [KK90].

A king, wishing to determine which of his three wise man is the wisest, puts a white spot on each of their foreheads, and tells them at least one of the spots is white. The king arranges the wise men in a circle, so that they can see and hear each other (but cannot see their own spot) and asks each wise man in turn what is the color of his spot. The first two say they do not know, and the third says that his spot is white.

The puzzle consists in explaining how the third wise man can reach the conclusion.

Instead of using common knowledge, we assume that, in order to answer a query about the color of his spot, the third wise man can ask questions to both the first and the second one, and the second wise man can ask the first one.

We use the operator 'v', with no predefined logical meaning: it is just an abbreviation for the use of a binary function symbol or(...,...).

The predefined predicate name is used for referentiation/dereferentiation: name($X,$Y) holds if $X=↑$Y. For instance, name("'wiseman1'","wiseman1") holds. The anonymous variable '_' has the same use as in Prolog.

The predicate spot("wiseman*i*",white*i*) means that the i–th wise man has a white spot on his forehead. We call white1, white2 and white3 the three white spots in order to distinguish among them. Below we report first those clauses which are the same for all wise men, and then the different knowledge of each wise man.

wiseman1,wiseman2,wiseman3:
solve(<spot>("'wiseman1'","white1") v
       <spot>("'wiseman2'","white2") v
       <spot>("'wiseman3'","white3")).   (ws1)

Clause ws1 states that at least one of the wise men has a white spot on his forehead. Notice that we need to state this at the metalevel, via solve, since 'v' is just an operator which takes the place of a function symbol, rather than the logical disjunction connective, which is unavailable in a Horn–clause language. Below are two auxiliary predicates.

diff($X,$Y).   (ws2)
solve_not(<diff>($X,$X)).   (ws3)
undefined($Y).   (ws4)
solve_not(<undefined>($X)):–name($X,$Y),solve($Y).   (ws5)
solve_not(<undefined>($X)):–name($X,$Y),solve_not($Y).   (ws6)

ws2–3 state that two propositions are different unless they are unifiable. By the definition of metalevel negation, the assertion solve_not(<diff>($X,$X)) excludes all couples diff($X,$X) from the consequences of the theory. ws4–6 state that a proposition $Y is undefined unless the agent can either prove or disprove it.

```
solve($P):-                                                    (ws7: reason)
    diff($P,_v_v_),
    can_ask($T),
    spot($T,W),
    name($TT,$T),name($W,W),
    name($TTT,$TT),name($WW,$W),
    told($T,<spot>($TT,$W) v $P),
    told($T,<undefined>("<spot>"($TTT,$WW))).
```

This is the main inference rule of the agents. It is based on the general principle that an agent can conclude $P whenever another agent is able to prove $P1 v $P, while it does not know about $P1. This is not a sound inference rule in general, but it is correct in absence of disjunctive information. Actually, the only disjunctive information in this database is ws1, which is explicitly excluded by the first condition. This rule implicitly contains two forms of confidence in the other agent. First, that it tells what it can prove (it tells *the truth*). Second, that it is logically reliable. This is consistent with the implicit assumption in the specification, that the wise men are logically omniscient and loyal. Since we mean to query the wise men about the color of their spots, in order to avoid a lot of useless backtracking, the rule is explicitly specialized to spots, and literally says: if $T can prove that either the color of its spot is W or $P is true, and it does not know the color of its spot, then $T can prove $P.

Observation 2. The invocation of told (like in conditions of ws7) implies a deliberate change of the context of the proof, by "asking" another agent about something. In a metalevel representation, this requires (as recognized since [BK82]) referentiation of the arguments. Presently, this referentiation is performed (like in ws7) by the predefined predicate name. This is expressively awkward, and also inefficient. Work is under way to suitably generalize unification, so as to perform referentiation automatically whenever necessary. ♦

```
tell($A,$P):-solve($P).                                        (ws8)
solve_not(<tell>($NA,"<spot>"($NNA,$C)):-name($NNA,$NA).        (ws9)
```

This is the specification of modes of communication. Every wise man tells the others everything it can prove, except the color of their spots. For instance, if wiseman1 can prove
spot("wiseman2",white2)
(which means wiseman2 has a white spot on its forehead) it cannot tell wiseman2. This because
wiseman1:tell("wiseman2",<spot>(" "wiseman2" ", "white2") fails.
In fact,
solve_not(<tell>(" "wiseman2" ","<spot>"(" " "wiseman2" " ", " "white2" ")))
is implied by
name(" " "wiseman2" " ", " "wiseman2" ").

Observation 3. In the above example many terms are referenced two or three times, looking rather awkward because of the many quotes. The reader should consider however that the extra quotes (generated by implicit upward reflection or by the application of name) are taken care of by the language interpreter, and not by the programmer/user. The interpreter needs them for avoiding ambiguities, but the user can ignore them. ♦

```
solve($P):-solve($Pv$Q),unprovable($Q).                        (ws10)
unprovable(_).                                                 (ws11)
solve_not(<unprovable>($QQ)):-name($QQ,$Q),solve($Q).          (ws12)
```

Axiom ws10 allows agents to exploit metalevel disjunctive information in reasoning. It states that a proposition $P holds if PvQ can be proved, and $Q does not hold. If for instance the only white spot were on wiseman1's forehead, by ws10 it could derive the color of its spot. ws11–ws12 are very similar to RP's conceptualization of Negation as Failure, except that they apply not only to base–level atoms like NAF, but also to metalevel disjunctions. For instance, unprovable($Q1v$Q2) holds (ws11) unless (ws12) $Q1v$Q2 is provable (possibly via ws10).

Below is the private knowledge of the three wise men. Object–level knowledge concerns other wise men's spots. Metaknowledge concerns the possibility of asking the others.The first wise man cannot ask anybody. The second one can ask only the first one, the third one can ask both the first and the second one.

wiseman1:
spot("wiseman2",white2).   (ws1.1)
spot("wiseman3",white3).   (ws1.2)

wiseman2:
spot("wiseman1",white1).   (ws2.1)
spot("wiseman3",white3).   (ws2.2)
can_ask("wiseman1").   (ws2.3)

wiseman3:
spot("wiseman1",white1).   (ws3.1)
spot("wiseman2",white2).   (ws3.2)
can_ask("wiseman2").   (ws3.3)
can_ask("wiseman1").   (ws4.4)

Observation 4. The generality of the above solution is attested from the fact that it does not need modifications to cope with different situations, i.e. different placements of the spots. The solution is also basically independent of the number of wise men. The generalization to N wise men only requires adapting ws1 and (the first subgoal of) ws7. ♦

6. Future Developments

One point that needs further refinement is full reversibility of told. Presently, told is reversible in the second argument, i.e. it is possible to use it with an unbound variable as second argument. The first argument instead must be instantiated when told is called. It might also be interesting to exploit reversibility in the first argument, like in the example below:

t1:p(X):–told($T,<q>("X")).

where p(X) holds in t1 if there exists any agent $T that asserts q(X). Work is under way for a suitable generalization.

The presented approach differs from those approaches in the literature where theories are used for structuring programs, since there is no predefined way of introducing/combining theories. It is not our intention to pursue this objective. It differs from approaches concerned with representation of attitudes like belief and knowledge, since no such concepts *per se* are involved in the representation of agents and their communication. In fact, there is no predicate like demo(t,A) meaning derivability of A from (the axioms in) t. The predicates tell('t,'A), told('t,'A) do not imply that A is provable in t, but only that A is communicated by t. In RP, the predicate solve remains a monadic predicate representing derivability in the single theory in which it appears.

Many different connections among these predicates may be established in a program (some of which are exemplified at the beginning of Section 5), giving rise to several different representations of propositional attitudes. The investigation of such possibilities is one main stream of the future work we foresee on this track. Presently, we just feel that reflective agents may be a basis for a very flexible syntactical rendering of modalities and attitudes. The utility of this approach would be to obtain all these possibilities without requiring separate semantic extensions for each of them.

References

[ANS88] Aiello L., Nardi D. and Schaerf M., *Reasoning about Knowledge and Ignorance*, in: Proceedings of FGCS, 1988.

[BK82] Bowen K. and Kowalski R., *Amalgamating Language and Metalanguage in Logic Programming*, in: Clark K.L. and Tarnlund S.-A. (eds.), *Logic Programming*, Academic Press, 1982.

[Br90] Bruynooghe M. (ed.), Proceedings of the Second Workshop on Metaprogramming in Logic (Meta90), Leuven, April 4–6, 1990.

[CL89] Costantini S. and Lanzarone G.A., *A Metalogic Programming Language*, in: Levi G. and Martelli M. (eds.), *Logic Programming*, Proceedings of the Sixth International Conference, MIT Press, 1989.

[CL91] Costantini S. and Lanzarone G.A., *Metalevel Negation and Non-Monotonic Reasoning*, to appear on the Journal on *Methods of Logic in Computer Science*, abstract in: Proceedings of the Workshop on Non-Monotonic Reasoning and Logic Programming, Austin, TX, November 1–2, 1990.

[CL92] Costantini S. and Lanzarone G.A., *On Procedural Semantics of Metalevel Negation*, in: Proceedings of the 2nd Russian Conference on Logic Programming (San Petersburg, September 1991), Lecture Notes in Artificial Intelligence, Springer–Verlag, 1992.

[Co90] Costantini S., *Semantics of a Metalogic Programming Language*, International Journal of Foundations of Computer Science, Vol. 1, N. 3, Sept. 1990 (draft version in [Br90]).

[Fe62] Feferman S., *Transfinite Recursive Progressions of Axiomatic Theories*, Journal of Symbolic Logic 27, 1962.

[KK90] Kim J.S. and Kowalski R., *An Application of Amalgamated Logic to Multi-Agent Belief*, in [Br90].

[Ko82] Konolige K., *A First-Order Formalization of Knowledge and Action for a Multiagent Planning System*, Machine Intelligence 10, Hayes J.E., Michie D., and Pao J.H. (eds.), Ellis Horwood, 1982.

[Ll87] Lloyd J.W., *Foundations of Logic Programming*, (Second, Extended Edition), Springer–Verlag, Berlin, 1987.

Logic Meta-Programming facilities in 'LOG

I. Cervesato[*], G.F. Rossi[**]

(*) Dip. di Informatica, Univ. di Torino - C.so Svizzera, 185 - Torino (I)
(**) Dip. di Matematica, Univ. di Bologna - P.zza Porta S.Donato, 5 - Bologna (I)

Abstract. A meta-level extension of a logic programming language is presented. The resulting language, called 'LOG (read *quote-log*), provides meta-programming facilities similar to those of Prolog while preserving a declarative logical semantics. It also offers new meta-programming opportunities as compared with Prolog due to its ability to treat whole programs, i.e. sequences of clauses, as data objects. The extension basically consists in defining a suitable *naming scheme*. It associates *two* different but related meta-representations with every syntactic object of the language, from characters to programs. The choice of the double meta-representation is motivated by both the user and the implementation viewpoints. All Prolog built-in meta-predicates can be redefined as 'LOG programs by exploiting the new naming scheme. Then some syntactic sugar is added to make the language more concrete. Some examples are given, in particular to show the ability of the language to deal with programs as data.

1 Introduction

The problem of meta-programming in the context of logic programming was systematically faced for the first time by Bowen and Kowalski in [3]. Since that time, a large number of researchers have carried this idea on in many directions. Relatively few efforts, however, have been devoted to the design of an *effective* logic programming language equipped with meta-programming capabilities similar to those usually available in Prolog but defined in a *cleaner* way. Among them, we must mention MetaProlog [2, 4], and, more recently, the Gödel language [6].

This paper moves along these lines and leads to the definition of an extended logic programming language - called 'LOG - which provides meta-programming facilities similar to (or, possibly, better than) those of Prolog. It has the very same aims as *Gödel*, at least as far as the introduction of meta-programming facilities is concerned: "... to have functionality and expressiveness similar to Prolog, but to have greatly improved declarative semantics compared with Prolog" [6]. 'LOG is also similar in aims to Barklund's proposal [1]: defining "a naming of Prolog formulas and terms as Prolog terms to create a practical and logically appealing language for reasoning about terms, programs, ...".

Also the applications we have in mind are mostly the same as those of the mentioned proposals, namely the development of software tools (the meta-programs) that manipulate other programs (the object programs) as data, such as debuggers,

Work partially supported by M.U.R.S.T. 40% and C.N.R. - Progetto Finalizzato Sistemi Informatici e Calcolo Parallelo", grant no. 890002369.

compilers, program-transformers, etc.. We do not consider as part of our language any reflection mechanism which would allow a meta-representation to be obtained from the object it denotes or vice versa. This differentiates (both in aims and in nature) our proposal from others, such as Reflective Prolog [5] and R-Prolog* [15], that, on the contrary, assume a reflection mechanism to be available, though not visible at the user level.

The main problem is that of defining a suitable *naming scheme* by which the syntactic entities of the language can be referred to and manipulated at the meta-level. Here we stress the fact that naming should apply to *every* syntactic entity of the language, from characters to programs. In addition, we want the naming scheme to be *effective*, that is to burden not too much the user with an heavy notation, and to allow *efficient implementations* of the language to be devised.

The key idea underlying our proposal is to provide *two* different but *related* meta-level *representations* for each syntactic object of the language. Precisely, the meta-level representations consist of a constant *name* and a structured ground term, called the *structural representation*. The name describes an entity as a whole, while the structural representation describes the structure of the entity in terms of the *names* of its components, thus allowing one to explore its internal structure. Moreover, for each composite syntactic object, it is possible to relate its name to its structural representation by means of the predefined predicate <=>. While neither meta-representation is especially original on its own, using them together seems to offer quite interesting possibilities.

The idea of a double meta-representation was already applied in a more pragmatical sense and to a limited extent (programs only) to the definition of the meta-logical facilities of the *EnvProlog* language, an *extended Prolog* aimed at building Prolog programming environments [10, 11, 12]. In this paper, we start instead with a pure logic programming language and we apply the naming scheme to every syntactic entity of the language. Then we show that a more concrete version of the language embodying this naming scheme can be obtained by the addition of a suitable syntactic level; this makes the language easier to be handled both for the user and the implementation. We show also that the resulting language exhibits higher meta-programming attitudes than Prolog (in particular, as regards the ability to deal with programs as first-class data objects) while preserving a logical reading.

Section 2 presents the main features of the naming scheme provided by 'LOG: names, structural representations and the <=> operator used to relate the name and the structural representation of each syntactic object. The usage and motivations of the double meta-representation from the user viewpoint are discussed in Section 3. Section 4 discusses how usual Prolog built-in meta-predicates can be redefined in 'LOG. Section 5 presents the syntactic additions and conventions we assume for the concrete version of 'LOG. The ability of our language to deal with programs as data is highlighted in Section 6 by showing some simple examples. Finally, Section 7 briefly discusses the implementation issue, pointing out some motivations for the use of the double meta-representation also from the language implementation viewpoint.

2 Meta-representations

'LOG syntax is mostly the usual syntax of logic programming languages (cf. for instance [9]) and will be skipped here, except for those parts concerning the meta-representations.

We start with an ordinary Horn clause language and we conservatively extend it to one in which every syntactic entity is named by ground terms of the language. Precisely, each 'LOG syntactic object has *two meta-representations* associated with it, called the *name* and the *structural representation* of the object.

2.1 Names

The *name* of an object is a *constant* symbol which is isomorphic in structure to the object it refers to. If e is a syntactic expression of the language, 'e' is its name. For example,

> 'append([],X,X).
> append([A|X],Y,[A|Z]) :- append(X,Y,Z)'

is the name of a program defining the usual append predicate which concatenates two lists. As another example, 'f(a,g(X))' is the name of the term f(a,g(X)).

Objects having a name in 'LOG are programs, clauses (including goals), terms, symbols and characters. Accordingly, names are partitioned into five different classes: program names, clause names, term names, symbol names and character names. Notice that these classes are not necessarily disjoint. The same name, in fact, can denote different syntactic expressions depending on the context where it is used; for instance, 'alpha' can represent either a symbol or a term with no arguments or a clause with no body or a single clause program. Also notice that we do not consider atomic formulas as a syntactic class of the language. Indeed it seems more appropriate to the meta-programming paradigm we are considering here to treat atoms simply as terms.

2.2 Structural representations

Every composite syntactic object (i.e., symbols, terms, clauses and programs) has a second meta-representation associated with it, called the *structural representation*. This meta-representation is a ground term which describes the structure of the object it denotes in terms of the *names* of its components.

If $e = e_1e_2...e_n$ is a syntactic expression where e_1, e_2, ..., e_n are its component sub-expressions then the structural representation of e is $['e_1','e_2',...,'e_n']$.

For instance, if $P = C_1.C_2. .... .C_n$ is a program then $['C_1','C_2',...,'C_n']$ is the *program structure* of P where 'C_i' is the clause name of the clause C_i. Similarly, if f(a,g(X,b)) is a term, the corresponding *term structure* is ['f','a','g(X,b)']. The only exception is the structural representation of clauses. *Clause* structures (other than goal clause structures) rely on the reserved symbol clause for distinguishing the head from the body part (e.g. clause('p',['q','r']) for p:- q,r). In Section 5 we will introduce a *synthetic notation* for structural representations which is more convenient for the user (in contrast with the list notation, also called explicit notation, presented here). Since it is simply syntactic sugar it can be ignored for the moment.

While names are constant symbols (hence atomic entities), structural

representations are compound ground terms. Therefore, one can easily define terms
similar to structural representations apart from the occurrence of *meta-level variables*
in place of some of the names 'e_i' of its component sub-expressions. Such terms
will be considered as *partially specified* structural representations. For instance,
['f',X] is not a term structure; however, if the meta-level variable X is instantiated to
the name of some term we get a complete term structure, e.g. ['f','a'], ['f','g(b)'],
and so on.

    Meta-variables in an incomplete structural representation are dealt with as real
variables in contrast with object level variables that are frozen inside the names that
constitute the structural representation. Therefore the two term names 'f(X)' and
'f(a)' cannot unify at the meta-level, whereas ['f',X] and ['f','a'] unify, yielding
the substitution X = 'a'.

    Names and structural representations are syntactic entities; therefore they have a
name and a structural representation too. For instance, the name of the term name
'Alpha' is ''Alpha''. Thus, 'LOG supports the definition of an infinite tower of
meta-levels. Anyway, meta-levels are strictly separated: at each level, the syntactic
entities of the lower levels are visible through their names only; variables do not
make an exception to this rule. No reflection mechanism is supported by 'LOG.

2.3 Relating names and structural representations

The name and the structural representation of an object can be related to each other by
the use of the predefined predicate <=> (written infix), called the *destructuring* or
simply the *double arrow* operator.

    The *informal semantics* of <=> is: a goal N <=> S is true if N is the name of an
object o and S is the ground structural representation of the same object o. Thus,
<=> simply defines a binary relation, called the *destructuring* relation, between names
and structural representations, i.e. between syntactic expressions of the language.

    Actually, we have distinguished five different classes of name symbols. It
follows that we must distinguish among different forms of the *generic operator* <=>,
accordingly to the different types of its arguments. We will use the four different
operators <=p=>, <=c=>, <=t=> and <=s=> for programs, clauses, terms and
symbols respectively, still using the generic double arrow operator when speaking of
its properties in general and no ambiguities arise (we will see in Section 5 that these
differences can be hidden by an upper syntactic level). Here are two simple examples
of goals involving <=>:

    ?- 'p :- q,r. q. r.' <=p=> ['p :- q,r','q','r'].
    yes.
    ?- 'f(g(a),b,C)' <=t=> ['f',A,'b','C'].
    A = 'g(a)'.

The second goal succeeds provided the meta-variable A is instantiated to 'g(a)'.

2.4 Semantics

The main differences in the semantics of 'LOG w.r.t. the standard case (as described for instance in [9]) are due to the presence of the *double arrow* operators.

As regards the *declarative semantics* of 'LOG, first a privileged interpretation domain resulting from suitable modifications to the classical Herbrand universe is defined then the privileged interpretation of <=> is given as a relation over this domain.

The modified *Herbrand universe* H is defined in almost the same way as usual except that it is built out of the set of characters composing names besides the set of function and constant symbols that occur in the program in such a way to include all the names and ground structures which can be constructed in that program.

The *privileged interpretations* of the <=> operators are defined as *binary relations* over such H. In particular, for any TN, TS $\in$ H, whether TN <=t=> TS holds or not can be established by: TN has the form 't', TS has the form ['f','t$_1$',...,'t$_n$'], t is a term, and t = f•(•t$_1$•,•...•,•t$_n$•) where • is the usual *string concatenation* relation and = is the usual syntactic equality. Similar definitions can be given for <=p=>, <=c=> and <=s=>.

Procedurally, a goal N <=> S succeeds from a program P if either N is a name and there exists a ground instance S' of S such that <N,S'> is in the destructuring relation, or N is a variable, S is ground and there exists an instance N' of N such that <N',S> is in the destructuring relation. If, on the contrary, N is a variable and S is not ground then the goal is unsolvable and its proof is delayed till either one of the above cases occurs.

A *refutation* of P $\cup$ {G} is a finite derivation G, G$_1$,...,G$_n$ of P $\cup$ {G} such that the last derived goal only contains destructuring goals in unsolvable form and there exists a substitution θ which makes all them true simultaneously. A *computed answer* for a refutation of P $\cup$ {G} is now a pair <σ,C> where σ is a substitution for the variables in G computed as in the standard case and C is the (possibly empty) set of destructuring goals in unsolvable form.

Delaying the solution of a goal containing the <=> operator allows a *declarative* reading of programs to be preserved. The order of literals in a clause or in a goal is immaterial. For example, the goal

?- S <=s=> ['a','l'|X], X = ['p','h','a'].

succeeds with computed answer substitution S = 'alpha' and no destructuring goal left unsolved. If, on the contrary, at the end of the computation a goal of the form N <=> S cannot be solved because N is a variable and S is not ground then N <=> S is returned as part of the computed answer: it will be considered as a *constraint* on values the not yet instantiated meta-variables occurring in it can assume. For example, in

?- N <=t=> ['f'|A], A = ['a'|B].

A = ['a'|B],

N <=t=> ['f','a'|B].

there are obvious valid instances of the meta-variables B and N, but not all of them are viable. Actually the way the double arrow operators are dealt with in our proposal can be viewed as a simple form of Constraint Logic Programming [7, 8].

3 Using the meta-representations

Names and structural representations are two descriptions of a syntactic object at two different levels of abstraction. There are circumstances in which the inner structure of the object we want to refer to is not important at all. For instance, when writing a procedure for appending two programs we just need to know that we have to append lists of clauses, without getting into their internal details. In other cases, on the contrary, it is important to access the inner components of the syntactic object we have to deal with.

The two different meta-representations 'LOG supplies are intended to satisfy these two different uses. The structural representation of an entity describes the structure of the entity in terms of the names of its components, which, on the contrary, are viewed as monolithic entities. Thus, for instance, the clause

$$p(X,f(a)) :- q(X),r(a,b)$$

can be represented in 'LOG as

$$clause('p(X,f(a))',['q(X)','r(a,b)'])$$

while in other proposals using structural descriptive names only (e.g. [1, 5]) also sub-components are represented as structured terms. For example, according to Barklund's proposal [1], the above clause should be represented as

$$clause(atom(p,[var(0),compound(f,[const(a)])]),$$
$$conj(atom(q,[var(0)]),atom(r,[const(a),const(b)])))$$

Notice that if symbols had a name too then they should be replaced by the structured terms, e.g. lists of characters, representing them.

Having the structural representation only, it may result quite cumbersome for the user to represent such entities as programs and clauses at the meta-level, and, on the other hand, it may result quite expensive for the implementation to maintain the structural representation of low level entities, such as symbols. Actually, proposals which use a structural meta-representation only usually do not cover the naming of all the syntactic entities of the language: they usually exclude the two extremes, namely programs and symbols. The use of a synthetic notation as a shorthand for complex structured names, such as the one proposed in [5], solves only the problem of notational conciseness but still leaves the implementation problems unsolved (the implementation issue will be briefly addressed in Section 7).

Whenever a deeper detail level is needed, 'LOG provides the user with the <=> operator. Given the name of an entity, one can obtain its structure by applying the proper <=> operator to the name. Thus, for instance, if we want to know which is the name of the predicate defined by the above clause we can go inside the clause structure by unification and then apply <=t=> to its first argument. The goal

$$? - clause('p(X,f(a))',['q(X)','r(a,b)']) = clause(H,\_),$$
$$H <=t=> [N|\_]$$

will instantiate N to 'p'.

As a more comprehensive example, which makes use of the full power of the double arrow operators, we show the definition of a predicate psort which is able to sort clauses of a program according to the names of their head predicates. The arguments of psort are two program names, namely the object program and its sorted version.

```
psort(Prog,Sorted_Prog) :-
        Prog <=p=> ProgStruct,
        sort(ProgStruct,Sorted_ProgStruct),
        Sorted_Prog <=p=> Sorted_ProgStruct.
sort(L1,L2) :- ...
        %true if list L2 is L1 sorted w.r.t. the order
        %relation defined by the predicate order/2
order(Cl1,Cl2) :-        %true if Cl1 precedes Cl2
        Cl1 <=c=> clause(Head1l_),
        Head1 <=t=> [PName1l_],
        Cl2 <=c=> clause(Head2l_),
        Head2 <=t=> [PName2l_],
        string_comp(PName1,PName2).
```

where the predicate string_comp(PName1,PName2) tests if the atomic symbol S1 precedes S2 w.r.t. a standard order of characters as defined by the list ['a','b',...,'z']. It employs the <=s=> operator to obtain the lists of the characters composing the given symbols and then compares these lists.

It is important to realize that this program does not use any extra-logical feature. To obtain something similar in C_Prolog one should use such extra-logical built-in predicates as clause, =.. and @< (the latter used in place of our string_comp).

4 "Reconstructing" Prolog built-in meta-predicates

All the meta-predicates Prolog usually supplies in the form of built-in predicates are definable in 'LOG, at least in principle, within the language itself. In particular, the object level provability relation of a goal from a program can be defined, even if quite inefficiently, as a 'LOG program, similarly to the definition of the *demo* predicate given in [3]. In this way it is possible, on the one hand, to give these predicates a logic semantics, and, on the other hand, to ignore them while performing a formal analysis of the language.

Actually, some of Prolog built-in meta-predicates, such as =.., name and ==, become unnecessary in 'LOG, since explicit representations of terms and symbols are directly available. For instance, the Prolog clause

```
p(X) :- X =.. [F,a1lArgs],q(F).
```

can be replaced in 'LOG by the clause

```
p(X) :- X <=t=> [F,'a1'lArgs],q(F).
```

where the argument of p is assumed to be a term name. Now, assume q is defined as q(f) (resp., q('f') in 'LOG). In Prolog the goal p(f(a1)) succeeds while, unfortunately, the goal p(X) fails. In 'LOG, on the contrary, also the goal p(X) succeeds yielding the constraint X <=t=> ['f','a1'lY]. This establishes that X is constrained to be the name of a term of the form f(a,...). In particular, the solution X = 'f(a1)' can be obtained from this constraint by instantiating Y to [].

Some other simple Prolog meta-predicates, such as var, atom, etc., can be defined quite easily in 'LOG. var, in particular, is concerned with a crucial point, that is *variable naming*. Let us briefly comment upon this point. The name of an object level variable Alpha is 'Alpha'. Its structural representation instead is ['Alpha'] that is a list of a single element which is a symbol name. Given the symbol name, its structural representation can be easily obtained via the <=s=> predicate. Thus we can inspect the internal structure of the symbol and check whether it is a variable or not (we assume the syntactic conventions of most Prolog systems where variables are symbols with initial capital). Therefore, the Prolog meta-predicate var can be redefined in 'LOG as follows:

```
var(TN) :- TN <=t=> [SN],
           SN <=s=> [CN|_],
           upperAlphabet(U), member(CN,U).
upperAlphabet(['_','A','B',...,'Z']).
```

where TN is intended to be instantiated to a term name. The goal ?-var('Alpha') clearly succeeds with this definition. The goal ?-var('f(a)'), on the contrary, fails since the call to <=t=> in var fails. Notice that if TN is not instantiated yet when var is called then the solution of the destructuring goals in var is simply postponed. Thus, the goal ?-var(X) succeeds with computed answer X<=t=>[SN],SN<=s=>['_'|_]; and then, through backtracking, with computed answer X<=t=>[SN],SN<=s=>['A'|_], and then X<=t=>[SN],SN<=s=>['B'|_], and so on. This result establishes that a variable is a non-structured term whose first character is any of '_', 'A', 'B',... . The goal ?-var(X),X='f(a)' clearly fails, whereas the same goal erroneously succeeds in conventional Prolog.

Using 'LOG meta-programming facilities it is also possible to define the *unification* procedure between two object level terms and then use it to define other typical Prolog meta-predicates. The unification procedure can be implemented as a predicate unify(T1,T2,Subs) where T1 and T2 are the names of the terms to be unified and Subs encodes the computed object level variable substitutions as a list of pairs X/t where X is the name of a variable occurring in T1 or T2 and t is a term name. Thus, for instance, the two term names 'f(X,b)' and 'f(a,Y)' do not unify at the meta-level but they unify at the object level: unify('f(X,b)','f(a,Y)',S) succeeds with S = ['X'/'a','Y'/'b'].

Using unify it is easy to define, for instance, extended versions of the call and clause built-in predicates of ordinary Prolog, where the program to work with and the generated substitutions are handled explicitly as new arguments of the predicates. Following EnvProlog [10], we call these meta-predicates ecall and eclause, respectively. In particular, the ecall predicate is defined as follows:

```
ecall(PN,GN,Subs,C)
```

holds if the goal represented by GN can be proved in the object level program represented by the program name PN. Subs is a (possibly empty) list of pairs X/t representing the substitutions of the object level variables occurring in G and C is a (possibly empty) list of constraints, that is destructuring goals N<=>S in unsolvable form, generated by the proof of G. For example:

```
?- ecall('p(a,Y) :- q(Y). q(f(b))', ':- p(X,Y)',S,C).
S = ['X'/'a', 'Y'/'f(b)'], C = [].
```

5 Towards a concrete language

Some syntactic sugar can be added to the language described so far to make its implementation more efficient and simplify its use.

First, we extend the syntax of names so to be able to determine for each name which kind of name it is, i.e. which syntactic class the named objects belong to, by simply looking at the name itself. This ability can be advantageously exploited by the language implementation to select *at compile time* the most adequate *internal representation* for each different kind of name (see Section 7 for a discussion of the implementation issue). Furthermore, the appropriate instance of the generic operator <=> can now be selected automatically, accordingly to the kind of its arguments; so the user must be concerned with only a single overloaded <=> operator, letting the language implementation have the task of disambiguating it.

The syntactic conventions we will use for names are summarized in Figure 1. Assuming these conventions, each name uniquely identifies an object of a precise syntactic class. For instance, '{alpha}', '.alpha.', 'alpha' and '/alpha/' represent a program, a clause, a term and a symbol, respectively. Also ground *structural meta-representations* of different kinds can easily be distinguished each others accordingly to the different kinds of their components. For instance, a list of clause names is necessarily a program structure, a list of term names is necessarily a goal clause and so on. As an example, the three similar structures ['b','e','t','a'], ['/b/','e','t','a'] and [%b,%e,%t,%a] can be easily mapped on to the goal clause :- b,e,t,a, the term b(e,t,a) and the symbol beta, respectively.

objects		names
program:	c1.c2. ... .cn	'{c1.c2. ... .cn}'
clause:	h:-b1, ... ,bn	'.h:-b1, ... ,bn.'
term:	f(t1, ... ,tn)	'f(t1, ... ,tn)'
symbol:	abc	'/abc/'
character:	c	%c

Figure 1

A further step towards a more concrete programming language is introducing a *synthetic notation* for structural representations which is more convenient for the programmer than the list-like explicit notation used so far. The synthetic notation for the structural representation of an expression e closely resembles the corresponding name of e, except that double quotes are used instead of single quotes. For instance, the synthetic structural representation of the program c1.c2. ... .cn is "{c1.c2. ... .cn}", whereas the synthetic structural representation of the term f(t1,...,tn) is "f(t1,...,tn)".

Meta-variables can be easily handled by the explicit notation of structural representations. However, it would be desirable to use meta-variables in the synthetic notation as well. In order to allow object level variables to be easily distinguished from meta-variables also when using the synthetic notation we admit the former to be enclosed in quotes whenever ambiguities might arise. Thus, for instance, the term structures ['/f/',X,'Y'] and [F,'g(X)','X'] can be represented unambiguously in synthetic notation as "f(X,'Y')" and "F(g(X),'X')", respectively.

Moreover, the usual Prolog notation used to represent the rest of a list is easily extended to structural representations in synthetic form. For instance, the (incomplete) term structure ['/f/','a'|R] can be rewritten in synthetic notation as "f(a|R)". As another example, a predicate that concatenates two program structures can be defined as follows (cf. [12]):

```
appendPS("{}",P,P).
appendPS("{C|P1}",P2,"{C|P3}") :- appendPS(P1,P2,P3).
```

Finally, a synthetic notation is introduced also to represent *nested term structures* in a more convenient way. Nested term structures are lists of term structures (rather than of term names) which can be constructed by repeated applications of <=t=>. For instance, given the term name 'f(a,g(b))' the corresponding nested term structure is ['/f/',['/a/'],['/g/',['/b/']]] or, in synthetic form, "*f(a,g(b))". Meta-variables in partially specified nested term structures may occur at any depth in the term. For instance, in ['/g/',['/h/',X]], i.e. "*g(h(X))", X is clearly a meta-variable. Notice that nested term structures where all object level variables are replaced by meta-level variables closely correspond to *non-ground representations* in Gödel [7]. Also notice that, as a special case, "f($X_1,...,X_n$)" and "*f($X_1,...,X_n$)", $n \geq 0$ and $X_1,...,X_n$ (meta-)variables, are equivalent synthetic notations for the partially specified term structure ['/f/',$X_1,...,X_n$].

Nested term structures can be advantageously exploited to give an alternative definition of the ecall meta-predicate which provides some form of communication from the object level to the meta-level which turn out to be very useful in practice.

```
ecall(PN,NGS)
```

holds if there is an instance NGS' of the list of partially specified nested term structures NGS such that the conjunction of goals represented by NGS' can be derived, at the object level, from the program represented by the program name PN. For example

```
?- ecall('{p(X) :- q(X). q(a)}', ["p(X)","q(X)"]).
X = 'a'.
```

Similarly, it would be possible to define also a version of eclause working with lists of partially specified nested term structures instead of lists of term names and then use it to define the vanilla meta-interpreter in almost the same way as in ordinary Prolog.

6 Programs as data

One of the most peculiar feature of our language is the ability to deal with whole programs, i.e. finite sequences of clauses, as data objects. In this section, we briefly point out two classes of problems for which program names may result particularly useful. A wider discussion about this topic can be found in [12].

6.1 Program structuring

A program can contain an assertion a(PN) about another program designated by the program name PN. Therefore, the set of clauses in a program can be partitioned into separate smaller subsets defined as *inner programs*. Inner programs can be dealt with as data by the enclosing program, i.e. by the program at the meta-level. For example, the program

```
alpha prog '{p(X) :- q(X),r. q(a). r}'.
beta prog '{p(b). q(a)}'.
p(X) :- q(X).
q(c).
```

where prog is a user-defined infix operator, contains the definition of two inner programs. The three definitions of the predicate p occurring in the three different programs are dealt with as definitions of three distinct predicates, i.e., predicate names in a program are *local* to that program. Predicates in an inner program can be accessed only using meta-predicates such as ecall and eclause. For example, adding the clause

```
demo(N,G) :- N prog P,ecall(P,G).
```

to the above program, we can issue the goal

```
?- demo(alpha,"p(X)").
X = "a".
```

where simple mnemonic names, such as alpha and beta in the example, can be used instead of program names to refer to programs.

The use of *structural-descriptive* names for representing programs at the meta-level is a major difference with respect to MetaProlog [2, 4]. In MetaProlog theories (i.e., sets of clauses) are named via *simple constant* names with no resemblance of the structure of the object they denote. Thus, "All MetaProlog program databases ... are set up either by reading them in from files or by dynamically constructing them using system predicates" [4]. In 'LOG, on the contrary, a program name lists explicitly clauses that compose it. Thus our solution is well suited to support *program modularization*. Program names can be statically nested at any depth. The global program database can be split into a number of smaller program units, possibly nested, which can be accessed only via meta-predicates such as ecall.

In addition, 'LOG allows meta-variables to occur in program structures whereas the same is not feasible in MetaProlog. As a consequence, it is not possible in MetaProlog to define for instance a predicate like the predicate appendPS of Section 5: such a predicate could be implemented in a rather awkward way as a series of add_to calls. On the other hand, the use of structural descriptive names is not adequate to support the construction of self-referential sentences.

6.2 Clausal representation of data structures

Programs can be used also to collect clauses defining some complex data structure so that it can be managed as any other term (e.g. passed to a procedure as a parameter) maintaining all the advantages of the clausal representation (e.g. access by unification). For example, the following assertion

g1 graph '{a(a,b).a(a,c).a(b,c).a(b,d).a(c,d)}'.

can be used to define a graph, named g1, where graph is a user-defined infix operator and, as usual, a(X,Y) represents an arc between two nodes X and Y.

With such a representation, it is easy to define general predicates dealing with graphs, such as, for instance, a predicate path(X,Y,G) for finding a path between two nodes X and Y in a given graph G:

path(G,X,Y) :- ecall(G,"a(X,Y)").
path(G,X,Y) :- ecall(G,"a(X,Z)"),path(G,Z,Y).

Thus, it is possible to solve, for instance, the goal:

?- g1 graph G,path(G,a,d).

Different graphs can coexist in the same program if each graph is defined as a separate program, possibly with its own mnemonic name. This is much more difficult, and less elegant, to obtain using standard Prolog.

7 Implementation issues

The naming scheme we have chosen for 'LOG and, in particular, the use of a double meta-representation is justified also by a number of implementation concerns we will try to summarize in this section.

First of all, we notice that the use of meta-level names which are *isomorphic in structure* to the named objects allows the internal representation of meta-level names to be used directly as the internal representation of the objects themselves. No explicit link between objects and object names must be maintained by the system as, on the contrary, it would be necessary if unstructured constant names were used (e.g. in MetaProlog).

Therefore, having two meta-representations for the same object actually amounts to having two different internal representations for the same object. When using the structural representation we are likely to inspect the structural composition of the named object. This requires a *list-like* internal representation to allow standard unification to be used to access components. Conversely, when using the constant name we want to deal with an object as a monolithic entity. No logical operation is allowed to access the internal structure of a constant. A name could be stored in main memory simply as a *string*.

However, an efficient implementation of the language could choose a different internal representation of names without interfering with their structural representation. Indeed, although all Prolog built-in meta-predicates are in principle definable in 'LOG, the concrete version of our language should provide also a low level implementation of most of them in order to obtain acceptable execution efficiency for practical applications. Of course this require an adequate internal representation of the object to be dealt with. In particular, an efficient implementation of meta-predicates dealing with *programs*, such as ecall and eclause, requires program names to be stored in such a way to make accessing clauses as fast as possible. Thus, for instance, program names can be represented internally as a tree-like structure with auxiliary pointers and indexes or hash tables for improving search operations on clauses.

If we had the *structural representation* only, and therefore only the list-like internal representation for every object then there would be an unacceptable decreasing in the overall efficiency of the language. Just think of the overhead (both in space and in time) caused by representing all symbols as character lists. If, on the other hand, we had only *names*, and therefore only special-purpose internal representations, as required by any real implementation to obtain a reasonable execution efficiency, then any attempt to access the structure of a syntactic entity would cause non-trivial difficulties. Ad-hoc operations should be provided in this case instead of standard unification.

With our solution, which internal representation of an object must be used can be established by the user by selecting the appropriate meta-level representation of the object. Given one representation, the other one can be built in a quite straightforward manner. To this regard, notice that a structural representation describes the structure of an object in terms of the *names* of its components. So it is possible to use directly the internal representation of names as the elements of the internal representation of the structured name of an object. Furthermore, one representation can be built from the other one only when it is really necessary, that is, whenever a goal N<=>S is encountered and one of its arguments is a variable while the other one is a ground term. If both the representations of the same object are present a double link is used to connect them to each other.

Finally, notice that the syntactic conventions established in Section 5 allows the language implementation to always select at compile time the appropriate internal representations for each different kind of names and ground structural representations.

8 Conclusions

In this paper we have presented the meta-logic facilities the language 'LOG supplies and briefly discussed their motivations and uses. In particular, we have described a *naming scheme* for logic programs which allows two different but related meta-representations (namely, names and structured representations) to be associated with each syntactic entity of the language, from programs to characters. The availability of such a naming scheme allows all the meta-predicates usually available in Prolog to be defined as 'LOG programs. Also interesting extended versions of some of these predicates (such as ecall, eclause, etc.) can be provided quite easily. Furthermore we have shown simple applications of program names which have no immediate correspondent in standard Prolog.

A simple though inefficient prototype of the language described in this paper has been implemented in C_Prolog. A more concrete and complete version of this language, called *Quote-Prolog*, is under development at present. In particular, the notion of program name, program structure and the <=> operator dealing with them had already been implemented in *EnvProlog* [10, 11, 12] and has been now successfully re-implemented (in a better structured way, and including clause names) in a new extended Prolog interpreter written in Modula 2. It is planned for the near future to continue this implementation to include all the meta-level features of 'LOG.

References

1. J. Barklund: What is a Meta-variable in Prolog?. In: H.D. Abramson and M.H. Rogers (eds.): *Meta-Programming in Logic Programming: Proceedings of the META88 Workshop*, Bristol. MIT Press, 1990, pp. 383-398.
2. K.A. Bowen: Meta-Level Programming and Knowledge Representation. *New Generation Computing* 3, 1985, pp. 359-383.
3. K.A. Bowen, R.A. Kowalski: Amalgamating Language and Metalanguage in Logic Programming. In: K.L. Clark, S.A. Tärnlund (eds.): *Logic Programming*. Academic Press, 1982, pp. 153-172.
4. K.A. Bowen, T. Weinberg: A Meta-Level Extension of Prolog. In: *IEEE Symposium on Logic Programming*, Boston, 1985, pp. 669-675.
5. S. Costantini, G.A. Lanzarone: A Metalogic Programming Language. In: G. Levi, M. Martelli (eds.): *Logic Programming: Proceedings of the 6th International Conference*, Lisbon. MIT Press, 1989, pp. 218-233.
6. P.M. Hill, J.W. Lloyd: The Gödel Report (Preliminary Version). Technical Report TR-91-02, Department of Computer Science, University of Bristol, March 1991.
7. J. Jaffar, J.L. Lassez: Constraint Logic Programming. In: *Proceedings of the 14th POPL Conference*, Munich. ACM, 1987, pp. 111-118.
8. P. Lim, P.J. Stuckey: Meta-Programming as Constraint Programming. In: S. Debray, M. Hermenegildo (eds.): *Logic Programming: Proceedings of the 1990 North American Conference on Logic Programming*, Jerusalem. MIT Press, 1990, pp. 416-430.
9. J.W. Lloyd: *Foundations of Logic Programming*. Springer Verlag, 2nd ed., 1987.
10. A. Martelli, G.F. Rossi: Enhancing Prolog to Support Prolog Programming Environments. In: H. Ganzinger (ed.): *ESOP'88: 2nd European Symposium on Programming*, Nancy. Lecture Notes in Computer Science 300, Springer Verlag, 1988, pp. 317-327.
11. G.F. Rossi: Meta-programming Facilities in an Extended Prolog. In: I. Plander (ed.): *Artificial Intelligence and Information-Control Systems of Robots-89*, North Holland, 1989
12. G.F. Rossi: Programs as Data in an Extended Prolog. To appear in: *The Computer Journal*, British Computer Society, 1992.
13. S. Safra, E. Shapiro: Meta-interpreters for Real. In H-J. Kugler, (ed.): *Information Processing 86*, North-Holland, 1986, pp. 271-278.
14. L. Sterling, A. Lakhotia: Composing Prolog Meta-interpreters. In: R.A. Kowalski, K.A. Bowen (eds.): *Logic Programming: Proceedings of the 5th International Conference and Symposium*, Seattle, MIT Press, 1988, pp. 386-403.
15. H. Sugano H.: Meta and Reflective Computation in Logic Programming and Its Semantics. In: M. Bruynooghe (ed.): *META90: 2nd Workshop on Meta-Programming in Logic Programming*, Leuven, 1990, pp. 19-34.

The Pandora Deadlock Handler Meta-Level Relation

Reem Bahgat

Dept. of Computer Science, City University
Northampton Square, London EC1V 0HB
e-mail: reem@cs.city.ac.uk

Abstract. Pandora was first introduced in [1] as a new parallel logic programming language which combines don't-know non-determinism and stream and-parallelism in a unified and efficient manner. In this paper, we present the Pandora *deadlock handler relation* whose input argument is a list of meta-level representations of suspended goals. One way of utilizing the deadlock handler relation is in implementing (application-dependent) heuristic search. More generally, the deadlock handler relation can be used to manipulate suspended goals in a flexible manner; redundant goals can be removed, new goals can be added, and a group of goals can be replaced by a simpler group. We illustrate the use of the deadlock handler relation by various application programs and outline the features of the Pandora abstract machine which are required to support its functionality.

1. Introduction

Most of the logic programming languages fall into one of two categories: (a) variants of Prolog, and (b) the concurrent committed-choice languages, such as Parlog [6]. The basic distinction is that Prolog features *don't-know non-determinism*, the ability to search through multiple solutions to a relation; the concurrent languages provide dependent or *stream and-parallelism*, whereby conjoined goals are evaluated concurrently and may communicate incrementally by bindings to shared variables.

In order to efficiently support stream and-parallelism, the committed-choice languages are based on *committed-choice non-determinism*: the commitment to one candidate clause for a goal when several are possible. Thus, at most one solution to a relation is produced.

To the best of our knowledge, the first successful attempt to efficiently combine don't-know non-determinism and stream and-parallelism in a unified manner is the *Basic Andorra model* for executing Prolog programs [7]. The essential idea is to delay the evaluation of non-deterministic goals while eagerly executing all deterministic ones in parallel. A goal is *deterministic* if and only if there is at most one clause with a unifiable head. When the computation deadlocks with all remaining goals being non-deterministic, one such goal (usually the leftmost) is selected and a choice point created for it, as in standard Prolog. The computation then proceeds in and-parallel for each or-branch.

Various considerations [1] suggested that the Basic Andorra model is insufficient to support all the applications that are currently expressible in committed-choice languages.

As a result, Pandora (Parlog + Andorra) was introduced as a new logic programming language which extends Parlog with deadlock handling capabilities and a non-deterministic fork. The operational semantics of Pandora is a generalization of the Basic Andorra model. A brief description of Pandora with its simple non-deterministic fork is given in Section 2, assuming the familiarity with Prolog and Parlog.

Moreover, Pandora provides a special-purpose meta-level relation, named the deadlock handler relation, by which the user can optionally program the behaviour on deadlock in a manner suited to the particular application. One way in which it can be used is in combination with the non-deterministic fork, to implement a heuristic search. This is a way to intelligently select a particular non-deterministic goal to execute when several are possible. Alternatively, the deadlock handler relation supports a non-forking deadlock breaking mechanism by removing redundant goals and/or adding new ones.

The deadlock handler relation is defined in Section 3 and examples of using it to program various application-dependent heuristics are illustrated in Section 4. Section 5 explains how to implement a process interaction discrete event simulation by using the non-forking deadlock breaking mechanism, while Section 6 discusses the development of constraint logic programming (CLP) systems in Pandora. In the last example, all the features of the deadlock handler relation are utilized. Namely, alternative sets of constraints are generated by forking the computation, and complex sets of constraints are replaced by simpler ones which are evaluated in a new and-parallel phase.

The data structures and operations of the Pandora abstract machine that are required to support the deadlock handler relation are outlined in Section 7, followed by our conclusions in Section 8.

2. Pandora with the Default Deadlock Handler

There are two kinds of relation in a Pandora program: don't-care relations and don't-know relations. A Pandora conjunction may contain goals for both kinds of relation. A *don't-care relation* is defined as a normal flat Parlog procedure [6], such as the clock/3 relation in Program 5.1. A *don't-know relation* is defined by an undeclared procedure comprising a sequence of (guarded) clauses, each of the form:

p(T1, T2, ..., Tn) :- D: B.

D is a *det-guard*: a conjunction of goals (possibly empty) for certain primitives, including at least unification (=); we also allow other primitives such as <, >, =<, >=. B, the clause body, is any Pandora conjunction. The qsort/5 relation in Program 4.1 is an example of a don't-know relation.

A Pandora computation alternates between two phases: the and-parallel phase and the deadlock phase. During the *and-parallel phase*, all goals in a parallel conjunction are evaluated concurrently. A goal for a don't-care relation may suspend on input matching, that is a goal is delayed if its arguments are insufficiently instantiated to match the clause head arguments. A goal for a don't-know relation P is reduced provided it is *deterministic*, i.e. if and only if at least k-1 clauses are non-candidates, where k is the number of clauses in P's procedure. A *non-candidate clause* for a goal is a clause which has false (unsatisfiable) head unification and/or a det-guard.

When all remaining goals are don't-care suspended goals and/or non-deterministic don't-know goals, a *deadlock phase* is begun. The default deadlock breaking mechanism in Pandora is to select an arbitrary goal for a don't-know relation and to create a choice point for it; a new and-parallel phase is then started for each or-branch. If no such goal exists, i.e. if all remaining goals being for don't-care relations, then a semantic error is reported.

In addition to the parallel conjunction operator ',', Pandora goals can be conjoined with the sequential conjunction operator '&'. Assume the following conjunction of goals:
   G1 , G2 & G3 , G4 & G5
This indicates that G1 and G2 are first evaluated concurrently until they both terminate successfully. Then, the evaluation of G3 and G4 is begun. G5 is evaluated when all the other goals in the conjunction have succeeded. Similarly, two kinds of clause search operators are supported: the parallel clause search operator '.', and the sequential one ';'. A ';' in a don't-care or a don't-know procedure specifies that the clauses following the ';' should not be checked for candidacy until all the clauses preceding the ';' are non-candidates. For a detailed definition of the Pandora language, the reader may refer to [3].

3. The Deadlock Handler Relation

The deadlock handler relation is a don't-care relation whose mode declaration is:
   mode deadlock_handler(Meta_Suspended?, Optimum_Goal^,
                          Removed_Goals^, New_Goals^).

As in Parlog, '?' stands for input argument while '^' stands for output argument. If deadlock_handler/4 is not defined in the program, the default mechanism for breaking deadlock is applied. Otherwise, its procedure is invoked at the time of deadlock with the input argument, Meta_Suspended.

Meta_Suspended is a list of ground terms, each representing a suspended goal in the current conjunction of goals. For instance, if the computation deadlocks with the following goals being suspended:
   (g1(X, a, Y) & g2(Y, Z)),   g3(f(X), b)

then the suspended goals are g1(X, a, Y) and g3(f(X), b) while g2(Y, Z) is not suspended; its evaluation is delayed, because of the sequential conjunction operator '&', until g1(X, a, Y) succeeds. Hence, the input argument for deadlock_handler will be:
   Meta_Suspended = [g1(X, a, Y), g3(f(X), b)]

where X, Y, a, and b are all constants (X and Y are meta-level representations of the corresponding variables in the arguments of the suspended goals).

A goal for the deadlock handler relation may either *succeed* or *fail* but must not deadlock. If it did deadlock, then the consequential re-entry of the deadlock phase will re-invoke the deadlock handler procedure which may in turn deadlock. This may lead to an infinite computation.

The second, third, and fourth arguments of deadlock_handler/4 are output arguments whose bindings depend on the particular application that is programmed. The second argument, Optimum_Goal, allows the programmer to select an input term which represents a particular suspended goal for a don't-know relation; a choice point will be created for the selected goal and a new and-parallel phase will then begin for each search branch.

Alternatively, the third and fourth arguments of deadlock_handler/4 provide means for breaking deadlock without forking the computation. Using the third argument, Removed_Goals, the user can specify a list of terms from its input argument. When the evaluation of deadlock_handler/4 terminates successfully, the goals corresponding to these terms will be removed from the existing conjunction of goals, effectively replaced by calls to true. For instance, if Removed_Goals is bound to [g1(X, a, Y), g3(f(X), b)] in the above example, then the computation will proceed in a new and-parallel phase evaluating g2(Y, Z).

The fourth argument, New_Goals, can be bound to a list of goals to be added to the current conjunction of goals, effectively replacing a call to true by the conjunction of the new goals. Adding new goals can also break deadlock by instantiating unbound variables on which goals are suspended. The computation will then proceed in a new and-parallel phase.

While the second argument of deadlock_handler/4 provides a forking means of breaking deadlock, the third and fourth arguments provide a non-forking deadlock breaking mechanism. The semantics of the deadlock handler relation restricts the use of exactly one of them at one time. That is, only one of the following cases can result from a successful evaluation of deadlock_handler/4:

(a) the second argument is bound to an input term, while both the third and fourth arguments are bound to empty lists.
(b) at least the third or the fourth argument is instantiated to a non-empty list while the second argument is unified with the constant no_choice to explicitly indicate a non-forking deadlock breaking mechanism.

4. Programming Heuristic Search

One way of utilizing the deadlock handler relation in a Pandora computation is by implementing a heuristic (guided) search, i.e. by selecting a particular don't-know non-deterministic goal to execute, when several are possible. The choice of a particular goal is usually specific to the application under consideration, so it clearly cannot be left to an implementation to decide.

Executing Pure Prolog Programs

Pure Prolog programs can be run unchanged on a Pandora system. Because of the eager execution of deterministic goals and lazy execution of non-deterministic ones, the Pandora computation may be in most cases more efficient than the Prolog one. However, it is not guaranteed that the solutions will be reported in the same order as reported by standard Prolog. If the same order of solutions is required, then the textual order of goals should be preserved during the computation and, when the computation deadlocks, the leftmost goal should be selected for forking. In order to achieve this behaviour, two extra arguments are added to each (don't-know) relation in the program, by which the goals in the query are connected in a chain. The left end of the chain is instantiated to a constant, say left, while the right end is a variable. For example, Program 4.1 below redefines a Prolog quick sort program by adding the extra arguments. Suppose that the Prolog query is:

    qsort(L,SortedL,[]),   produce(L),   consume(SortedL)

Then, its corresponding Pandora query would be:

    qsort(left, R1, L, SortedL, []),
    produce(R1, R2, L),   consume(R2, R3, SortedL)

When a goal is reduced, it replaces itself in the chain with the goals in the body of its clause, in their correct order (clause 2). If the body of the clause is empty (clause 1), the goal removes itself from the chain by unifying its left and right connections.

```
    qsort(Chain, Chain, [], R, R).                        % clause 1.
    qsort(Left, Right, [X|Xs], R0, R) :-
        partition(Left, Temp1, Xs, X, L1, L2),
        qsort(Temp1, Temp2, L1, R0, [X|R1]),
        qsort(Temp2, Right, L2, R1, R).                   % clause 2.
```
Program 4.1. The don't-know relation qsort/5.

At any time during the computation, the leftmost goal according to the textual order in the program will have its first argument instantiated to the constant left, while the other goals will have an unbound variable in their first argument position. However, the leftmost goal may be anywhere in the current conjunction of goals. In order to select the leftmost goal for forking in the deadlock phase, deadlock_handler/4 can be defined as in Program 4.2. When the computation deadlocks, the deadlock_handler goal searches the list of meta-level representations of the suspended goals until it finds the term whose first argument instantiated to the constant left. This term represents the leftmost suspended goal. deadlock_handler then succeeds after unifying its second argument with the selected term. Notice that because of the sequential clause search operator ';', the second clause of deadlock_handler/4 will not be tried for candidacy except when the guard in the first clause fails.

```
mode deadlock_handler(?, ^, ^, ^).
deadlock_handler([ Goal | Meta_Suspended], Goal, [], []) <-
    arg(Goal, 1, left): true;
deadlock_handler([_|Meta_Suspended], Optimal, Removed, New) <-
    deadlock_handler(Meta_Suspended, Optimal, Removed, New).
```
Program 4.2. Using the deadlock_handler/4 relation to simulate a Prolog system.

Making Choices in Fly-Pan

Fly-Pan is a Pandora program which solves a real-life resource allocation problem. The program accepts two input arguments: a list of scheduled flights and a list of available aircraft, and assigns to each flight an aircraft whilst taking into account a number of constraints. For instance, an aircraft cannot be assigned to two flights that are overlapping in their scheduled flying time. Also, an aircraft can be assigned to several flights as long as it does not exceed its maximum flying hours in one day. For a full list of the constraints that are considered in Fly-Pan, the reader may refer to [2].

Using Pandora to program Fly-Pan reduces the search space dramatically [2]. However, since the problem under consideration is computationally complex, it is most likely that the and-parallel phase of the computation will not be sufficient to assign an aircraft to each input flight and choices have to be made. In order to increase the efficiency of the system even further, several heuristics are defined in the Fly-Pan deadlock handler (Program 4.3) which are applied in the following order:

1. *Selecting a Flight*
There are two alternative methods for selecting a flight from the remaining unassigned ones:

(a) *Flights Priorities*
The user may optionally specify a priority for each input flight as one of its attributes. The unassigned flight with the highest priority is the one chosen to be assigned an aircraft. Since the flights priorities are input to the program, the flights can be given different priorities each time the program is invoked, reflecting the changes in the environment of the problem without the need to modify the program itself.

(b) *First-Fail Principle*
If the user does not specify priorities for the flights, then the flight with the least number of choices is selected since it is likely to be the most difficult one to assign an aircraft.

select_aircraft/2 in Program 4.3 accepts Meta_Suspended: the list of ground terms

representing the suspended goals and returns Aircraft: the list of all aircraft terms in
Meta_Suspended, representing the aircraft which can still be assigned to flights.
choose_flight/2 accepts Aircraft and returns Flight: a list of those aircraft terms
that represent the set of possible aircraft for the selected flight.

```
mode deadlock_handler(?, ^, ^, ^).
deadlock_handler(Meta_Suspended, Optimal, [], []) <-
    select_aircraft(Meta_Suspended, Aircraft),
    choose_flight(Aircraft, Flight),
    choose_aircraft(Meta_Suspended, Flight, Optimal).
```
Program 4.3. The deadlock_handler/4 relation in Fly-Pan.

2. *Selecting an Aircraft for the Selected Flight*
The aircraft with the smallest number of remaining flying hours is selected since
assigning it to the flight is more likely to make it unavailable for other flights, reducing
the search space and leading to a more efficient problem-solving behaviour. For each
input aircraft, a don't-care relation goal:

```
    current_max(Identity, _, RemainingHrs, _)
```

holds a counter of the remaining flying hours for the aircraft as its third argument while
the first argument for the goal is the identity of the aircraft itself. When the aircraft is
assigned to a flight, current_max/4 reduces the remaining flying hours for the aircraft
by the length of the flight and recursively spawns a current_max/4 goal with the new
counter of the flying hours then terminates successfully. When the computation
deadlocks, current_max/4 will be represented by a ground term in Meta_Suspended.

```
    mode choose_aircraft(?, ?, ^).
    choose_aircraft(Meta_Suspended, Flight, Aircraft) <-
        select_remaininghrs(Meta_Suspended, RemainingHrs),
        aircraft_cost(Flight, RemainingHrs, Costs),
        min_cost(Costs, Aircraft).
```
Program 4.4. Choosing an aircraft for the selected flight.

choose_aircraft/3 in Program 4.4 selects the aircraft term in Flight which represents
the aircraft with the smallest counter value. It spawns three concurrently executing
processes:
1. select_remaininghrs/2, which scans Meta_Suspended searching for all the
   current_max/4 terms and returns RemainingHrs: a list of pairs, each of the form
   (Identity,FlyingHours), representing the input aircraft with their remaining flying
   hours.
2. aircraft_cost/3, which accepts Flight as well as RemainingHrs and assigns a *cost
   value* to each possible aircraft for the selected flight. The cost value for each aircraft
   is its remaining flying hours. aircraft_cost/3 produces Costs: a list of all the
   aircraft terms in Flight together with the cost values of the corresponding aircraft.

3. min_cost/2, which selects the aircraft with the minimum number of remaining flying hours.

The deadlock handler of the Fly-Pan program illustrates the need to inspect the suspended don't-care relation goals (e.g. current_max/4) as well as the suspended don't-know relation goals in order to select an optimal don't-know relation goal for forking.

5. Non-Forking Deadlock Breaking Mechanism

In some applications, a more powerful control than a simple heuristic search is required. Goals should be inspected and manipulated in different ways; redundant goals should be removed, new goals should be added, existing goals should be replaced by alternative ones, and inconsistent goals should be detected and possibly fail the computation.

For instance in a process interaction discrete event simulation [5], a real system is simulated by a number of concurrent logical processes, each representing activities in the real system. The simulated time is represented by a *central clock process* which controls the times at which activities are simulated. A process simulating an activity at time T_i would send an alarm request to the clock process and receive an alarm signal in reply when the simulated time is T_i. It can then start its activity. Events that are scheduled at the same time can be processed simultaneously by the concurrent logical processes. After executing all the events at T_i, the logical processes suspend and the clock process advances the simulated time to T_{i+1}: the time of the next simultaneous events.

```
mode clock(?, ?, ?).
clock(T, [], []).                                    % clause 1.
clock(Time, Events, [hold(Delay, Alarm)|Requests]) <-
    EventTime is Time + Delay,
    order_insert(e(EventTime, Alarm), Events, Events1),
    clock(Time, Events1, Requests).                  % clause 2.
clock(EventTime, [e(EventTime, Alarm)|Events], Requests) <-
    Alarm = EventTime,
    clock(EventTime, Events, Requests).              % clause 3.
```
Program 5.1. The clock process in a process interaction simulation.

One way of implementing the central clock process in Pandora is illustrated in Program 5.1. The first argument of clock/3 is the current simulated time. The second argument, Events, is a chronologically ordered list of the alarm signals to be sent. The clock process receives alarm requests from other processes in the system, each of the form hold(Delay, Alarm). Delay is the delaying period, from the current simulated time, after which an event should take place. For each alarm request (clause 2), the process computes the time for the event to occur, EventTime, and adds the alarm signal e(EventTime, Alarm) to Events. If the scheduled time for the next event in Events is equal to the current simulated time (clause 3), the clock process instantiates Alarm to that time.

When all the scheduled events for the present time have been processed, all the processes suspend waiting for the simulated time to be advanced. A goal for the deadlock_handler/4 relation in Program 5.2 will then bind its third and fourth arguments to singleton lists. The member of the list in the third argument is the meta-level representation of the suspended clock process, while the member of the list in the fourth argument is a term representing a new clock process with the advanced simulated time in its first argument. After the successful evaluation of deadlock_handler/4, the suspended clock process will be replaced by the new clock process. An alarm signal will be then sent by the new clock process (clause 3 in Program 5.1) to start the next scheduled event. deadlock_handler/4 unifies its second argument with the constant no_choice in order to explicitly indicate that no forking is required.

```
mode deadlock_handler(?, ^, ^, ^).
deadlock_handler( Meta_Suspended, no_choice, [OldClock], [NewClock]) <-
    get_clock(Meta_Suspended, OldClock),
    advance_time(OldClock, NewClock).

mode advance_time(?, ^).
advance_time(clock(T, [e(EventTime, Alarm)| Events], Request),
             clock(EventTime, [e(EventTime,Alarm)|Events], Request).
```

Program 5.2. The deadlock handler in a process interaction simulation.

6. Implementing Constraint Logic Programming Systems

A common method for implementing experimental constraint logic programming (CLP) systems is to write meta-interpreters for them in logic programming languages. Among the systems written this way are CLP(Σ^*) [13], CAL [11], and CLP*(X) [10]. These systems are all built on top of Prolog but differ in the domains of computation they are handling. The domain of computation, in turn, affects the type of constraints provided by the system as well as the constraint solver which solves them. Since goals in a Prolog conjunction are executed sequentially, none of these systems supports stream and-parallelism.

Pandora provides a new programming paradigm for writing such meta-interpreters which facilitates the development of these systems and increases their efficiency. Constraints can be represented by a network of concurrent goals for don't-care relations, whereby the shared variables among these goals represent the objects whose values are constrained. A constraint goal is suspended if its arguments are not sufficiently instantiated. As soon as a variable gets bound to a value, its value is propagated to all the constraints that are sharing it. This *constant propagation* may wake up suspended constraints, to be concurrently evaluated during the and-parallel phase, and this may in turn result in binding more variables, and so on.

Constraints can be incrementally generated by a constraint generator goal. If this goal is for a don't-know relation, then alternative sets of constraints can be (lazily) produced.

Constant propagation is a well-known simple and efficient technique for solving various constraint problems. The and-parallel phase in Pandora provides a natural means for applying constant propagation. However, more specialized constraint solvers and optimization techniques are required for solving complex types of constraints. Suppose, for example, that the computation is deadlocked and the following goals are suspended, representing equations and inequations:

$1 < Y$, $Y < X$, $X \leq 3$, X is $5 - Y$

Since these goals are for don't-care relation primitives, the default Pandora implementation would report a deadlock situation without producing a solution to the query. However, the deadlock handler relation can be defined to inspect the meta-level representation of the suspended constraints and to apply specialized algorithms for solving these constraints. Assuming in this example that the domain of computation is the set of Integer numbers, then deadlock_handler/4 may infer from the three inequality goals that $1 < Y < 3$, $Y < X \leq 3$, and hence Y is equal to 2 and X is equal to 3. As a consequence, deadlock_handler/4 should succeed after binding its third argument, Removed_Goals, to [$1 < Y$, $Y < X$, $X \leq 3$] and its fourth argument, New_Goals, to [$Y = 2$, $X = 3$]. A new and-parallel phase will then begin after removing the goals corresponding to the terms in Removed_Goals, and adding the goals corresponding to the terms in New_Goals. In this particular example, Y will be unified with 2, X will be unified with 3, the "3 is 5 - 2" goal will succeed, and the computation will terminate successfully.

Program 6.1 illustrates an abstract definition of the deadlock handler in a Pandora constraint program. The saggregate/3 goal scans Meta_Suspended and produces:

1. Dont_Know: a list of the meta-level terms representing suspended goals for don't-know relations,
2. Constraints: a list of the meta-level terms representing the suspended constraints.

constraint_solver/2 checks the satisfiability of the constraints and fails if inconsistent constraint goals are suspended. For example, suppose the following constraint goals are suspended: $X < 2$, $X > 4$. Since there is no value for X which would satisfy both inequalities, the constraint_solver goal should fail. Otherwise, constraint_solver applies a specialized algorithm to solve the constraints according to their domain of computation, and produce a new specified set of constraints, NewConstraints, which may include equality constraints that assign values to unbound variables (such as $Y = 2$ in a previous example).

decide_on_behaviour/6 decides on the mechanism for breaking deadlock based on the bindings of Dont_Know, Constraints and NewConstraints. If the set of constraints could not be further simplified by the constraint solver, then a term representing a don't-know relation goal is selected so that the corresponding goal is non-deterministically reduced (clause 2). This goal either generates alternative sets of constraints or alternative values for a constrained variable(s).

If no such goal exists (clause 3), a symbolic solution is produced as a set of existing simplified satisfiable constraints. However, if the set of constraints have been simplified (clause 4), then the new set of constraints is then compared with the old one in order to determine which of the old constraints should be removed and what are the constraints which should be added to the current conjunction of goals.

```
mode deadlock_handler(?, ^, ^, ^).
deadlock_handler(Meta_Suspended, Optimal, Removed, New)<-
    saggregate(Meta_Suspended, Dont_Know, Constraints),
    constraint_solver(Constraints, NewConstraints),
    decide_on_behaviour(Constraints, NewConstraints, Dont_Know,
                        Optimal, Removed, New).              % clause 1.

mode decide_on_behaviour(?, ?, ?, ^, ^, ^).
decide_on_behaviour( Constraints, Constraints, [Generator|Dont_Know],
                     Optimal, [], []) <-
    select_optimal([Generator|Dont_Know], Optimal).          % clause2.
decide_on_behaviour( Constraints, Constraints, [], no_choice, Removed,
                     Solution) <-
    symbolic_solution(Constraints, Removed, Solution);       % clause3.
decide_on_behaviour(Old, New, _, no_choice, Removed, Added) <-
    compare(Old, New, Removed, Added).                       % clause4.
```
Program 6.1. The deadlock handler for implementing a CLP constraint solver.

7. Supporting the Deadlock Handler Relation in PAM

An abstract machine for Pandora (PAM) is designed and implemented. PAM is based on a process-oriented execution model. There are two kinds of processes: don't-care processes and don't-know processes, corresponding to the two kinds of relation in a Pandora program. A process is responsible for reducing a goal using a candidate clause for it and spawning child processes to evaluate the goals in the body of the clause.

PAM is designed for a shared memory multi-processor architecture. A processor in PAM is either a master or a slave. There is only one master processor; the others are slaves. Each processor has its own set of data structures which are accessed by the other processors. For a detailed description of the Pandora abstract machine, the reader may refer to [4].

In order to be able to manipulate suspended goals in the deadlock phase, each processor maintains a *deadlock list*. The union of all processors deadlock lists forms a distributed data structure by which suspended processes can be accessed in the deadlock phase.

When processes suspend on an unbound variable, a *suspension list* attached to that variable is formed. It is a linked list which starts at the variable itself. Instead of adding each suspended process to the deadlock list, an optimization is employed: when a process is the first to suspend on a variable, the processor executing it adds a pointer to that variable at the end of its deadlock list. Other suspensions on the same variable do not affect the deadlock list. As a result, each element in the deadlock list represents a list of processes that are suspended on the same variable.

If a pointer to a variable is added to the deadlock list and before creating a choice point the variable gets bound, then the pointer can be deallocated ([3] explains how to remove such elements from the deadlock list).

When the computation deadlocks, each processor scans its deadlock list and *Suspended_Processes*: a list of all the processes that are still suspended on variable(s), is generated. Then, a slave processor searches for work while the master processor loads its first argument register with the Suspended_Processes list and creates a process to evaluate a goal for the predefined deadlock/1 relation, which the master executes itself.

If deadlock_handler/4 is not defined in the program, deadlock/1 searches for a don't-know process in Suspended_Processes and sets a global register, **choice-point flag**, to point to that process. A *choice point event* is then signalled so that processors synchronize and create a choice point for the selected (goal) process.

If deadlock_handler/4 is defined, *Meta_Suspended*: a list of meta-level representation of Suspended_Processes is generated. While creating the terms representing the goals of the suspended processes, the tag field of each unbound variable of these terms is (conditionally/unconditionally) changed to meta_var; these variables are then treated as constants by the unification algorithm in PAM. Additionally, a dictionary, *Meta_Var_Dictionary*, of all the variables in the arguments of the suspended goals is generated. Then, deadlock/1 creates a child (don't-care) process for deadlock_handler/4, whose input argument is Meta_Suspended, and suspends on children.

Failure of deadlock_handler/4 is the same as failure of any other (goal) process. If deadlock_handler/4 succeeds, the deadlock/1 process resumes to inspect the bindings of the deadlock_handler/4 output arguments.

deadlock/1 may remove processes in Suspended_Processes. A process is removed by marking it to be dead. When a variable on which the process is suspended gets bound, the process will not be executed since it is dead. If new goals should be added to the conjunction of goals, deadlock/1 spawns sibling processes for the new goals. Finally, the tags of the variables that are accessible by Meta_Var_Dictionary are changed back to unbound, Meta_Var_Dictionary is deallocated, and deadlock/1 terminates successfully.

Alternatively, a don't-know process may be selected for a non-deterministic execution; **choice-point flag** is then set to point to that process, the tags of the meta-variables are changed back to unbound, Meta_Var_Dictionary is deallocated, and a choice point event is signalled.

8. Conclusions

Pandora provides a programming paradigm of non-deterministic concurrent, communicating processes which appears to be well suited to many applications as well as being easy to implement. In addition to the simple non-deterministic fork, Pandora supports a more powerful mechanism for handling deadlock in which deadlocked goals can be accessed and manipulated by a meta-level program. Sections 4, 5, and 6 demonstrate the use of the Pandora meta-level deadlock handler relation in various applications. Namely: (1) controlling the order in which solutions are reported when running Pure Prolog programs as a Pandora system; (2) applying heuristic search to find an optimal solution for a resource allocation problem; (3) implementing process interaction discrete event simulation; and (4) implementing constraint logic programming prototype systems.

In all these examples, the definition of the deadlock handler relation is *sound* with respect to the intended meaning of the program. Selecting a don't-know relation goal for a non-deterministic reduction always preserves the soundness property. However, if the state of computation is changed by removing existing goals and/or adding new ones, then the new state of computation should be equivalent to the deadlocked state of computation for the intended interpretation of the program. At present, it is the responsibility of the programmer to ensure that the definition of the deadlock handler relation preserves the soundness property of the program. However, we are investigating possible algorithms for checking the equivalence of the old and new states of computation, given a formal specification of the intended meaning of the program.

The Pandora abstract machine PAM with the simple non-deterministic fork has been implemented [12]. The extensions for supporting the deadlock handler relation has been designed, as described in Section 7, but are not yet implemented. However, a prototype implementation of full Pandora has been developed on top of Parlog [3] and has been used to run the example programs in this paper, as well as others.

Related Work

Kernel Andorra Prolog [9] is a language closely related to Pandora and, like Pandora, it is inspired by the Basic Andorra model. It extends Prolog with a commit operator to commit to any clause whose guard is satisfied. An extended form of the Basic Andorra model, called the *Extended Andorra model* [14], is used to run Kernel Andorra Prolog programs. In the Basic Andorra model, a non-deterministic reduction of a goal is delayed until all deterministic computations take place. However, the Extended Andorra model allows and-parallel execution of non-deterministic goals but delays the propagation of their bindings to shared variables. If the execution of a conjunction of goals deadlocks, the non-deterministic bindings to shared variables are propagated.

Unlike Pandora, Kernel Andorra Prolog does not support a meta-level manipulation of suspended goals when the computation deadlocks. Moreover, the extended Andorra execution model is relatively complex and it remains to be seen how efficiently it can be implemented.

Future Research

In the immediate future, we will extend the current implementation of PAM with the requirements of the Pandora meta-level programming facilities. These extensions will not have any run-time overhead on programs which do not include the definition of the deadlock_handler/4 relation except for a test in the deadlock phase to check whether it is defined. Removing and/or adding processes is straightforward and has no significant extra cost. The price paid by supporting deadlock_handler/4 is in creating its input argument list as well as modifying the tags of the unbound variables before and after executing deadlock_handler/4. We intend to measure this cost and possibly investigate alternative implementations for supporting meta-level programming in the deadlock phase.

At the language level, we will investigate other application areas suited to Pandora. Another interesting subject is to design higher-level languages as front ends to Pandora; we are investigating the design of special-purpose constraint languages with primitives for expressing objects and constraints in specific domains. The same objective, but with a different approach, is taken by Gregory and Yang [8]. In their paper, they describe extensions to the Andorra-I abstract machine which support some primitive constraint handling techniques over finite domain variables.

References

[1] Bahgat, R.M.R. and Gregory, S. 1989. "Pandora: non-deterministic parallel logic programming". In *Proceedings of the 6th International Conference on Logic Programming*, Lisbon, pp.471-486.

[2] Bahgat, R.M.R. 1990. "Symbolic constraint-based reasoning in Pandora". In *Proceedings of the 2nd International Conference of the OUG Artificial Intelligence SIG*, London (October).

[3] Bahgat, R.M.R. 1991. "Pandora: non-deterministic parallel logic programming". PhD Thesis, Dept. of Computing, Imperial College, London University.

[4] Bahgat, R.M.R. 1991. "The Pandora abstract machine: an extension of JAM". In *Proceedings of the ICLP91 Pre-Workshop on Parallel Execution of Logic Programs*, 569 Lecture Notes in Computer Science, Springer-Verlag.

[5] Broda, K. and Gregory, S. 1984. "Parlog for discrete event simulation". In *Proceedings of the 2nd International Conference on Logic Programming*, Uppsala, Uppsala University Press, pp.301-312.

[6] Clark, K.L. and Gregory, S. 1986. "Parlog: parallel programming in logic". *ACM Transactions on Programming Languages and Systems*, 8(1), pp.1-49.

[7] Costa, V.S.; Warren, D.H.D. and Yang, R. 1990a. "The Andorra-I preprocessor: supporting full Prolog on the basic Andorra model". In *Proceedings of the 8th International Conference on Logic Programming*, Paris, pp.443-456.

[8] Gregory, S. and Yang, R. 1991. "Parallel Constraint solving in Andorra-I". Submitted to *FGCS92*.

[9] Haridi, S. and Janson, S. 1990. "Kernel Andorra Prolog and its computational model". In *Proceedings of the 7th International Conference on Logic Programming*, Jerusalem, pp.31-46.

[10] Hickey, T. 1989. "CLP* and constraint abstraction". In *Proceedings of Principles of Programming Languages*, Chicago, pp.125-133.

[11] Sakai, K. and Aiba, A. 1987. "Introduction to CAL". ICOT Technical Report, Tokyo.

[12] Schneider, L. 1991. "An Implementation of Pandora". MSc. Thesis, Dept. of Computing, Imperial College, London University.

[13] Walinsky, C. 1989. "CLP(Σ*): Constraint logic programming with regular sets". In *Proceedings of the 6th International Conference on Logic Programming*, Lisbon, pp.181-198.

[14] Warren, D.H.D. 1990. "The extended Andorra model with implicit control". Presented in *the Workshop of the 7th International Conference on Logic Programming*, Eilat, Israel.

Object-oriented Programming in Gödel: An Experiment

K. Benkerimi* and P.M. Hill**

(*) Department of Computer Science, University College London, London WC1E 6BT
(**) School of Computer Studies, University of Leeds, Leeds LS2 9JT

Abstract. This paper describes the results of an experiment in the use of the Gödel logic programming language for object-oriented programming. An object-oriented program is implemented in Gödel at two levels. First, at the base or object level, the static features such as object identity, the classification of objects, and message passing between objects are implemented using the basic Gödel language. Secondly, at the top or meta-level, the dynamic features such as changing the attributes of an object or creating a new object are implemented using the meta-programming facilities provided by Gödel. This experiment highlights the advantages of using Gödel for such a task as well as showing its limitations.

1 Introduction

Object-oriented programming [23] is a methodology that has been well researched and widely used in knowledge-base systems, artificial intelligence, and software engineering. As a result, programming languages such as Smalltalk [11] have been developed to directly support object-oriented programming.

Logic programming [17] is based on a subset of first order logic, namely Horn clauses, together with SLD-resolution as its inference strategy. The most popular and best known programming language using this formalism is Prolog, and a number of object-oriented environments have already been built on top of Prolog (for example, BIM_Probe [8] and Explore/L [9]). However, as it has been pointed out in [10], this does not represent an integration of the object-oriented and logic programming paradigms. Furthermore, since Prolog has no type system or declarative meta-programming facilities and few compilers allow for the modularization of a program, the object-oriented languages are implemented using the non-logical features of Prolog.

By contrast, Gödel [12] is a language with a parametric type system, simple module structure, and special declarative meta-programming facilities. Since Gödel is intended to be a declarative replacement to Prolog, it is worthwhile investigating how the above features of Gödel can be used for object-oriented programming. Furthermore, by modelling the object-oriented concepts directly in Gödel, we can provide these concepts both a model theoretic semantics in first order logic and an executable procedural semantics based on SLDNF-resolution.

The structure of the paper is as follows. In the next section, we outline the basic concepts of object-oriented programming and discuss related work. In Section 3, the main features of the Gödel programming language are described. In Section 4, we explain, with the help of an example, how the features of Gödel can be used to implement the concepts of the object-oriented programming methodology. In the last section, we discuss why the Gödel type and module systems cannot be used to model objects directly and indicate directions for future research.

2 Object-Oriented Programming

In this section, we describe briefly our own view of object-oriented programming. This has been sifted from many sources, the most important of which are [11] and [23].

The increasing interest and use of object-oriented techniques reflect a growing realisation that the design of a computer system must parallel a user's own abstraction of the underlying real world. The object-oriented method relies on a representation of the world in which the primary entity is an object. Such an object may be complex: that is, not only does an object have an identifier, but also properties and attributes. We regard properties as general rules for computing values whereas attributes define specific values. In the logic programming context, properties are defined by rules whereas attributes are usually defined as facts. The object identifier together with its properties and attributes need to be grouped in some way as a complete entity. This can be done by textually bracketing them to delimit the object, or (as we do in this paper) by labelling each of the properties and attributes with the identifier.

The objects in a system are collected into classes. An object can be a member of more than one class so that the classes may not be disjoint. Usually, a class is also regarded as an (abstract) object so that classes can be grouped into meta-classes. Since the meta-classes are also viewed as classes, this process can iterate, producing a hierarchy of classes. Thus, the class structure is assumed to be in the form of a directed graph without cycles. The properties and attributes of a class can be inherited by its subclasses. Furthermore, it is usually accepted that subclasses may override the inherited attribute values by declaring exceptions to the default value.

Objects communicate with other objects by message passing. The messages can be simple queries or instructions to the object to perform some action. A message must include the name of the receiving object, a selector, which is used to determine which action is required, and, preferably, the name of the sending object, to provide a context for the request. The procedure that carries out the action is called a method. The action may be to send further messages or to actually change the receiving object. The messages may have to be between classes and their subclasses, or between objects unconnected in the class hierarchical network.

Objects can change with time. A message to an object can request the object to update the values of certain of its attributes, demand the default value for an attribute inherited from its meta-class to be overridden, or create a new object. In such cases, it is desirable that the object has built-in integrity constraints and these should be checked before making any permanent changes.

Thus it can be seen that an object-oriented programming environment needs to provide support for complex objects, classification of objects, communication between objects, and changing objects. It is these features that we look at in relation to logic programming and, in particular, the logic programming language Gödel.

The research on object-oriented logic programming languages divides into two main areas in the same way that logic programming divides into the non-deterministic languages such as Prolog, and the concurrent languages [20]. In this paper, we concentrate on the non-deterministic object-oriented languages. In our approach, we separate the representation of the static aspects of object-oriented programming from the dynamic ones. This is reflected in the papers on this subject, where they either concentrate on the representation of dynamic objects (see, for example, [2], [3], [5], [6], [19]) or the representation of the classification of objects and inheritance of their properties (see, for example, [1], [4], [7], [13], [14], [16], and [22]).

In most of the literature, object identifiers are normally represented as constants but, in [16], arbitrary terms are allowed to represent an object identifier. The classification of objects and the inheritance networks connecting the classes of objects are either expressed as in [1] (LIFE), [18] (O-Logic) and [22] using order-sorted logic or by means of logical implication as in [7] and [16]. In this paper, we use the latter approach. Both [7] and [16] use negative literals in the heads of clauses to define the overriding of inheritance. The intended semantics of these negative literals is described in [7] to be that of three-valued logic and there does not appear to be a simple mapping of this to first order logic. The programming language LOCO [15] is based on the logic defined in [16].

The representation of complex objects, which includes encapsulation and inheritance, is the subject of [4], [13], and [14] and is based on Maier's O-Logic [18]. In [4], C-Logic is defined as a corrected version of O-Logic that supports object identity and encapsulation. It is shown how C-Logic can be mapped to first order logic. In [14], a revised and corrected version of O-Logic that allows for set-valued attributes is presented and, in [13], this is further extended to F-Logic which incorporates the overriding of inherited attributes using frame-based reasoning.

The modelling of dynamic objects has been approached in two distinct ways. The first approach models the state of an object by a first order theory and defines an action to be the process by which the state is changed. The intensional logic, defined in [3], associates the complete history of the state of an object with the object. This uses a frame assumption so that only changes to the database need be recorded. It is discussed in [19] how an event calculus might be used to model the changing states. The second approach extends logic programming to allow for disjunctive heads in the clauses although only a restricted form of resolution is used. In all the papers that use this approach, a subset of the atoms in the goal must unify (with the same substitution) with all the atoms in the head of a clause. This means that an atom cannot be selected for resolution before there is a complete set of atoms in the goal whose predicates match those occurring in the head of a clause. In [2], there is no limit on the number of atoms that can occur in the head of the clause. However, in [5] and [6], there are two kinds of predicates, those that name objects and those that name properties. A head of a clause can have at most two atoms, one with an object name and the other with a property name. There is no restriction on the atoms that can occur in the bodies of the clauses.

3   A Brief Overview of the Language Gödel

In this section, we only describe those features of Gödel that are used in this paper. For a complete specification of the language, see [12]. The language Gödel is intended to be a declarative successor to Prolog. However, it is also intended for practical programming in the large providing the support for good software engineering paradigms. Gödel is a strongly typed language with a simple module system. The type and module systems are explained by means of an example which is given below.

```
EXPORT Travel.
IMPORT Lists.
BASE Country.
CONSTANT Belgium,France,Germany,Netherlands : Country.
PREDICATE Journey : Country * Country * List(Country).
DELAY Journey(_,y,_) UNTIL NONVAR(y).

LOCAL Travel.
PREDICATE Neighbour : Country * Country.
Journey(x,x,[x]).
Journey(x,y,[y|w]) <-
        (Neighbour(z,y) \/ Neighbour(y,z)) & Journey(x,z,w).
Neighbour(France,Belgium).
Neighbour(Netherlands,Belgium).
Neighbour(France,Germany).
Neighbour(Germany,Netherlands).
```

A possible query to this program is

```
<- Journey(France,Netherlands,k).
```

A Gödel program is a set of modules, closed with respect to the import statements. There must be exactly one module in the set (called the *main* module) which does not occur in any import statements in the set. The example program here consists of a main module Travel together with the imported module Lists and the module Integers imported by Lists. The module Travel is in two parts. Symbols declared in the *export* part are available for use by other modules that import Travel. Symbols declared in the *local* part are not visible outside the module Travel. In addition, predicates declared in either part of Travel must be completely defined in the local part. The module structure of Gödel provides a means of defining abstract data types since a type symbol can be defined in the export part (and thus can be used by importing modules), but the functions used to construct the type may be declared in the local part, effectively hiding these functions and hence terms with this function at the top-level from other users.

Lists is a *system* module and, by importing Lists, we can use the standard Prolog notation [...|...] for lists together with a number of list processing predicates such as Append and Member. The module Lists imports the system module Integers. This makes the integers together with the usual integer arithmetic functions and predicates available for use in Lists and all modules that import Lists. Another useful system module is Strings that exports a type String. Thus, by importing the module Strings, the notation "..." can be used for a string of characters so that, for example, "ABC" denotes the character sequence A,B,C and, as \ is an escape character, "A\"BC\"" denotes the character sequence A,",B,C,". The module also exports a number of useful predicates for manipulating such strings. The system modules such as Integers, Lists, and Strings are *closed* modules. This means that the local part is not available for inspection and may be implemented (for efficiency reasons) in another language. However, all Gödel system predicates except for those declared and defined in input/output modules can be defined in first order logic. Thus the declarative semantics is not affected by using closed modules.

All symbols used in a Gödel module must be declared or imported into that module. In Travel, we declare one type constructor Country to be a base (that is a type constructor with arity 0). Lists exports the type constructor List with arity 1 and this can be used to construct types such as List(Country). The types, Country and List(Country), are needed by the constant and predicate declarations. The constant declaration defines the symbols such as France that are required to name individual countries. The predicate declarations define the two symbols Journey and Neighbour. Note that Journey is declared in the export part of Travel and Neighbour is declared in the local part. This means that, if another module imports Travel, it can use the predicate Journey but not Neighbour.

The predicates are always defined in the local part of a module. The syntax is straightforward. Variables begin with lower case letters. The connectives <-, &, and \/ denote backward implication, conjunction, and disjunction, respectively.

Finally, the execution of a program can be controlled by means of delay declarations. Any call to Journey will be delayed until the second argument is bound to a non-variable. There are a number of other facilities provided by Gödel, such as the pruning operator. These are defined and explained in [12].

Since Gödel is based on first order logic, it does not provide predicates such as *assert* and *retract* in Prolog. Instead, special meta-programming facilities are provided. This includes a utility program_compile that transforms an object level program to a term representing it, an input/output module ProgramIO that defines predicates for reading and writing to files containing such a term, and a system module Program which defines predicates needed to manipulate this term. The module Program exports the type OProgram. This is the type of the ground representation of a Gödel program used by the utility program_compile as well as the modules ProgramIO and Program. The directive

program_compile Travel.

will transform the program {Travel, Lists, Integers} to a ground term of type OProgram and store the result on the file Travel.prm. There is a similar utility program_decompile that reverses the process. The goal

<- FindInput("Travel",In(t)) & GetProgram(t,travel_prog).

for the module ProgramIO opens the input stream from file Travel.prm and binds the term on the file to the variable travel_prog. The module ProgramIO also exports the predicates FindOutput and PutProgram: these can be used to put a term of type OProgram back on a file, making any changes to a program persistent. In addition to the type OProgram, the module Program makes available base types including OName, OTerm, and OFormula. The base OName, (resp., OTerm, OFormula) is the type of the representation of a symbol (resp., term, formula) in the object level language. These base types are abstract data types. That is, the terms of these types can only be constructed using the implementation part of Program. However, Program exports a large number of predicates that provide access to the constituent parts of such terms and facilitate their modification. For example, there are predicates that take the string representing a symbol, term, or formula and convert it to the internal representation of type `OName`, `OTerm`, or `OFormula`, respectively. There are predicates that update the object program by adding or deleting language declarations and program statements. The predicate `Succeed` finds the computed answer for a representation of a goal and program. For a complete list of predicates exported by `Program`, see [12].

4   Object-Oriented Programming in Gödel

In this section, we show how an object-oriented system might be defined in Gödel. We assume that the static features can be modelled using sentences in first order logic of the form $A \leftarrow W$, where A is an atom. We illustrate our ideas with a program consisting of a bibliographic database. An object-oriented database has to be defined in a single module. (The reason for this is given in Section 5.) In the example, this is called `Bibliographies`. This module exports the base types `ClassType` (for the object identifiers) and `ValueType` (for the objects' attributes and properties).

So that the subclass relationship between objects is visible to all modules that import the database module, the object identifiers are declared in the export part of the module. In the `Bibliographies` example, the constants declared with type `ClassType` are either bibliographies (such as `Bibliography`, `OOBib`, `LPBib1`) or individual entries within a bibliography (such as `Key1`). The root class (that is an object which is a class with no parent class) has identifier `Bibliography` and subclasses `LPBib` (for logic programming bibliographies) and `OOBib` (for object-oriented bibliographies). The logic programming bibliography is in two parts. These are subclasses of `LPBib` and have identifiers `LPBib1` and `LPBib2`. The bibliographies `OOBib`, `LPBib1`, and `LPBib2` have subclasses `Key1`, `Key2`, and `Key3` that identify the individual entries.

An object-oriented database also needs to export the selectors that define the type `ValueType` needed for querying the database. In `Bibliographies`, these selectors are the functions `Entry`, `Subject`, and `Author`. For the arguments of these selectors, we import the type `String` via the module `Strings`.

Finally, the predicates `SubClass` and `Query` are exported. `SubClass` defines the hierarchical structure of the objects in the database. The predicate `Query` is intended for finding the attribute and property values of these objects.

In the local part of the module the predicates `SubClass` and `Query` are defined. The definition of `Subclass` depends on the particular object-oriented database. For example, in `Bibliographies`, there is the fact

`SubClass(OOBib,Bibliography).`

The predicate `Query` is defined by

`Query(x,y) <- Send(x,x,y).`

where `Send` is a message passing predicate. The difference between `Send` and `Query` is that `Send` keeps a record of the object that sent the message in its middle argument. Thus `Send`(s,t,u) is true if s is the identifier for the object receiving the message, t is the identifier of the (possibly the same) object that is sending the message and u is either an attribute of s or a property for s but where the rule for this property is attached to t. The definition of `Send` is as follows.

`Send(x,_,z) <- Attribute(x,z).`
`Send(x,y,z) <- Property(x,y,z).`
`Send(x,y,z) <- Inherit(x,z) & SubClass(x,w) & Send(w,y,z).`

The predicates Attribute and Property define the attributes and properties of objects, respectively. Attribute has two arguments. The value of the second argument is the attribute of the object named in the first argument. Thus, in the Bibliographies module, there is the fact

Attribute(LPBib,Subject("logic programming")).

Property has three arguments. The value of the third argument is a property of the object named in the second argument although the rule for this property is attached to the object named in the first argument. Thus, in Bibliographies there is the rule

```
Property(Bibliography,self,Author(author)) <-
          Send(self,Bibliography,Entry(x,y)) &
          Send(self,Bibliography,Order(order)) &
          Select(Entry(x,y),order,author).
```

The attribute selector Order is declared to be a function with one argument of type OrderType. Hence the selector Order and its argument cannot be referenced from outside this module. Select is a predicate required just for this rule. Select(s,t,u) is true if s is an attribute with selector Entry, t is the constant AT or TA of type OrderType, indicating whether the order of the arguments to Entry is author/title or vice versa, and u is the first argument of Entry in s if t is AT and the second if t is TA.

An object normally inherits an attribute or property value defined in an ancestor class. This is defined by the predicate Inherit. An object may need to override the default attribute values with local values. This is defined by the predicate Override. For both Inherit and Override, the second argument is the attribute or property which is inherited or overridden by the object identified in the first argument. In the Bibliographies example, we have a clause of the form

Inherit(x,Subject(_)) <- ~Override(x,Subject(_)).

for each attribute selector such as Subject. (It might be thought that the definition Inherit(x,y) <- ~Override(x,y) for Inherit would be adequate, but Gödel has safe negation which means that, apart from anonymous variables, the selected literal must be ground.) There is also the fact

Override(LPBib,Subject(_)).

so that the subject's value "logic programming" of LPBib overrides the default value "general" of Bibliography. To avoid complicating the examples, we do not allow properties to be overridden. Hence there is a fact

Inherit(x,Author(_)).

Thus all the static features of an object-oriented system can be represented in a pure logic programming language such as Gödel. It is assumed that the object-oriented system is represented in the form described above so that it is completely contained within a single module. We call such a module an *object-oriented database*.

For dynamic or changing objects such as a bibliography, we need the ability to modify existing objects and create new objects in the database. Changes to an object-oriented database could cause unwanted consequences to hold. For example, in Bibliographies, no bibliography should have both attribute values Order(AT) and Order(TA). As a result, it is envisaged that together with an object-oriented database, there will be one or more integrity constraints in the form of goals. It is intended that these must be true in any version of the database. Thus it will be necessary to check that these goals succeed in an updated database before making the changes persistent. The predicates defined in the meta-programming input/output module ProgramIO can be used to make database changes persistent. Other predicates, exported by the system module Program, are fully declarative so that any changes made to the database by these predicates is undone on backtracking. Thus, in Gödel, non-persistent changes can be made and then only if the integrity constraints are satisfied would they be made persistent.

An object-oriented database module is represented at the meta-level as part of a ground term of type OProgram. This term has to be manipulated using predicates in the meta-programming module Program. As described in Section 2, only four forms of updating need to be fully supported.

1. Updating the attribute values.

2. Updating the override rules.

3. Updating the subclass hierarchy.

4. Creating or removing an object.

It is envisaged that, having designed an object-oriented database, all interaction will be at the meta-level. Thus, not only do we need to support the updating of an object-oriented database, but also the different questions that might be asked of such a database. These are assumed to include the following.

5. Querying an object's property and attribute values.

6. Querying the existence of an object and its attributes.

7. Querying the inheritance hierarchy of the objects.

Thus, we define a meta-level module Oo that imports Program and defines a number of predicates for querying and updating an object-oriented database. The export part of Oo is given in the appendix. We do not give the local part, but the predicates in the export part can all be defined in Gödel using predicates exported by Program.

We now explain the predicates in Oo. The first two arguments of each predicate in the module Oo have the same types OProgram and String. The first argument should contain the representation of a complete program and the second argument the module name of an object-oriented database in the program.

The predicate OoObject provides a means of converting an object constant to the corresponding term of type OTerm. Thus OoObject is true if the fourth argument is the meta-level representation of the constant represented by the string in the third argument. Similarly, the predicate OoSelector provides a means of converting an attribute or property selector to the corresponding term of type OName.

The predicates OoUpdateAttribute, OoUpdateOverride, OoUpdateSubClass, and OoUpdateObject provide the means of changing the database and correspond to items 1 to 4 above. Finally, the predicates OoQueryValue, OoQueryObject, and OoQueryInherit provide the ability to query the database and correspond to items 5, 6, and 7 above.

We illustrate the use of Oo using Bibliographies as the object level module. For this we need to define a module ProcessBib.

```
EXPORT ProcessBib.
IMPORT ProgramIO, Oo.
```

We assume that ProcessBib is compiled and loaded, so that all the predicates defined in Oo, IO, and ProgramIO are available for use in a goal. The directive

```
program_compile Bibliographies.
```

will take the program whose main module is Bibliographies and write the ground representation of it in the file Bibliographies.prm. Next, the goal

```
<- FindInput("Bibliographies",In(bib)) &
   GetProgram(bib,bib_prog) &
   EndInput(new_bib) &
   OoObject(bib_prog,"Bibliographies","OOBib",oobib) &
   OoObject(bib_prog,"Bibliographies","Key4",key4) &
   OoSelector(bib_prog,"Bibliographies","Entry",2,entry) &
   OoUpdateObject(bib_prog,"Bibliographies",key4,bib_prog1) &
   OoUpdateSubClass(bib_prog1,"Bibliographies",oobib,key4,bib_prog2) &
   OoUpdateAttribute(bib_prog2,"Bibliographies",key4,entry,
              [["\"Davies\"","\"What is an Object\""]],bib_prog3) &
   FindOutput("Bibliographies",Out(new_bib)) &
   PutProgram(new_bib,bib_prog3) &
   EndOutput(new_bib).
```

will bind bib_prog to the ground representation of the program whose main module is Bibliographies and is on file Bibliographies.prm, add a bibliographic item to the OOBib bibliography with key Key4 and entry Entry("Davies","What is an Object"), and put the modified bibliography back on file Bibliographies.prm. Finally, the directive

```
program_decompile Bibliographies.
```

will take the file Bibliographies.prm and create the new Bibliographies module.

Note that until the call to PutProgram these changes are not persistent. Thus we may insert after OoUpdateAttribute a call that checks the integrity constraints. If this call fails, then the update fails and the database is unchanged.

5 Discussion

In exploring the use of Gödel to represent the object-oriented concepts described in Section 2, three questions arise.

- Can the type system of Gödel be used to model the classification of objects?

- Can the module system of Gödel be used to model the encapsulation of objects?

- Can the meta-programming facilities provided in Gödel be used to model the dynamic behaviour of objects?

We consider each of these in turn. First, if an object is identified as a type, the classification of objects is most easily modelled by means of subtypes. However, the Gödel type system does not allow for subtyping so that such a representation is not possible. Thus, in the bibliography example, `ClassType` is the type of all the object identifiers and `ValueType` is the type of all the attributes and properties.

Secondly, we consider the possible use of the Gödel module system for object encapsulation. If an object is to be encapsulated in a module, then all the attributes and properties of that object should be defined in that module. Furthermore, if an object is the subclass of another class, then, as it will inherit values from this class, it must import the module identified with this class. For good software engineering reasons, Gödel has a number of module conditions which restrict both the importation of modules and the use of symbols in a module. First, Gödel does not allow a symbol to have more than one declaration in the set of modules defining a program. Thus each symbol is declared in a unique module. Secondly, a module can only use symbols that it declares or imports. Thirdly, a Gödel module system must be hierarchical, that is, if module A imports a symbol from module B, then module B cannot import any symbols from module A. Thus the only way that both A and B can use the same symbol is by declaring it in a common module and importing this module into both A and B. This means that the base types `ClassType` and `ValueType` and the message passing predicate `Send` must be declared in a module imported by all the object modules. Fourthly, the range types in constant and function declarations in a module must be either bases that are declared in the same module or types whose top-level constructors are declared in the same module. Thus, the common module would have to declare all the constants and functions (such as `Bibliography`, `OOBib`, `Key1`, `Subject`, and `Author`) whose range types were `ClassType` or `ValueType`. Finally, every predicate declared in a module must be defined in the local part of that module. Hence, `Send` must be defined in the common module and all the predicates such as `Property` and `Attribute` used in the definition of `Send` must be declared or imported into the common module. As a result, the encapsulation breaks down and all the objects, their attributes and properties for an object-oriented system such as `Bibliographies` are defined in a single module. Research on extending the Gödel type and module systems for object-oriented programming is required.

Finally, we have shown how the meta-programming facilities of Gödel can be used in modelling the dynamic behaviour of objects. In Gödel (as in first order logic), meta-level programming is strictly separated from the object level that is being manipulated. Thus an object cannot modify another object in the same database and the traditional view of object-oriented programming in which one object is able to send a message to another object to update itself is not definable. However, a meta-level object can query and change objects in the object level database. By defining the special meta-level module `Oo` for manipulating object-oriented databases, we have shown how such an object might be supported. Furthermore, by separating the input/output predicates in `ProgramIO` from the updating predicates in `Program`, it is possible to check any updating against the integrity constraints before making the changes to the database persistent. Further work on the precise form of the module `Oo` together with its implementation is also required.

It is shown in [21] that the basic concepts of object-oriented programming can be modelled naturally in Concurrent Prolog. Since Gödel has a pruning operator of which a special case (called the *bar commit*) is similar to the commit of the concurrent logic programming languages and has delay declarations that force variables to be instantiated before a literal can be selected, it should be possible to use the same approach to model the concepts of object-oriented programming in Gödel. Such an approach to object-oriented programming in Gödel should also be explored.

References

[1] H. Aït-Kaci and P. Lincoln. LIFE, a natural language for natural language. Technical Report ACA-ST-074-88, MCC, Austin, USA, 1990.

[2] J. Andreoli and R. Pareschi. Linear objects: Logical processes with built-in inheritance. In *The Seventh International Conference on Logic Programming*, Jerusalem, Israel, pages 495–510. MIT Press, 1990.

[3] W. Chen and D.S.Warren. Objects as intensions. In K. Bowen and R. Kowalski, editors, *The Fifth International Conference and Symposium on Logic Programming*, Seattle, USA, pages 404–419. MIT Press, 1988.

[4] W. Chen and D.S.Warren. C-logic of complex objects. In *The Eighth ACM Symposium on Principles of Database Systems*, Philadelphia, USA, pages 369–378, 1989.

[5] J.S. Conery. Object oriented programming with first order logic. Technical Report CIS-TR-87-09, Dept. of Computer and Information Science, University of Oregon, USA, 1987.

[6] J.S. Conery. Logical objects. In R.A. Kowalski and K.A. Bowen, editors, *The Fifth International Conference and Symposium on Logic Programming*, Seattle, USA, pages 420–434. MIT Press, 1988.

[7] G. David and A. Porto. Semantics of inheritance in hierarchic systems. Position paper in the ICLP91 pre-conference workshop on merging object-oriented and logic programming, June 1991.

[8] I. de Zegher and M. Baudinet. BIM-Probe : An object oriented language on top of BIM-Prolog. In *The EUREKA Project PROTOS*, Zürich, April 1990.

[9] M.P.J. Fromhertz. Explore/L: An object-oriented logic language. Technical Report Nr. 91.06, Institut für Informatik der Universität Zürich, 1991.

[10] P. Gailly and J. Binot. Position paper. In the ICLP91 pre-conference workshop on merging object-oriented and logic programming, June 1991.

[11] A. Goldberg and D. Robson. *SmallTalk-80, The Language and its Implementation*. Addison-Wesley, 1983.

[12] P.M. Hill and J.W. Lloyd. The Gödel report. Technical Report TR-91-02, Dept. of Computer Science, University of Bristol, UK, 1991. Revised Feb 1992.

[13] M. Kifer, G. Lausen, and J. Wu. Logical foundations of object-oriented and frame-based languages. Technical Report 90/14, Dept. of Computer Science, State University of New York at Stony Brook, USA, 1990.

[14] M. Kifer and J. Wu. A logic for programming with complex objects. To be published in the Journal of Computing and System Sciences, 1992.

[15] E. Laenens, B. Verdonk, D. Vermeir, and D. Sacca. The LOCO language: Towards an integration of logic and object oriented programming. Technical Report 90-09, Dept. of Mathematics and Computer Science, Universitaire Instelling, Antwerp, Belgium, 1990.

[16] E. Laenens and D. Vermeir. Object oriented logic programming using ordered logic. Position paper in the ICLP91 pre-conference workshop on merging object-oriented and logic programming, June 1991.

[17] J.W. Lloyd. *Foundations of Logic Programming.* Springer-Verlag, 2nd edition, 1987.

[18] D. Maier. A logic for objects. In J. Minker, editor, *Workshop on Foundations of Deductive Databases and Logic Programming,* Washington D.C., USA, 1986.

[19] M.J. Sergot and F.N. Kesim. On the dynamics of objects in a logic programming framework. Position paper in the ICLP91 pre-conference workshop on merging object-oriented and logic programming, June 1991.

[20] E. Shapiro. The family of concurrent logic programming languages. *ACM Computing Surveys,* 21(3):412–510, 1989.

[21] E. Y. Shapiro and A. Takeuchi. Object-oriented programming in Concurrent PROLOG. *New Generation Computing,* 1(1):25–48, 1983.

[22] G. Smolka and H. Aït-Kaci. Inheritance hierarchies: Semantics and unification. *Journal of Symbolic Computation,* 7:343–369, 1989.

[23] M. Stefik and D.G. Bobrow. Object-oriented programming: Themes and variations. *The AI Magazine,* 6(4):40 – 62, 1984.

Appendix

Database Example: Bibliographies

```
EXPORT Bibliographies.

%The bases and predicates required by all object-oriented programs.
BASE       ClassType, ValueType.
PREDICATE  Query    : ClassType * ValueType;
           SubClass : ClassType * ClassType.
DELAY      Query(x,y) UNTIL NONVAR(x) & NONVAR(y).


%Classes used in Bibliographies.
CONSTANT   Bibliography, LPBib, OOBib,LPBib1, LPBib2,
           Key1, Key2, Key3                          : ClassType.
```

```
% The module Strings and selectors Subject, Entry, Author are used
% in the attributes and properties of Bibliography.
IMPORT     Strings.
FUNCTION   Subject, Author : String -> ValueType;
           Entry : String * String -> ValueType.

LOCAL Bibliographies.

% The main predicates for the module.
PREDICATE Override, Inherit, Attribute : ClassType * ValueType;
          Property, Send : ClassType * ClassType * ValueType.

Query(x,y)  <- Send(x,x,y).              % The definition of Query.

% We can use any attribute or property in the current class.
% Unless overridden use the selector for the superclass.
Send(x,_,z) <- Attribute(x,z).
Send(x,y,z) <- Property(x,y,z).
Send(x,y,z) <- Inherit(x,z) & SubClass(x,w) & Send(w,y,z).

% The default attributes for Bibliography.
Attribute(Bibliography,Subject("general")).
Attribute(Bibliography,Order(AT)).

% The property rule for Bibliography and all its subclasses.
Property(Bibliography,self,Author(author)) <-
                Send(self,Bibliography,Entry(x,y)) &
                Send(self,Bibliography,Order(order)) &
                Select(Entry(x,y),order,author).

% Order is locally defined and can only be used within the module.
BASE       OrderType.
CONSTANT   TA,AT : OrderType.
FUNCTION   Order : OrderType -> ValueType.

% Local declaration and definition for Select
PREDICATE Select : ValueType * OrderType * String.
Select(Entry(x,y),AT,x).
Select(Entry(x,y),TA,y).

% The attributes are inherited if not overridden.
Inherit(x,Subject(_)) <- ~Override(x,Subject(_)).
Inherit(x,Entry(_,_)) <- ~Override(x,Entry(_,_)).
Inherit(x,Order(_)) <- ~Override(x,Order(_)).

% The property Author is always inherited.
```

```
Inherit(x,Author(_)).

SubClass(LPBib,Bibliography).      % LPBib: a subclass of Bibliography.
Override(LPBib,Subject(_)).
Attribute(LPBib,Subject("logic programming")).

% There is one object-oriented bibliography.
SubClass(OOBib,Bibliography).      % OOBib: a subclass of Bibliography.
Override(OOBib,Subject(_)).
Attribute(OOBib,Subject("object oriented")).

% There are two logic programming bibliographies.
SubClass(LPBib1,LPBib).            % LPBib1: a subclass of LPBib.
Override(LPBib1,Order(_)).
Attribute(LPBib1,Order(TA)).
SubClass(LPBib2,LPBib).            % LPBib2: a subclass of LPBib.

SubClass(Key1,LPBib1).             % Key1 is a key in LPBib1.
SubClass(Key2,LPBib2).             % Key2 is a key in LPBib2.
SubClass(Key2,OOBib).             % Key2 is also a key in OOBib.
SubClass(Key3,OOBib).             % Key3 is a key in OOBib.

% The bibliography entries.
Attribute(Key1,Entry(" Programming in Logic","Jones")).
Attribute(Key2,Entry("Thomas","Logic in Objects")).
Attribute(Key3,Entry("Evans","Objects in Programming")).
```

A Module for Dynamic Object-oriented Programming

```
EXPORT Oo.
% This program requires the object-level program to contain an
% object-oriented database module.

IMPORT Program.

% In the following predicates, the first argument is the
% representation of a program P and the second argument is
% the name of an object-oriented database module O in P.
PREDICATE  OoObject : OProgram * String
* String        % String representation of an object constant I in O.
* OTerm;        % Representation of I.

           OoSelector : OProgram * String
* String        % String representation of a selector M in O.
* Integer       % Arity of M.
* OName;        % Representation of M.
```

```
             OoQuery : OProgram * String
* OTerm        % Representation of class C in O.
* OName        % Representation of a selector M with arity n for C.
* List(String) % List of arguments T1,...,Tn for M.
* String;      % String representation of computed answer for the goal
               % "<- Query(C,M(T1,...,Tn))" for P.


             OoClass : OProgram * String
* OTerm        % Representation of a class C in O.
* List(OName)  % List of representations of attribute selectors in C.
* List(OName); % List of representations of property selectors in C.


             OoSubClass : OProgram * String
* OTerm        % Representation of a class C in O.
* OTerm        % Representation of a subclass S of C in O.
* List(OName); % List of representations of attribute selectors for C
               % overridden in S.


             OoUpdateObject : OProgram * String
* OTerm        % Representation of an object constant S which is not a
               % class in O.
* OProgram;    % Representation of the program which differs from P
               % only in module O where S is now a class.


             OoUpdateSubClass : OProgram * String
* OTerm        % Representation of a class C in O.
* OTerm        % Representation of an object constant S which
               % is not a subclass of C in the module O in P.
* OProgram;    % Representation of the program which differs from P
               % only in module O where S is now a subclass of C.


             OoUpdateOverride : OProgram * String
* OTerm        % Representation of a class S in O.
* OTerm        % Representation of the parent class C of S.
* OName        % Representation of an attribute selector M for C
               % which is not overridden in S in the module O in P.
* OProgram;    % Representation of the program which differs from P
               % only in module O where M is now overridden in S.


             OoUpdateAttribute : OProgram * String
* OTerm        % Representation of a class C in O.
* OName        % Representation of a selector M with arity n for C.
* List(List(String)) % List of lists of arguments T1,...,Tn for M.
* OProgram.    % Representation of the program which differs from P
               % only in O where all attributes for C with selector M
               % are replaced by attributes with values M(T1,..,Tn).
```

A Sensible Least Herbrand Semantics for Untyped Vanilla Meta-Programming and its Extension to a Limited Form of Amalgamation

Danny De Schreye and Bern Martens

Department of Computer Science, Katholieke Universiteit Leuven
Celestijnenlaan 200A, B-3001 Heverlee, Belgium
e-mail: {dannyd,bern}@cs.kuleuven.ac.be

Abstract. We study a semantics for untyped, vanilla meta-programs, using the non-ground representation for object level variables. We introduce the notion of language independence for definite programs, which generalises range restriction. For language independent, definite object programs, we prove that there is a natural one-to-one correspondence between atoms $p(t_1, \ldots, t_r)$ in the least Herbrand model of the object program and atoms of the form $solve(p(t_1, \ldots, t_r))$ in the least Herbrand model of the associated vanilla meta-program. Thus, for this class of programs, the least Herbrand model provides a sensible semantics for the meta-program. The main attraction of our approach is that the results can be further extended – in a straightforward way – to provide a sensible semantics for a limited form of amalgamation.

1 Introduction

Meta-programming has become increasingly important in logic programming and deductive databases. Applications in knowledge representation and reasoning, program transformation, synthesis and analysis, debugging and expert systems, the modeling of evaluation strategies, the specification and implementation of sophisticated optimisation techniques, the description of integrity constraint checking, etc. are constituting a significantly large part of the recent work in the field (see e.g. [1], [2], [14], [10], [24], [9], [22], [21], [3], [25], [4]). In the last few years, theoretical foundations for meta-programming in logic programming have been developed in [12] and [23]. As a result, the programming language Gödel ([13]) now provides – not merely in the context of meta-programming – a fully declarative successor of Prolog, supporting the sound development of further meta-programming applications.

It should be clear however, that it can not have been the sound semantics for meta-programming, nor the existence of Gödel, that attracted so much interest into meta-programming in logic programming to start with. (Although they have clearly accelerated the activity in the area). One attraction certainly is the desire to extend the expressiveness of Horn clause logic augmented with negation as failure. Meta-programming adds extra knowledge representation and reasoning facilities ([14]). A second attraction is related to practicality. In applicative languages – both pure functional and pure logical –, data and programs are syntactically indistinguishable. This is an open invitation to writing programs that take other programs as input. We believe that the practical success of – in particular – vanilla-type meta-programming has resulted from this.

There are two basic motivations for the work reported in this paper. The first is based on the observation that, given the amount of work that has been spent on the development of practical applications using untyped, non-ground representation vanilla-type meta-programming, insufficient effort was made to provide it with a reasonable supporting semantics.

The second is our ultimate goal of evaluating the merits of a language such as Gödel for the development of a meta-programming approach to integrity checking in deductive databases – including the application of partial deduction in this context. However, we feel that no sensible evaluation is possible if the results of our experiments can not be compared with the results of matching experiments performed in a "competing" language. In particular, since untyped, non-ground representation vanilla-type meta-programming has been standard practice for many years, this would be the perfect candidate for a comparison. Clearly, such a comparison is completely unfeasible in the absence of a reasonable, formal semantics for the untyped, non-ground representation vanilla approach.

In [12], the possibility of providing such a semantics is rejected immediately, on the basis that – under the usual semantics for untyped logic programs – the intended interpretations of vanilla meta-programs can never be models. Now, this statement is somewhat inaccurate. The intended *meaning* of a vanilla-type meta-theory (in which different variables range over different domains) can simply not be captured within the formal notion of an interpretation, as it is defined for untyped, first order logic. So, a more precise statement would be that *the intended meaning can not be formalised as an untyped interpretation*. However, we argue that this problem is not typical for untyped vanilla programs, but that, in general, it is a problem for the semantics of most untyped logic programs. Any untyped logic program in which a functor is used to represent a partial function suffers from the same semantical problem and, in practice, total functions seldom occur in real applications. (See [6] for a thorough discussion of this issue.)

Whether this (and other) argument(s) in favour of typed logic programs should convince us to abandon the notational simplicity of untyped logic programs all together, is an issue we do not want to address in this paper. From here on, we will assume that the semantics of an (untyped) program is captured by the alternative notion of its (least/minimal/perfect) Herbrand model(s), avoiding the problems with intended interpretations. Even in this more restricted context, there remain problems with the semantics of untyped vanilla programs.

Consider the object program P:

$$p(x) \leftarrow$$
$$q(a) \leftarrow$$

Let M denote the standard *solve* interpreter:

$solve(empty) \leftarrow$
$solve(x\&y) \leftarrow solve(x), solve(y)$
$solve(x) \leftarrow clause(x, y), solve(y)$

In addition, let M_P denote the program M augmented with the following facts:

$clause(p(x), empty) \leftarrow$
$clause(q(a), empty) \leftarrow$

Although the least Herbrand model of our object program is $\{p(a), q(a)\}$, the least Herbrand model of the meta-program M_P contains completely unrelated atoms, such as $solve(p(empty))$, $solve(p(q(a)))$, etc..

This is certainly undesirable, since we, in general, would like at least that the atoms of the form $solve(p(t))$ in the least Herbrand model of M_P correspond in a one-to-one way with the atoms of the form $p(t)$ in the least Herbrand model of P.

In this paper, we introduce the notion of *language independence* of a program and show that it generalises range restrictedness. As one of our main results, we prove that for definite, language independent programs, the least Herbrand model of the program corresponds in a one-to-one way with a natural subset of the least Herbrand model of the corresponding vanilla theory. In addition, we show how this approach can be extended to provide a semantics for a limited form of amalgamation. This extension is rather interesting, since it reflects one of the main advantages the untyped approach may have over the typed one.

The paper is organised as follows. In the next section, we present our main results for definite, language independent object programs and their straightforward untyped vanilla meta-version. In Section 3, we briefly discuss overloading of symbols and its effect on the semantics of amalgamated programs. We propose a semantics for a limited form of amalgamation in Section 4, using the foundations laid out in the previous two sections. We end with a discussion.

2 A Semantics for Untyped Vanilla Meta-Programming

As mentioned in the introduction, in this paper, we only consider definite logic programs. We suppose the reader to be familiar with the basic concepts of predicate logic (see e.g. [8]) and logic programming (see e.g. [15]).

2.1 Language Independent Programs

We start by introducing the concept of a *language independent* definite program.

Definition 1. Let $\mathcal{L}$ be a (first order) language and $\mathcal{R}$, $\mathcal{F}$ and $\mathcal{C}$ its sets of predicate, function and constant symbols respectively. We call a language $\mathcal{L}'$, determined by $\mathcal{R}'$, $\mathcal{F}'$ and $\mathcal{C}'$ an *extension* of $\mathcal{L}$ iff $\mathcal{R} \subseteq \mathcal{R}'$, $\mathcal{F} \subseteq \mathcal{F}'$ and $\mathcal{C} \subseteq \mathcal{C}'$.

When considering a logic program, it is customary to speak about the *language underlying the program*. In this language, $\mathcal{R}$, $\mathcal{F}$ and $\mathcal{C}$ are the sets of predicate, function and constant symbols occurring *in the program*. Notice that this implies that we assume the set of constants in the underlying language to be empty if the program does not contain any constants. (We return to this below.) Although this is not imposed as a limitation in e.g. [15], Herbrand interpretations of the program are usually constructed with this underlying language in mind. For our purposes in this paper, however, we need more flexibility. Therefore, the language in which the interpretations are constructed, is made explicit in the following definition.

Definition 2. Let P be a definite program with underlying language $\mathcal{L}$. A Herbrand interpretation of P in a language $\mathcal{L}'$, extension of $\mathcal{L}$, is called an $\mathcal{L}'$-*Herbrand interpretation* of P.

We are now in a position to introduce the notion of *language independence*.

Definition 3. A definite program P with underlying language $\mathcal{L}$ is called *language independent* iff for any extension $\mathcal{L}'$ of $\mathcal{L}$, its least $\mathcal{L}'$-Herbrand model is equal to its least $\mathcal{L}$-Herbrand model.

Notice that this definition immediately entails that no atom in any least Herbrand model can contain any terms with function and/or constant symbols not occurring in P. In particular, propositions are the only possible elements of any least Herbrand model for a language independent program without constant symbols. This justifies our choice of a language without constant symbols as the one underlying such programs, instead of the more usual approach of picking one with a single arbitrary constant symbol.

Before concluding this subsection, we point out that the notion of language independence generalises the well-known concept of *range restriction* for definite logic programs.

We first repeat the definition of the latter.

Definition 4. A clause in a definite program P is called *range restricted* iff any variable that appears in its head also appears in its body.
A definite program P is called *range restricted* iff all its clauses are range restricted.

Range restriction has been defined for more general formulas and/or programs and was used in other contexts. See e.g. [18] and [5] for its use in the context of integrity checking in relational and deductive databases. Two equivalent notions are *safety*, used by Ullman in [26] and *allowedness* (at the clause level), defined in [15].

The following proposition shows that this important class of logic programs is a subclass of the language independent ones.

Proposition 5. *A range restricted definite program is language independent.*

Proof. Let P be a range restricted definite program with underlying language $\mathcal{L}$ and let $\mathcal{L}'$ be an extension of $\mathcal{L}$. Let T_P be the immediate consequence operator applied in the context of $\mathcal{L}$ and T_P' the corresponding operator applied in the context of $\mathcal{L}'$. We prove that for each $n \geq 0 : T_P{\uparrow}n = T_P'{\uparrow}n$.

The proof is through induction on n. The base case is trivial. Furthermore, if the body literals in a ground instance of a range restricted clause are instantiated with terms in $\mathcal{L}$, then so is the head. The induction step now follows immediately. □

2.2 A Meta-Programming Semantics

In this subsection, we show that there is a natural correspondence between the least Herbrand model of a language independent definite program and the least Herbrand model of its vanilla meta-program. We claim that this provides a sensible semantics for untyped non-ground representation vanilla-type meta-programming.

Definition 6. Suppose $\mathcal{L}$ is a first order language and $\mathcal{R}$ its finite (or countable) set of predicate symbols. Then we define $\mathcal{F}_\mathcal{R}$ to be a *functorisation* of $\mathcal{R}$ iff $\mathcal{F}_\mathcal{R}$ is a set of function symbols such that there is a one-to-one correspondence between elements of $\mathcal{R}$ and $\mathcal{F}_\mathcal{R}$ and the arity of corresponding elements is equal.

We introduce the following notation: Whenever A is an atom in a first order language $\mathcal{L}$ with predicate symbol set $\mathcal{R}$, A' denotes the term produced by replacing in A the predicate symbol by its corresponding element in $\mathcal{F}_\mathcal{R}$.

Definition 7. Let P be a definite program. Then, M_P, the *vanilla meta-program associated to* P, will be the following definite program:

$solve(empty) \leftarrow$
$solve(x \& y) \leftarrow solve(x), solve(y)$
$solve(x) \leftarrow clause(x, y), solve(y)$

together with a fact of the form

$clause(A', B_1' \& \ldots \& B_n') \leftarrow$

for every clause $A \leftarrow B_1, \ldots, B_n$ in P and a fact of the form

$clause(A', empty) \leftarrow$

for every fact $A \leftarrow$ in P.

Notice that if $\mathcal{L}_P$, the language underlying P, is determined by $\mathcal{R}_P$, $\mathcal{F}_P$ and $\mathcal{C}_P$, then $\mathcal{L}_{M_P}$, the language underlying M_P, is determined by

$\mathcal{R}_{M_P} = \{solve, clause\}$
$\mathcal{F}_{M_P} = \mathcal{F}_P \cup \mathcal{F}_{\mathcal{R}_P} \cup \{\&\}$
$\mathcal{C}_{M_P} = \mathcal{C}_P \cup \{empty\}$

where $\mathcal{F}_{\mathcal{R}_P}$ is a functorisation of $\mathcal{R}_P$.

In the sequel, when we refer to Herbrand interpretations and/or models (of a program P or M_P) and related concepts, this will be in the context of the languages $\mathcal{L}_P$ and $\mathcal{L}_{M_P}$, as defined above, unless stated explicitly otherwise.

Before continuing, we introduce the following notation:

1. U_P: the Herbrand universe of a program P
2. $U_P{}^n = U_P \times \ldots \times U_P$ (n copies)
3. p/r: a predicate symbol with arity r in $\mathcal{R}_P$
   p'/r: its associated function symbol in $\mathcal{F}_{M_P}$

The following proposition, which will implicitly be used in the sequel, is immediate.

Proposition 8. *Let P be a definite program and M_P its vanilla meta-program. Then $U_P \subset U_{M_P}$.*

We are now finally in a position where we can formulate and prove the main result of this paper. This will be done in Theorem 11. First, we present two lemmas.

Lemma 9. *Let P be a definite program and M_P its vanilla meta-program. Then the following holds for every $p/r \in \mathcal{R}_P$:*
$$\forall t \in U_P{}^r, \forall n \in \mathbb{N} : p(t) \in T_P \uparrow n \implies \exists m \in \mathbb{N} : solve(p'(t)) \in T_{M_P} \uparrow m$$

Proof. The proof is through induction on n. The base case ($n = 0$; $T_P \uparrow 0 = \emptyset$) is trivially satisfied. Now suppose that $p(t) \in T_P \uparrow n, n > 0$. Then there must be at least one clause C in P such that $p(t) \leftarrow C_1, \ldots, C_k$ ($k \geq 0$) is a ground instance of C and $C_1, \ldots, C_k \in T_P \uparrow (n-1)$. Consider first the case that we have one with $k = 0$. In other words, $p(t)$ is a ground instance of a fact in P. In that case Definition 7 immediately implies that $solve(p'(t)) \in T_{M_P} \uparrow 2$.
Suppose now $k \geq 1$. The induction hypothesis guarantees for every C_i the existence of an $m_i \in \mathbb{N}$ such that $solve(C_i') \in T_{M_P} \uparrow m_i$. Let mm denote the maximum of these m_i. It takes only a completely straightforward proof by induction on l to show the following:
$$\forall\, 1 \leq l \leq k : solve(C_{k-l}' \& \ldots \& C_k') \in T_{M_P} \uparrow (mm + k - l).$$
From this, it follows that
$$solve(C_1' \& \ldots \& C_k') \in T_{M_P} \uparrow (mm + k - 1)$$
and therefore
$$solve(p'(t)) \in T_{M_P} \uparrow (mm + k). \qquad \square$$

Lemma 10. *Let P be a definite, language independent program and M_P its vanilla meta-program. Then the following holds for every $p/r \in \mathcal{R}_P$:*
$$\forall t \in U_{M_P}{}^r, \forall n \in \mathbb{N} : solve(p'(t)) \in T_{M_P} \uparrow n \implies$$
$$t \in U_P{}^r \ \& \ \exists m \in \mathbb{N} : p(t) \in T_P \uparrow m$$

Proof. We first define $\mathcal{L}'$ to be the language determined by $\mathcal{R}_P$, $\mathcal{F}_{M_P}$ and $\mathcal{C}_{M_P}$. $\mathcal{L}'$ is an extension of $\mathcal{L}_P$.
The proof proceeds through an induction on n. The base case ($n = 0$; $T_{M_P} \uparrow 0 = \emptyset$) is trivially satisfied. Suppose that $solve(p'(t)) \in T_{M_P} \uparrow n$ where $n > 0$. Then either there is a *clause*-fact in M_P of which $clause(p'(t), empty) \leftarrow$ is a ground instance or this is not the case. Suppose first there is. Then P must contain a fact

of which $p(t) \leftarrow$ is a ground instance in $\mathcal{L}'$. This means that $p(t) \in T_P\uparrow 1$ in $\mathcal{L}'$. But, since P is language independent, this implies that $t \in U_P{}^r$ and $p(t) \in T_P\uparrow 1$. If there is no such *clause*-fact in M_P, then there must be one with a ground instance $clause(p'(t), C_1' \& \ldots \& C_k')$ where $k \geq 1$, such that $solve(C_1' \& \ldots \& C_k') \in T_{M_P}\uparrow(n-1)$. A simple induction argument on k shows that we obtain the following:

$$\forall\, 1 \leq i \leq k : \exists\, n_i < n \in \mathbb{N} : solve(C_i') \in T_{M_P}\uparrow n_i.$$

Through the induction hypothesis, we get:

$$\forall\, 1 \leq i \leq k : \exists m_i \in \mathbb{N} : C_i \in T_P\uparrow m_i \;\&\; t_i \in U_P{}^{r_i}$$

(where r_i is the arity of the predicate symbol in C_i). From the above and the fact that P is language independent, the desired result follows. $\qquad\square$

Theorem 11. *Let P be a definite, language independent program and M_P its vanilla meta-program. Let H_P and H_{M_P} denote the least Herbrand model of P and M_P respectively. Then the following holds for every $p/r \in \mathcal{R}_P$:*

$$\forall t \in U_{M_P}{}^r : solve(p'(t)) \in H_{M_P} \Longleftrightarrow t \in U_P{}^r \;\&\; p(t) \in H_P$$

Proof. The theorem follows immediately from Lemmas 9 and 10. $\qquad\square$

3 A Justification for Overloading

In the next section, we extend some of the results of Section 2 to provide a semantics for a limited form of amalgamation. The most trivial example of the programs we will consider is the (textual) combination $P + M_P$ of the clauses of an object program P with the clauses for its associated vanilla program M_P. A more complex example is obtained by further (textually) extending $P + M_P$ with one additional *clause*/2-fact for each of the (three) clauses in M_P defining the *solve*/1-predicate. In the most general case, we allow – in addition – the occurrence of *solve*/1-calls in the bodies of clauses of P. As a final difference with Section 2, we here impose the use of one particular functorisation $\mathcal{F}_{\mathcal{R}_P}$, namely the one in which all functors in $\mathcal{F}_{\mathcal{R}_P}$ are identical to their associated predicate symbols in $\mathcal{R}_P$.

Postponing the discussion on the generalisation of our results, we first address the more basic problem with the semantics of these programs, caused by overloading the symbols in the language. Clearly, the predicate symbols of P occur both as predicate symbol and as functor in $P + M_P$ (and in any further extensions). Now – although this was not made explicit in e.g. [15] – an underlying assumption of first order logic is that the class of functors and the class of predicate symbols of a first order language $\mathcal{L}$, are disjoint (see e.g. [19]). So, if we aim to extend our results to amalgamated programs – without introducing any kind of naming to avoid the overloading – we need to verify whether the constructions, definitions and results on the foundations of logic programming are still valid if the functors and predicate symbols of the language overlap. In particular, from discussions on this issue that we had with a number of researchers in the field, we found that it is generally believed that overloading may cause paradoxes to become expressible in the language (or that, at the least, the resulting semantics would be completely unclear).

We have checked the formalisation and proofs in [15] in detail, starting from the assumption that the set of functors and the set of predicate symbols may overlap. We found that none of the results become invalid. Of course, under this assumption, there is in general no way to distinguish well-formed formulas from terms. They as well have a non-empty intersection. But, this causes no problem in the definition of pre-interpretations, variable- and term-assignments and interpretations (see [15], p.12). It is clear however, that a same syntactical object can be both term and formula and can therefore be given two different meanings, one under the pre-interpretation and variable-assignment, the other under the corresponding interpretation and variable-assignment. But this causes no confusion on the level of truth-assignment to well-formed formulas under an interpretation and a variable-assignment ([15], p.12–13). This definition performs a complete parsing of the well-formed formulas, making sure that the appropriate assignments are applied for each syntactic substructure. In particular, it should be noted that no paradoxes can be formulated in these languages, since each formula obtains a unique truth-value under every interpretation and variable-assignment.

On the level of declarative semantics (for definite programs), the main results – the existence of a least Herbrand model and its characterisation as the least fixpoint of T_P ([15], Prop.6.1, Th.6.2, Th.6.5) – remain valid in the extended languages. Thus, the amalgamated programs we aim to study can in any case be given a unique semantics. Again, we postpone the discussion on whether it is a sensible semantics.

4 Amalgamation

In this section, we build on the results of the previous two sections and present a semantics for different kinds of amalgamated programs. We will *fix the choice of the functorisation* for a given set of predicate symbols: It will contain *exactly the same symbols*. A justification for this practice was given above. It leads to an increased flexibility in considering meta-programs with several layers. In fact, we can now deal with an unlimited amount of meta-layers.

The first extension we consider is completely straightforward: We include the object-program in the resulting meta-program.

Definition 12. Let P be a definite program and M_P its associated vanilla meta-program (see Definition 7). Then we call the textual combination $P + M_P$ of P and M_P the *amalgamated vanilla meta-program associated to* P.

It is clear that $P + M_P$ is also a definite program. Notice that $\mathcal{L}_{P+M_P}$ is determined by

$$\mathcal{R}_{P+M_P} = \mathcal{R}_P \cup \{solve, clause\}$$
$$\mathcal{F}_{P+M_P} = \mathcal{F}_P \cup \mathcal{F}_{\mathcal{R}_P} \cup \{\&\}$$
$$\mathcal{C}_{P+M_P} = \mathcal{C}_P \cup \{empty\}$$

where $\mathcal{F}_{\mathcal{R}_P}$ is equal to $\mathcal{R}_P$.

We immediately have the following:

$$U_{P+M_P} = U_{M_P}$$

We can now prove the following theorem through an immediate extension of the proof for Theorem 11.

Theorem 13. *Let P be a definite, language independent program, M_P its vanilla meta-program and $P+M_P$ its amalgamated vanilla meta-program. Let H_P, H_{M_P} and H_{P+M_P} denote their least Herbrand models. Then the following holds for every $p/r \in \mathcal{R}_P$:*

$$\forall t \in U_{P+M_P}^r : \text{solve}(p(t)) \in H_{P+M_P} \Longleftrightarrow p(t) \in H_{P+M_P}$$

$$\forall t \in U_{P+M_P}^r : \text{solve}(p(t)) \in H_{P+M_P} \Longleftrightarrow t \in U_P^r \ \& \ p(t) \in H_P$$

$$\forall t \in U_{P+M_P}^r : \text{solve}(p(t)) \in H_{P+M_P} \Longleftrightarrow t \in U_{M_P}^r \ \& \ \text{solve}(p(t)) \in H_{M_P}$$

Theorem 11 and Theorem 13 are interesting because they provide us with a reasonable semantics for vanilla and amalgamated vanilla meta-programming for a large class of definite programs. However, they also show that we do not seem to *gain* much by this kind of programming. Indeed, (the relevant part of) the meta-semantics can be *identified* with the object semantics. So, why going through the trouble of writing a meta-program in the first place ? The answer lies of course in useful *extensions* of the vanilla interpreter. (See e.g. [22] and further references given there.) The following definition captures many such extensions.

Definition 14. Let P be a definite program. Let E be a definite program of the following form – called *extended* meta-interpreter – :

$$\text{solve}(empty, t_{11}, \ldots, t_{1n}) \leftarrow C_{11}, \ldots, C_{1m_1}$$

$$\text{solve}(x\&y, t_{21}, \ldots, t_{2n}) \leftarrow \text{solve}(x, t_{31}, \ldots, t_{3n}), \text{solve}(y, t_{41}, \ldots, t_{4n}),$$
$$C_{21}, \ldots, C_{2m_2}$$

$$\text{solve}(x, t_{51}, \ldots, t_{5n}) \leftarrow \text{clause}(x, y), \text{solve}(y, t_{61}, \ldots, t_{6n}), C_{31}, \ldots, C_{3m_3}$$

where the t_{ij}-terms are extra arguments of the *solve*-predicate and the C_{kl}-atoms extra conditions in its body, together with defining clauses for any other predicates occurring in the C_{kl}. Then we define E_P, the E-*extended meta-program associated to* P, to be the program E together with the usual *clause*-facts, one for each clause in P (as in Definition 7).

The following proposition can now be proved. (The reasoning is largely similar to the one underlying the proofs above.) It ensures us that working with extended meta-interpreters is "safe".

Proposition 15. *Let P be a definite, language independent program, Let M_P be the vanilla and E_P be an E-extended meta-program associated to P. Then the following holds for every $p/r \in \mathcal{R}_P$:*

$$\forall t \in U_{E_P}^r : (\exists s \in U_{E_P}^n : \text{solve}(p(t), s) \in H_{E_P}) \Longrightarrow$$
$$t \in U_{M_P}^r \ \& \ \text{solve}(p(t)) \in H_{M_P}$$

Observe that from Theorem 11, we have that the right hand side of the implication is equivalent to: $t \in U_P^r \ \& \ p(t) \in H_P$

Example 1. As an example, we include the following program E, adapted from [22]. It builds proof trees of object level queries.

$solve(empty, empty) \leftarrow$
$solve(x\&y, proofx\&proofy) \leftarrow solve(x, proofx), solve(y, proofy)$
$solve(x, x \ if \ proof) \leftarrow clause(x, y), solve(y, proof)$

As is illustrated in [22], the proof trees thus constructed can be used as a basis for explanation facilities in expert systems.

Definition 14 and Proposition 15 do not involve amalgamation. But, it is of course perfectly possible to give a definition analogous to Definition 12 for extended meta-interpretation. We will not do this explicitly and only illustrate by an example the extra programming power one can gain in this way.

Example 2. In applications based on the proof tree recording program given in the previous example, it may be the case that users are not interested in branches for particular predicates. To reflect this, clauses of the form:

$solve(p(x), some\_info) \leftarrow p(x)$

can be added (combined with extra measures to avoid also using the standard clause for these cases). Notice that analogues of Theorem 13 and Proposition 15 still hold for such meta-programs.

Further extensions of the framework remain possible. A very interesting one is adding *clause*-facts for the *solve*-clauses themselves. The formal definition is as follows.

Definition 16. Let P be a definite program. Let the vanilla meta-interpreter M as before be defined by the following clauses:

$solve(empty) \leftarrow$
$solve(x\&y) \leftarrow solve(x), solve(y)$
$solve(x) \leftarrow clause(x, y), solve(y)$

Then, $M^2{}_P$, the *vanilla meta2-program associated to* P, will be the definite program consisting of M together with the following clause:

$clause(clause(x, y), empty) \leftarrow clause(x, y)$

and a fact of the form

$clause(A, B_1\& \ldots \&B_n) \leftarrow$

for every clause $A \leftarrow B_1, \ldots, B_n$ in P or in M and a fact of the form

$clause(A, empty) \leftarrow$

for every fact $A \leftarrow$ in P or in M.
If we also include P, we speak about the *amalgamated vanilla meta2-program associated to* P, which we denote by $P + M^2{}_P$.

We have the following theorem:

Theorem 17. *Let P be a definite, language independent program and $M^2{}_P$ its vanilla meta2-program. Let H_P and $H_{M^2{}_P}$ denote the least Herbrand model of P and $M^2{}_P$ respectively. Then the following holds:*

$\forall t \in U_{M^2{}_P} : \ solve(solve(t)) \in H_{M^2{}_P} \Longleftrightarrow solve(t) \in H_{M^2{}_P}$

Moreover, the following holds for every $p/r \in \mathcal{R}_P$:

$$\forall t \in U_{M^2{}_P}{}^r : \; solve(p(t)) \in H_{M^2{}_P} \Longleftrightarrow t \in U_P{}^r \; \& \; p(t) \in H_P$$

In other words, for any "layer" of meta-interpretation, we obtain the same correspondence with the object program as before. Formulating and proving a theorem similar to Theorem 13 for amalgamated vanilla meta2-programs is not difficult: We will not do this explicitly.

Yet another step can be taken. We can consider amalgamated meta2-programs in which the "object" clauses contain meta-calls. It is clear that we can in such cases no longer discern between an object- and a meta-level. Results similar to what we obtained before make no sense. But we can, along lines of reasoning analogous to those above, establish the following:

Proposition 18. *Let P be a definite, language independent program and $P + M^2{}_P$ its amalgamated vanilla meta2-program. Let PM be a program textually identical to $P + M^2{}_P$, except that an arbitrary number of atoms A in the bodies of clauses in the P-part of it have been replaced by $solve(A)$. Then the following holds:*

$$H_{P+M^2{}_P} = H_{PM}$$

Example 9. In this framework, we can formulate examples similar to some of the more interesting ones in [2]. Consider e.g. the following clause, telling us that a person is convicted for a crime when he is found guilty of it:

$$convicted(x, y) \leftarrow person(x), crime(y), solve(guilty(x, y)))$$

Of course, such possibilities only become really interesting when using extended meta-interpreters – involving e.g. an extra *solve*-argument limiting the resources available for proving a person's guilt.

Further extensions are possible, but we believe that the above sufficiently illustrates the flexibility, elegance and power of our approach. The essence of this section being that in each of these extensions, true atoms of the form $p(t)$ or $solve(p(t), \ldots)$ or even such atoms involving more layers of *solve* correspond to true atoms of the original object theory.

5   Discussion

First, it should be noted that Prolog meta-predicates, such as $var/1$, $assert/1$, $retract/1$ and $call/2$ are not included in our language. A thorough discussion of the problems related to these predicates is given in [12], [11] and [16].

Not having these predicates in our language certainly puts some limitation on the obtained expressiveness. Observe, however, that in the typed non-ground representation proposed in [12], no alternative for $var/1$ was introduced either and that the declarative $var/1$ predicate introduced in the ground representation approach of [12] provides no direct support for the sort of functionalities (e.g. control and coroutining facilities) that the $var/1$ predicate in Prolog is typically used for. For the $assert/1$ and $retract/1$ predicates, the solution of [11], to represent dynamic theories as terms in the meta-program, can as well be applied in our approach.

Next, observe that the condition of range restriction – which is the practical, verifiable, sufficient condition for language independence our approach was mostly designed for – is strongly related to typing. Conversely, typing can often be expressed by means of additional atoms that are added in the bodies of clauses, expressing the range of each variable. (See e.g. [7] and [15].) In this perspective, our approach is not so different from the typed non-ground representation in [12], except that we stick to the untyped logic programming syntax. Also, in view of the applications in deductive databases we have in mind, the condition of range restriction is very natural and commonly imposed.

With respect to the extension to amalgamated programs, we should point out that our use of overloading is very strongly related to the approach proposed in [20]. The main difference we can see at this time is that Richards adds each sentence as a new constant to the language, where we represent it as a new term. Here as well, more work is needed to clarify the relative merits of both approaches. The relation to the framework in [23] is much harder to investigate, since here a totally different approach based on a – much more powerful, but also more complex – explicit naming mechanism for formulas is proposed.

Finally, the presentation here has been restricted to definite programs. An extension of our work to normal programs is described in [17].

Acknowledgements

This work was partially supported by ESPRIT BRA COMPULOG (project 3012). Both authors are supported by the Belgian National Fund for Scientific Research.

References

1. K. A. Bowen. Meta-level programming and knowledge representation. *New Generation Computing*, 3(3):359–383, 1985.
2. K. A. Bowen and R. A. Kowalski. Amalgamating language and metalanguage in logic programming. In K. L. Clark and S.-A. Tärnlund, editors, *Logic Programming*, pages 153–172. Academic Press, 1982.
3. F. Bry. Query evaluation in recursive databases: Bottom-up and top-down reconciled. *Data & Knowledge Engineering*, 5(4):289–312, 1990.
4. F. Bry, H. Decker, and R. Manthey. A uniform approach to constraint satisfaction and constraint satisfiability in deductive databases. In *Proceedings EDBT'88*, March 1988.
5. F. Bry, B. Martens, and R. Manthey. Integrity verification in knowledge bases. In *Proceedings 2nd Russian Conference on Logic Programming*, pages 114–139, St-Petersburg, September 1991. Springer, LNAI 592.
6. M. Denecker, D. De Schreye, and Y. D. Willems. Terms in logic programs : a problem with their semantics and its effect on the programming methodology. *CCAI, Journal for the Integrated Study of Artificial Intelligence, Cognitive Science and Applied Epistemology*, 7(3 & 4):363–383, 1990.

7. H. B. Enderton. *A Mathematical Introduction to Logic.* Academic Press, 1972.
8. M. Fitting. *First-Order Logic and Automated Theorem Proving.* Springer-Verlag, 1990.
9. J. Gallagher. Transforming logic programs by specialising interpreters. In *Proceedings ECAI'86*, pages 109–122, 1986.
10. H. Gallaire and C. Lasserre. Metalevel control for logic programs. In K. L. Clark and S.-A. Tärnlund, editors, *Logic Programming*, pages 173–185. Academic Press, 1982.
11. P. M. Hill and J. W. Lloyd. Meta-programming for dynamic knowledge bases. Technical Report CS-88-18, Computer Science Department, University of Bristol, Great-Britain, 1988.
12. P. M. Hill and J. W. Lloyd. Analysis of meta-programs. In H. D. Abramson and M. H. Rogers, editors, *Proceedings Meta'88*, pages 23–51. MIT Press, 1989.
13. P. M. Hill and J. W. Lloyd. The Gödel report. Technical Report TR-91-02, Computer Science Department, University of Bristol, Great-Britain, March 1991. (Revised September 1991).
14. R. A. Kowalski. Problems and promises of computational logic. In J. W. Lloyd, editor, *Proceedings of the Esprit Symposium on Computational Logic*, pages 1–36. Springer-Verlag, November 1990.
15. J. W. Lloyd. *Foundations of Logic Programming.* Springer-Verlag, 1987.
16. J. W. Lloyd. Directions for meta-programming. In *Proceedings FGCS'88*, pages 609–617. ICOT, 1988.
17. B. Martens and D. De Schreye. A perfect Herbrand semantics for untyped vanilla meta-programming. In *Proceedings JICSLP'92*, Washington, November 1992. MIT Press.
18. J.-M. Nicolas. Logic for improving integrity checking in relational databases. *Acta Informatica*, 18(3):227–253, 1982.
19. A. Ramsay. *Formal Methods in Artificial Intelligence.* Cambridge University Press, 1988.
20. B. Richards. A point of reference. *Synthesis*, 28:431–445, 1974.
21. L. Sterling and R. D. Beer. Metainterpreters for expert system construction. *Journal of Logic Programming*, pages 163–178, 1989.
22. L. Sterling and E. Shapiro. *The Art of Prolog.* MIT Press, 1986.
23. V. S. Subrahmanian. A simple formulation of the theory of metalogic programming. In H. D. Abramson and M. H. Rogers, editors, *Proceedings Meta'88*, pages 65–101. MIT Press, 1989.
24. A. Takeuchi and K. Furukawa. Partial evaluation of Prolog programs and its application to metaprogramming. In H.-J. Kugler, editor, *Information Processing 86*, pages 415–420, 1986.
25. J. L. Träff and S. D. Prestwich. Meta-programming for reordering literals in deductive databases. In *Proceedings Meta'92*, Uppsala, June 1992. Springer-Verlag.
26. J. D. Ullman. *Database and Knowledge-Base Systems, Volume 1.* Computer Science Press, 1988.

A complete resolution method for logical meta-programming languages

Henning Christiansen

Roskilde University, P.O. Box 260, DK-4000 Roskilde, Denmark. E-mail: henning@dat.ruc.dk

Abstract. A resolution method for logical meta-programming languages with reflexive power corresponding to the binary demo predicate is presented. The method is complete in the sense that it answers also queries with uninstantiated variables which represent arbitrary fragments of the program currently being executed.

1 Why search for completeness?

The motivation for this work is a desire for a more declarative style of meta-programming than the present-day sort of meta-programs which appear as detailed recipes for synthesis or analysis of programs represented as data. As pointed out by Kowalski (1979), a binary demo-predicate can be used for specifying in an abstract way properties of desired programs. A program is represented by a ground term and a variable here will thus represent an unknown piece of program text. A fully logical implementation of binary demo will naturally generate the unspecified program parts.

In this paper we formalize a procedural semantics in the form of a resolution method. It is shown to be sound and complete which implies that it answers correctly also queries with variables which in this way stand for parts of the program being executed.

To mention a few potential applications we may suggest automatic database update or belief revision so that the modified database or belief set (the program argument) conforms with new observations (given as goals to be provable in the program). Various program generation tasks seem also obvious.

Bowen and Kowalski (1982) present a procedural semantics by a sketch of an interpreter which suffers from the assumption that the program argument must be completely specified when the demo call is reached by execution. In Hill and Lloyd's (1989) study of such interpreters the program clauses are represented by global facts and thus there is no way to have variables to stand for unknown program text. Brogi et al (1990) use a version of demo which allows expressions over program names where each name refers to an existing set of clauses. A straightforward implementation may then generate new such expressions. However, in this way no novel clauses can arise.

In our approach, variables can stand for arbitrary program phrases and these variables may be further constrained by syntactic requirements expressed in a conventional way.

An overview

Our method is described for a language called generative clause programs introduced in (Christiansen, 1990). It corresponds to definite clause programs extended with a demo predicate, but it has a more uniform structure which simplifies the formal expositions.

Consider, as an example, the following clause whose body contains a call of the demo predicate.

p :– q, demo(*Prog*, r).

It specifies that goal p can be proved if q and r can be proved, q in the same program as p, r in the program represented by the term *Prog*. In our language, this clause is written as follows.

p in Var :– q in Var, r in *Prog*.

When this rule is chosen for the execution of p, the variable Var is bound to a representation of the current program which, then, is carried over to q — whereas r is executed in the program represented by *Prog*. The constellation of goal and program representation is called a meta-goal. In section 3 we give the syntax of the language and its declarative, model-based semantics which is formulated in terms of Herbrand models of ground meta-goals.

Our unification algorithm described in section 4 is a direct generalization of Martelli and Montanari's (1982) algorithm. It is based on rewriting of systems of equations which in our case contain delayed function calls. The functions represent the translation of ground terms to terms composed with variable renaming. The presence of these functions implies a close relationship with the generally undecidable semi-unification problem (Kfoury, Tiuryn, Urcyczyn, 1990). It is actually easy to see that any semi-unification problem can be written as a simple call of binary demo. However, this theoretical discussion will be deferred to a forthcoming article.

Section 5 describes the resolution method. It consists of the same steps as ordinary SLD-resolution (van Emden, Kowalski, 1976, Lloyd, 1987). However, we must be more careful when selecting meta-goals from queries, or clauses from programs which appear as delayed function calls. In section 6, we show that Bowen and Kowalski's (1982) interpreter can be formalized as an incomplete restriction of our resolution method.

Acknowledgment

Part of this work was made at INRIA, Rocquencourt, with support from Eureka Software Factory Project (ESF) and Danish Natural Science Research Council.

2 Related work

Model-theoretic semantics similar to ours has been described earlier for languages with implications as goals, e.g. (Nait Abdallah, 1986, 1989, Miller, 1986, Subrahmanian, 1989, Giordano, Martelli, Rossi, 1991). The direct use of implication yields static extensions of the given program and corresponds thus to modules or blocks in traditional programming languages. Giordano, Martelli (1991) have shown that different interpretations of the implication symbol by modal operators correspond to different visibility rules.

For a more dynamic style of meta-programming, i.e., treating programs as data, we need some sort of Gödel encoding or naming relation to avoid semantical problems (cf. Bowen, Kowalski, 1982). However, we cannot take over directly the mathematicians' "Gödel-brackets", ⌈···⌉. This corresponds to quotation without un-quoting and we are back to the static modules. A representation of terms as ground terms provides the necessary flexibility to have variables to stand for pieces of program text.

Bowen and Kowalski (1982) presented an incomplete interpreter for this sort of language followed by implementations of the MetaProlog language (Bowen, 1985, Bacha, 1987, 1988), all with the "semi-dynamic" view that variables standing for program fragments must be grounded before reached by execution. The only approach we are aware of which allows a limited form of partially specified programs is the work by Brogi *et al* (1990) commented on above.

What we intend to show here is that the declarative as well as the procedural semantics can be specified almost as simply as for plain, definite clause programs.

3   The language

The syntax is given as follows. The special constants defined below are used as ground names for variables. The symbol "in" is a particular binary functor, TRUE a constant.

Definition 3.1. The notions of *constants, variables, functors, terms, instances* of a term, and *ground terms* are defined as usual. For any variable v assume a family of constant symbols, called *special constants*, $*v, **v, ***v$, etc. A *meta-goal* is a term of the form $(G \operatorname{in} P)$ or the constant symbol TRUE. The subterms G and P are called *goal* and *program parts*, respectively. A *generative clause* is a term of the form

$$M_0 :- M_1, \ldots, M_n, \quad n \geq 1,$$

where $M_0, \ldots, M_n$ are meta-goals. A *generative clause program* is a list of generative clauses. $\square$

The ground representation for programs as data is given by the following function.

Definition 3.2. The *syntactic denotation function* is the function $\mathcal{D}$ from the set of ground terms to the set of terms defined inductively as follows.

– $\mathcal{D}[\![c]\!] = c$,   for any non-special constant c,

– $\mathcal{D}[\![*^n v]\!] = *^{(n-1)} v$,   for any variable v and $n \geq 1$,

– $\mathcal{D}[\![f(t_1, \ldots, t_n)]\!] = f(\mathcal{D}[\![t_1]\!], \ldots, \mathcal{D}[\![t_n]\!])$.

Whenever $\mathcal{D}[\![t]\!] = t'$, we say that t *denotes* t'. $\square$

Note in the second clause that the right hand side is a variable for $n = 1$, a special constant otherwise. Special constants with more that one asterisk are useful in clauses which create new clauses which in turn, when they are applied, generate yet other clauses and so on.

The meta-level power is provided by the way clauses are selected. In order to rewrite a given meta-goal, we use an instance of a clause found in its particular program part — as opposed to plain logic programs where the clause is selected from *the* one and only, unalterable program.

Example 3.1. Let $\mathcal{P}_0$ be an abbreviation for the following ground term

$$[(p(*z) \operatorname{in} *x :- r \operatorname{in} *x, q(*z) \operatorname{in} \mathcal{P}_1), (r \operatorname{in} *x :- \text{TRUE})]$$

where again, $\mathcal{P}_1$ abbreviates

$[q(a)$ in $**x :-$ TRUE$]$.

We will go through the arguments which show that the meta-goal

$(p(a)$ in $\mathcal{P}_0)$

is true. Its program part denotes the program

$[(p(Z)$ in $X :- r$ in $X, q(Z)$ in $[q(a)$ in $*x :-$ TRUE$]), (r$ in $X :-$ TRUE$)]$

from which we select the first clause and instantiate Z to a and X to $\mathcal{P}_0$ to get the clause instance

$p(a)$ in $\mathcal{P}_0 :- r$ in $\mathcal{P}_0, q(a)$ in $[q(a)$ in $*x :-$ TRUE$]$.

The head is identical to the meta-goal we want to prove, so in order to do this, we may prove the two meta-goals in the body. The first one is true since its program part denotes a program (the same as above) which has a clause with an instance

r in $\mathcal{P}_0 :-$ TRUE.

The second meta-goal,

$(q(a)$ in $[q(a)$ in $*x :-$ TRUE$])$,

is seen to be true in a similar way. $\square$

The idea of selecting a clause in the meta-goal's program part and instantiating it properly is formalized in the following notion of self-contained clause instances. Such clauses are, as shown in the example, exactly those that can be applied a in proof.

Definition 3.3. A ground instance

$C = ((G$ in $P) :- B)$

of a generative clause is *self-contained* if $\mathcal{D}[\![P]\!]$ has an instance of the form $[\cdots, C, \cdots]$.
$\square$

The clause instances in the example above are self-contained and illustrate, thus, that there is no circularity problem in the definition. A self-contained instance contains a representation of a pattern of which it is an instance — which is different from containing a copy of itself.

The formalization of a declarative semantics is based on the following concepts generalized from the usual notions for plain logic programs (cf. Lloyd, 1987).

Definition 3.4. An *Herbrand interpretation* is a set of ground meta-goals. An *Herbrand model* is an Herbrand interpretation $\mathcal{M}$ such that TRUE $\in \mathcal{M}$ and for any self-contained clause instance

$M_0 :- M_1, \ldots, M_n$

it holds that

$\{M_1, \ldots, M_n\} \subseteq \mathcal{M}$ implies $M_0 \in \mathcal{M}$.

The intersection of all Herbrand models is called the *least* Herbrand model and will be denoted $\mathcal{M}^{\text{GCP}}$. $\square$

We quote the following fixed-point characterization from (Christiansen, 1990).

Definition 3.5. The immediate consequence operator for generative clause programs is the mapping T^{GCP} from Herbrand interpretations to Herbrand interpretations defined as follows.

$T^{GCP}(I) = \{M \mid M = \text{TRUE}$ or there exists a self-contained clause instance

$$M :\!\!- M_1, \ldots, M_n$$

such that

$$\{M_1, \ldots, M_n\} \subseteq I\}.$$

□

Theorem 3.1. $\mathcal{M}^{GCP} = lfp(T^{GCP}) = T^{GCP}\!\uparrow\!\omega.$ □

This motivates our usage of the notion of truth. We will say that a ground meta-goal M is true if and only if $M \in \mathcal{M}^{GCP}$. A ground query, which is a sequence $M_1, \cdots, M_n$, is true if and only if each M_i is a true meta-goal. For the proof of completeness of the resolution method we note the following reformulation of the theorem.

Corollary 3.1. A ground query Q_0 is true if and only if the following holds. There exists a sequence of ground queries $Q_0, \cdots, Q_n$ with $Q_n = (\text{TRUE}, \cdots, \text{TRUE})$ such that for $i = 0, \cdots, n - 1$,

$$Q_i = M_{i,1}, \cdots, M_{i,k_i}, \cdots, M_{i,n_i}, \quad \text{and}$$

$$Q_{i+1} = M_{i,1}, \cdots, M_{i,k_i-1}, B_i, M_{i,k_i+1}, \cdots, M_{i,n_i}$$

where $M_{i,k_i} :\!\!- B_i$ is a self-contained clause instance. □

Example 3.2. Let $\mathcal{P}$ stand for the following non-ground term.

$$[(p(*y) \text{ in } *x :\!\!- r \text{ in } *x, Z), (r \text{ in } *x :\!\!- \text{TRUE}), (q(a) \text{ in } *x :\!\!- \text{TRUE})]$$

The declarative semantics tells that the meta-goal $(p(X) \text{ in } \mathcal{P})$ is true when, e.g., X takes the value a and Z the value $(q(*y) \text{ in } *x)$. A *complete* resolution method must provide such answers. □

4   Extended terms and unification

We extend the notion of terms with function symbols $\mathcal{R}_i \mathcal{D}[\![-]\!]$ which stand for the composition of the syntactic denotation function $\mathcal{D}$ and a renaming function for the i'th resolution step. The following definitions are standard for idempotent substitutions (e.g., Ko, Nadel, 1991) although a few extensions are necessary in order to cope with the unusual functions.

Definition 4.1. The set of *extended terms* is defined inductively as follows.

– a variable is an extended term,

– a constant is an extended term,

– whenever f is a functor of arity n and $t_1, \ldots, t_n$ are extended terms,

$$f(t_1, \ldots, t_n)$$

is an extended term,

– for any integer i and extended term t,

$$\mathcal{R}_i \mathcal{D}[\![t]\!]$$

is an extended term.

A *structure* is either a constant or an extended term whose first symbol is a functor. A *term* is an extended term without $\mathcal{R}_i \mathcal{D}[\![-]\!]$ function symbols. A *ground term* is a term without variables. A *generalized variable* is an extended term of the form

$$\mathcal{R}_{i_1} \mathcal{D}[\![\cdots \mathcal{R}_{i_n} \mathcal{D}[\![v]\!] \cdots]\!], \quad \text{for some } i_1, \ldots, i_n, n \geq 0, v \text{ a variable.}$$

An extended term is *reduced* if the function symbols $\mathcal{R}_i \mathcal{D}[\![-]\!]$ in it (if any) appear only as part of generalized variables. □

For example, the extended term

$$\mathcal{R}_1 \mathcal{D}[\![f(\mathcal{R}_4 \mathcal{D}[\![x]\!], y, a)]\!]$$

is not reduced. A distribution of the function symbols will reduce it into

$$f(\mathcal{R}_1 \mathcal{D}[\![\mathcal{R}_4 \mathcal{D}[\![x]\!]]\!], \mathcal{R}_1 \mathcal{D}[\![y]\!], a);$$

this is made precise in the following definition. Note also that in the extended term $\mathcal{R}_8 \mathcal{D}[\![\mathcal{R}_7 \mathcal{D}[\![x]\!]]\!]$, three different, generalized variables can be identified, one of which is a variable.

We assume a family of functions $\mathcal{R}_i [\![-]\!]$ which map variables to new, unique variables.

Definition 4.2. The functions denoted $\mathcal{R}_i \mathcal{D}[\![-]\!]$ from ground terms to terms are defined inductively as follows. We use the notation $\overline{\mathcal{R}_i \mathcal{D}[\![t]\!]}$ to distinguish the value of the function application from the extended term $\mathcal{R}_i \mathcal{D}[\![t]\!]$.

– $\overline{\mathcal{R}_i \mathcal{D}[\![a]\!]} = a$, for any non-special constant a,

– $\overline{\mathcal{R}_i \mathcal{D}[\![*^n v]\!]} = *^{(n-1)} v$, for any special constant $*^n v$, $n > 1$,

– $\overline{\mathcal{R}_i \mathcal{D}[\![*v]\!]} = \mathcal{R}_i [\![v]\!]$, for any special constant of the form $*v$,

– $\overline{\mathcal{R}_i \mathcal{D}[\![f(t_1, \ldots, t_n)]\!]} = f(\overline{\mathcal{R}_i \mathcal{D}[\![t_1]\!]}, \ldots, \overline{\mathcal{R}_i \mathcal{D}[\![t_n]\!]})$ for any functor f of arity n and ground terms $t_1, \ldots, t_n$.

The overline notation is generalized to apply to any extended term by extending the last clause to allow extended terms for $t_1, \ldots, t_n$ and by the following two additional rewrite rules.

– $\overline{v} = v$ for any generalized variable v,

– $\overline{f(t_1, \ldots, t_n)} = f(\overline{t_1}, \cdots, \overline{t_n})$.

Whenever t is a generalized term, we say that $\overline{t}$ is a *reduced form* of t. □

Definition 4.3. For any mapping m from a set of variables to terms, let dom(m) be the set of variables for which m defines a value and let range(m) be the set of variables which can occur in a term $m(v)$ for $v \in$ dom(m); vars(m) is the union of dom(m) and range(m). A *substitution* is a mapping θ from a finite set of variables to terms, such that dom(θ) $\cap$ range(θ) = $\emptyset$. □

The restriction on the variables in a substitution gives the idempotent property.

Definition 4.4. The *value of* an extended term t under a substitution θ, denoted $t\theta$, is a term defined inductively as follows.

- $v\theta = \theta(v)$ for a variable $v \in \mathrm{dom}(\theta)$, otherwise $v\theta = v$,

- $f(t_1, \ldots, t_n)\theta = f(t_1\theta, \ldots, t_n\theta)$ for any extended term of the form $f(t_1, \ldots, t_n)$,

- $\mathcal{R}_i \mathcal{D}[\![t]\!]\theta = \overline{\mathcal{R}_i \mathcal{D}[\![t\theta]\!]}\theta$ for any extended term t such that $t\theta$ is ground.

The *composition* of two substitutions θ_1 and θ_2 is the mapping $\theta_1\theta_2$ defined as follows. However, if $\theta_1\theta_2$ is not a substitution, the composition is undefined. For any $v \in \mathrm{dom}(\theta_1) \cup \mathrm{dom}(\theta_2)$ let

- $(\theta_1\theta_2)(v) = s\theta_2$ provided $\theta_1(v) = s$, otherwise

- $(\theta_1\theta_2)(v) = \theta_2(v)$,

except if $(\theta_1\theta_2)(x)$ defined as above equals x for a variable x, we let $(\theta_1\theta_2)(x)$ be undefined.

Whenever, for substitutions σ_1 and σ_2, that $\sigma_2 = \sigma_1\varphi$ for some substitution φ, we say that σ_2 is a *specialization* of σ_1. If, furthermore, $\mathrm{vars}(\sigma_1) \cap \mathrm{dom}(\varphi) = \emptyset$, we call σ_2 an *extension* of σ_1. $\square$

The value of $\mathcal{R}_i \mathcal{D}[\![t]\!]\theta$ is undefined whenever $t\theta$ is non-ground. The definitions can be extended to handle this by the introduction of yet another level of generalized terms which include delayed substitutions. However, we do not need this generalization and we avoid the implied complications. So, for example, for $\mathcal{R}_1 \mathcal{D}[\![\mathcal{R}_2 \mathcal{D}[\![x]\!]]\!]\theta$ to be defined θ must provide ground values for x and the variables which arise in $\overline{\mathcal{R}_2 \mathcal{D}[\![x\theta]\!]}$. The repeated application of θ in the distribution rule for $\mathcal{R}_i \mathcal{D}[\![-]\!]\theta$ is necessary for the following standard properties to hold.

Proposition 4.1. For any substitutions θ, θ_1, θ_2, and θ_3 and extended term t, the following properties hold.

- $\theta\theta = \theta$,

- $(t\theta_1)\theta_2 = t(\theta_1\theta_2)$ provided both values are defined,

- $(\theta_1\theta_2)\theta_3 = \theta_1(\theta_2\theta_3)$ provided the indicated compositions are defined,

- $t\theta = \overline{t}\theta$ provided $t\theta$ is defined. $\square$

The proposition follows directly from the definitions. We will illustrate the distribution rule by a little example. Let

$$\theta = \{x \to {}^*a\} \text{ and } \varphi = \{A_7 \to const\}$$

where $\overline{\mathcal{R}_7 \mathcal{D}[\![{}^*a]\!]} = A_7$ and consider the following identity which is a consequence of the proposition.

$$(\mathcal{R}_7 \mathcal{D}[\![x]\!]\theta)\varphi = \mathcal{R}_7 \mathcal{D}[\![x]\!](\theta\varphi)$$

According to the distribution rule, this is equivalent to the following.

$$(\mathcal{R}_7 \mathcal{D}[\![x\theta]\!]\theta)\varphi = \overline{\mathcal{R}_7 \mathcal{D}[\![x\theta\varphi]\!]}(\theta\varphi)$$

With or without the repeated application of θ, the left hand side evaluates into $A_7\varphi = const$. Consider now the right hand side, $\overline{\mathcal{R}_7 \mathcal{D}[\![x\theta\varphi]\!]}$ evaluates into A_7 and the repeated application will trigger off the otherwise forgotten φ.

Definition 4.5. A *constraint* is a finite set of equations between extended terms in reduced form. A *unifier* for a constraint C is a substitution σ such that for any θ for which $C(\sigma\theta)$ is defined, that $C(\sigma\theta)$ consists of equations between identical terms. $\square$

We will now introduce the notion of constraints in normal form. For this purpose, we define the following ordering on extended terms. It should not be confused with subsumption ordering; its intuitive meaning is rather "less committed than" without actually considering the particular sort of commitments.

Definition 4.6. The partial ordering $\prec$ on extended terms is defined as follows. For any two distinct, generalized variables v_1 and v_2,

- if v_1 contains more function symbols than v_2, $v_1 \prec v_2$,

- otherwise, assume some standard ordering, (e.g. lexicographic) such that one and only one of $v_1 \prec v_2$ and $v_2 \prec v_1$ holds.

For any generalized variable v and structure s, let $v \prec s$. $\square$

In an equation $l = r$ with $l \prec r$, l will always be a generalized variable. If r occurs in l, the equation will have a unifier. Consider for example, the following archetypical equation,

$$\mathcal{R}_8 \mathcal{D}[\![\mathcal{R}_7 \mathcal{D}[\![x]\!]]\!] = \mathcal{R}_7 \mathcal{D}[\![x]\!].$$

The substitution $\{x \to *a, A_7 \to *b, B_8 \to *b\}$, where $\overline{\mathcal{R}_7 \mathcal{D}[\![*a]\!]} = A_7$ and $\overline{\mathcal{R}_8 \mathcal{D}[\![*b]\!]} = B_8$, is a unifier. This motivates the following definition and illustrates the construction in the proof of the lemma. The other way round, if l occurs in r, still with $l \prec r$, it can be seen that r must be of the form $f(\cdots l \cdots)$. In this case the equation $l = r$ has no unifier.

Definition 4.7. A constraint N is in *normal form* if

$$N = \{l_1 = r_1, \ldots, l_n = r_n\}$$

where $l_i \prec r_i$ for all i and each l_i occurs only once in N. $\square$

Lemma 4.1. A constraint in normal form has at least one unifier. $\square$

Proof. Let N be a constraint in normal form. We construct a unifier for N as follows. Let α_0 be the empty substitution and let $N_0 = N$. If N_{j-1} contains a generalized variable of the form $\mathcal{R}_{i_j} \mathcal{D}[\![x_j]\!]$ where x_j is a variable, let $\alpha_j = \alpha_{j-1}\{x_j \to *a_j\}$ where $*a_j$ is a special constant chosen such that the variable $\overline{\mathcal{R}_{i_j} \mathcal{D}[\![*a]\!]}$ does not occur in $N_0, \ldots, N_{j-1}$ and let $N_j = N_{j-1}\alpha_j$. Let k be largest k such that α_k is defined in this way; clearly k is finite. The constraint $N_k = N_{k-1}\alpha_k = N\alpha_k$ will contain only equations of the form $v = t$, v a variable and t a term, where each such v occurs only once. Let β be the substitution such that $\beta(v) = t$ for each such equation. Clearly β is a unifier for N_k and thus $\alpha_k\beta$ is a unifier for N. $\square$

In general, a given constraint may have many unifiers — and unifiers which are not directly comparable. Consider, for example, the following normal form constraint.

$$\{\mathcal{R}_7\mathcal{D}[\![v]\!] = (x := y)\}$$

The proof of lemma 4.1 may give the unifier

$$\{v \to {}^*a, A_7 \to (x := y)\},$$

assuming $\overline{\mathcal{R}_7\mathcal{D}[\![{}^*a]\!]} = A_7$. Alternatively, we may use a more specific pattern as the value of v and construct the following unifier,

$$\{v \to ({}^*a := {}^*b), A_7 \to x, B_7 \to y\}$$

with $\overline{\mathcal{R}_7\mathcal{D}[\![{}^*a]\!]} = A_7$ and $\overline{\mathcal{R}_7\mathcal{D}[\![{}^*b]\!]} = B_7$.

The following algorithm takes an arbitrary constraint and transforms it into an equivalent normal form if possible. Except for the use of the "$\prec$" relation and the reduction of functional terms, our algorithm is identical to that given by (Martelli, Montanari, 1982)

Normalization algorithm. Initially, let $C_0 = C$ and $i = 0$. Repeatedly do the following: Select any equation $l = r$ in C_i such that one of the rules 1 to 4 below applies. If no such equation exists, stop with C_i as result. Otherwise perform the corresponding action, i.e., stop with failure or construct C_{i+1} from C_i. Then increment i by 1.

1) If l and r are identical, remove the equation $l = r$.

2) $l \prec r$ and l occurs elsewhere in C_i. If l occurs in r, stop with failure, otherwise replace all other occurrences of l in C_i by r (i.e., leave $l = r$ unchanged). If any non-reduced, extended term t arises, replace t by $\overline{t}$.

3) If $r \prec l$, replace $l = r$ by $r = l$.

4) $l = f_1(l_1,\dots,l_n)$ and $r = f_2(r_1,\dots,r_m)$ where f_1 and f_2 are functors. If f_1 and f_2 are different or $n \neq m$, stop with failure, otherwise replace $l = r$ by $l_1 = r_1,\dots,l_n = r_n$.

□

Lemma 4.2. Assume the algorithm is given a constraint C as input and it outputs a constraint N, then N is in normal form and any unifier for N is a unifier for C and vice-versa. If the algorithm outputs failure, C cannot have a unifier. □

Proof. Firstly, we will show that each possible step made by the algorithm preserves the set of unifiers. Assume the algorithm in one step transforms a constraint C_i into C_{i+1}. Rule 1 removes equations which are satisfied under any substitution, so any unifier for C_i is a unifier for C_{i+1} and vice-versa. Rule 2 exchanges extended terms l with extended terms r which are indifferent under any unifier and thus does not affect the unifier set. Rules 3 and 4 clearly do not affect the unifier set either. By induction, any constraint output by the algorithm possesses the same set of unifiers as the input constraint.

If the algorithm returns failure, rule 2 has found an equation of the form $v = f(\cdots v \cdots)$ where v is a generalized variable or step 4 has found an equation $f(\cdots) = g(\cdots)$. In neither case, a unifier can exist and the already proved part gives that the input constraint has no unifier.

Finally, we show that if the algorithm outputs a constraint N, then N must be in normal form. The proof will be indirect, so assume that N is *not* in normal form. Then N has an equation $l = r$ where either not $l \prec r$ or l occurs somewhere else in N. If not $l \prec r$ then either $r \prec l$, which is not possible since rule 3 could have been applied, or l and r are not comparable by $\prec$. The only possibility for the latter is if both l and r are structures. This is not the case either, since rule 4 or perhaps rule 1 could have been applied. Is it possible, then, that l occurs somewhere else in N? No, because then rule 2 could have been applied. Hence it is not the case that N is not normal. □

The termination properties are expressed in the following conjecture based on experiences from the closely related semi-unification problem.

Conjecture. Assume the algorithm is given a constraint C as input. If C has unifiers, it terminates. If C has no unifier, the algorithm will terminate in most cases. □

It may seem a little disappointing, but the potential loops do not affect our completeness result.

Our algorithm is identical to Leiß' (1989) for semi-unification which is the problem of solving systems of term in-equations where inequality refers to subsumption ordering. Leiß has given a translation of semi-unification problems into equations with a sort of generalized variables which obey exactly the same formal laws as our objects of the same name. Intuitively, the relation between the two problems is clear, the sort of commitments made for the head of an unknown program clause will have the character of subsumption constraints. Coming back to the conjecture, it was shown by Kfoury, Tiuryn, and Urcyczyn (1990) that semi-unification in general is undecidable. Several decidable sub-classes of semi-unification are known and it is commonly believed that the algorithm terminates when there is a solution (cf. Leiß, Henglein, 1991, and private communication).

5 Resolution

A resolution state consists of a constraint describing the current set of commitments together with a query Q which stands for the meta-goals which are yet to be resolved. The query is embedded in an otherwise insignificant equation in order to simplify the presentation of the resolution method.

Definition 5.1. A *resolution state* is a constraint in normal form $S = C \cup \{\text{QUERY} = \text{query}(Q)\}$ where QUERY is a distinguished variable, "query" a functor. If C is empty, S is said to be an *initial state* and Q an *initial query*; in this case, we may write Q as an abbreviation for S. If Q takes the form $(\text{TRUE}, \ldots, \text{TRUE})$ we say that S is a *success state* and C a *success constraint*; in this case we may write C as an abbreviation for S.

A *correct answer substitution* for a query Q is a substitution σ such that for any γ for which $Q\sigma\gamma$ is ground, $Q\sigma\gamma$ is true. A *solution* for a resolution state $S = C \cup \{\text{QUERY} = \text{query}(Q)\}$ is a substitution which is a correct answer substitution for Q and a unifier for C. □

The normalization algorithm cannot change the overall appearance of the query equation but the query itself can be affected. The variable QUERY does not appear anywhere else and thus it will not be eliminated or distributed by the algorithm's rule 2 and the embracing functor "query" prevents step 3 from turning the equation around.

In each resolution step a meta-goal is selected from the current query and in this meta-goal's program part, a clause is chosen. There must be paid attention to possible, generalized variables which stand for unknown sub-structures of the objects from which these items are selected. We present the selection procedures as generators of sequences of skeletons, some of which may unify with the given object, i.e., skeletons which match the structure already given and which supply the remaining structure. The definition refs to the notion of new variables which are variables which do not appear in the implicit context of constraints and queries — or in the substitutions which are used in the correctness proofs to follow. We are also very careful in formalizing what it means to "replace the selected meta-goal by something".

Definition 5.2. The property

$$\text{query-skeleton}(v_{\text{select}}, Q_{\text{skel}}, Q'_{\text{skel}}, v_{\text{replace}})$$

is satisfied if $Q_{\text{skel}} = (v_1, \cdots, v_n), n \geq 1$ where each v_i is a new variable and $v_{\text{select}} = v_i$ for some i, $1 \leq i \leq n$. The structure Q'_{skel} is identical to Q_{skel} except that v_{select} is replaced by another new variable v_{replace}. □

The skeleton Q_{skel} will be unified with the query and v_{select} stands for the subterm which serves as the selected meta-goal. The derived query is given by Q'_{skel} in which v_{replace} represents the new meta-goals replacing the selected one.

Definition 5.3. The property

$$\text{program-skeleton}(v_i, [v_1, \cdots, v_i, \cdots, v_n])$$

is satisfied when $v_1 \cdots, v_n$ are distinct, new variables, $1 \leq i \leq n$ and $n \geq 1$. □

In a practical implementation we will probably prefer selection procedures which examine the actual structures given and which do not select a meta-goal already reduced to TRUE. Such procedures are given in our report (Christiansen, 1992). Furthermore, the possible set of choices in a given case should be represented by some finite constraint. We prefer to think of our definitions above as theoretically convenient abstractions over constraints of the sort "P is a list with cl as an element." A perhaps better solution would be to think of programs as sets of clauses and use the approach of Dovier *et al* (1991) for representing finite sets in logic programs. For our present purpose, however, we are satisfied by the observation that the inherent nondeterminism in each selection step is countable.

Definition 5.4. A state S' is *derived from* a state $S = C \cup \{\text{QUERY} = \text{query}(Q)\}$ at level i if

– $\text{query-skeleton}(v_{\text{select}}, Q_{\text{skel}}, Q'_{\text{skel}}, v_{\text{replace}})$,

– $\text{program-skeleton}(v_{\text{cl}}, P_{\text{skel}})$,

and if the normal form algorithm outputs S' when applied to one of the following two constraints; v_g and v_p are new variables.

$$C \cup \{Q = Q_{\text{skel}},$$
$$v_{\text{select}} = (v_g \text{ in } v_p),$$
$$\mathcal{R}_i \mathcal{D}[\![v_p]\!] = P_{\text{skel}},$$
$$v_{\text{cl}} = (v_{\text{select}} :- v_{\text{replace}})\}$$
$$\cup \{\text{QUERY} = \text{query}(Q'_{\text{skel}})\}$$

or

$$C \cup \{Q = Q_{\text{skel}},$$
$$v_{\text{select}} = \text{TRUE}, v_{\text{replace}} = \text{TRUE}\}$$
$$\cup \{\text{QUERY} = \text{query}(Q'_{\text{skel}})\}.$$

□

The first constraint represents a traditional proof step. In addition to the unification of the selected meta-goal with the head of the selected clause, the normalization algorithm will automatically place the body of the selected clause in the new query skeleton Q'_{skel}

in the position of the variable $v_{replace}$ and copy over the matching parts of Q. The second case is when a perhaps unknown meta-goal is committed to be TRUE. For simplicity, we do not sort out the useless cases in which the selected meta-goal is TRUE already.

The following two lemmas describe the properties of a single resolution step which imply the general soundness and completeness theorems.

Lemma 5.1. Let $S = C \cup \{\text{QUERY} = \text{query}(Q)\}$ be a resolution state and $S' = C' \cup \{\text{QUERY} = \text{query}(Q')\}$ a state derived from S and σ a solution for S'. Then σ is a solution for S. $\square$

Proof. Let S, S', and σ be as in the lemma and assume that S' in the derivation step appears as the normalization of the constraint

$$C \cup \{Q = Q_{skel}, \, v_{select} = (v_g \text{ in } v_p), \, \mathcal{R}_i \mathcal{D}[\![v_p]\!] = P_{skel}, \, v_{cl} = (v_{select} :- v_{replace})\}$$

$$\cup \{\text{QUERY} = \text{query}(Q'_{skel})\}.$$

where the conditions in definition 5.4 are satisfied. Lemma 4.2 gives that σ unifies these equations (but the last one mentioned) and hence σ is a unifier also for C.

To see that σ is a correct answer substitution for Q, we proceed as follows. Let γ be a substitution such that $Q\sigma\gamma$ is ground. Since $\sigma\gamma$ unifies the equations above and due to the defined relation between Q_{skel} and Q'_{skel}, we see that

$$Q\sigma\gamma = M_1, \cdots, M_{select}, \cdots, M_n$$

where $M_1, \cdots, M_n$, except M_{select}, recur in $Q'_{skel}\sigma\gamma = Q'\sigma\gamma$ and hence are true. About $M_{select} = v_{select}\sigma\gamma$, the equations above tell that there is a self-contained clause instance $M_{select} :- B$ where $B = v_{replace}\sigma\gamma$ is part of $Q'_{skel}\sigma\gamma = Q'\sigma\gamma$. By definition of model and truth, M_{select} is true. I.e., σ is a correct answer substitution for Q and thus a solution for S.

For the second sort of derivation with $v_{select} = \text{TRUE}$, the result follows in a similar way. $\square$

Lemma 5.2. Assume $S = C \cup \{\text{QUERY} = \text{query}(Q)\}$ is a state which has a solution σ and let γ be a substitution such that

$$Q\sigma\gamma = M_1, \cdots, M_n$$

is ground (and hence true). For any s, $1 \leq s \leq n$, either there exists a self-contained clause instance $M_s :- B$ or $M_s = \text{TRUE}$ (in which case we let $B = \text{TRUE}$). Furthermore, there exists a state

$$S' = C' \cup \{\text{QUERY} = \text{query}(Q')\}$$

derived from S and an extension of σ, σ' which is a solution of S' such that

$$Q'\sigma'\gamma = M_1, \cdots M_{s-1}, B, M_{s+1}, M_n.$$
$\square$

Proof. Let $S, \sigma, \gamma, M_1, \cdots, M_n$, etc. be as in the lemma. We can then find new variables, $v_1, \cdots, v_n$, and $v_{replace}$ such that

$$\text{query-skeleton}(v_s, Q_{skel}, Q'_{skel}, v_{replace})$$

with $Q_{skel} = (v_1, \cdots, v_s, \cdots, v_n)$, $Q'_{skel} = (v_1, \cdots, v_{replace}, \cdots, v_n)$.

Let us consider the case $M_s = \text{TRUE}$. Here we define σ' as an extension of σ with $\sigma'(v_l) = M_l$, $1 \leq l \leq n$, and $\sigma'(v_{\text{replace}}) = \text{TRUE}$. Then $\sigma'\gamma$ is clearly a solution for

$$C \cup \{Q = Q_{\text{skel}}, v_s = \text{TRUE}, v_{\text{replace}} = \text{TRUE}\} \cup \{\text{QUERY} = \text{query}(Q'_{\text{skel}})\}.$$

When the normalization algorithm is applied to this constraint, we get a state $S' = C' \cup \{\text{QUERY} = \text{query}(Q')\}$ derived from S with $\sigma'\gamma$ as a solution. To see this, we refer to lemma 4.2 and the conjecture about termination and to the fact that the algorithm only exchanges components of Q'_{skel} with parts indifferent under $\sigma'\gamma$. This gives also that $Q'\sigma'\gamma = M_1, \cdots M_{s-1}, B, M_{s+1}, M_n$.

We consider, now, the case where M_s is a true meta-goal of the form $(G \text{ in } P)$. By definition of model and truth, $M_s :- B$ exists as a self-contained instance of cl_k, where $\overline{\mathcal{R}_i \mathcal{D}[\![P]\!]} = [cl_1, \cdots, cl_k, \cdots, cl_m]$ for any i and where B is true. We assume i is such that the variables in $\overline{\mathcal{R}_i \mathcal{D}[\![P]\!]}$ do not occur elsewhere. We find new variables $w_1, \cdots, w_m$ such that

prog-skeleton(w_k, P_{skel}) where $P_{\text{skel}} = [w_1, \cdots, w_m]$.

Here we define σ' as an extension of σ with $\sigma'(v_l) = M_l$ for $1 \leq l \leq n$, $\sigma'(v_{\text{replace}}) = B$, $\sigma'(w_l) = cl_l$ for $1 \leq l \leq m$, and $\sigma'(v_g) = G$, $\sigma'(v_p) = P$ where v_g and v_p are new variables. Finally, for the variables in cl_k, let σ' determine the matching components of $M_s :- B$ such that $cl_k \sigma' = (M_s :- B)$. We can argue as above, that $\sigma\gamma$ is a solution for a constraint $S' = C' \cup \{\text{QUERY} = \text{query}(Q')\}$ derived from S (cf. definition 5.4) and that $Q'\sigma'\gamma$ is of the form stated in the lemma. $\square$

We can now formulate the resolution method which we call SLG-resolution (G for "generative" instead of D for "definite") and the corresponding theorems of soundness and completeness.

Definition 5.5. An SLG-*derivation* is a series of resolution states

$$S_0, S_1, S_2, \ldots,$$

such that S_0 is an initial state and S_{i+1} is derived from S_i at level i.

An SLG-*resolution* of a query Q is a finite SLG-derivation whose initial state is Q and whose last state is a success state. $\square$

Theorem 5.1. (Soundness) Assume Q is a query and that there exists an SLG-resolution

$$Q = S_0, S_1, \cdots, S_n = C.$$

Then any unifier for C is a correct answer substitution for Q. $\square$

Proof. Follows by induction from lemma 5.1. $\square$

Theorem 5.2. (Completeness) Assume Q is a query which has a correct answer substitution σ. Then there exists an SLG-resolution

$$Q = S_0, S_1, \cdots, S_n = C$$

and an extension of σ which is a unifier for C. $\square$

Proof. Follows by induction from corollary 3.1 and lemma 5.2. $\square$

Finally, we observe also the compactness property that the set of all success states for a given query can be given an effective enumeration. All possible derivation steps and steps made by the normalization algorithm can be organized in tree with an at most countable branching at each node.

6 The Bowen-Kowalski interpreter as a special case

If we restrict the selection step in the resolution procedure to handle only ground program parts, we get a resolution method which can be seen as a formalization of the interpreter given by Bowen and Kowalski (1982) which we criticized in the introduction.

Definition 6.1. A resolution state $S = C \cup \{\text{QUERY} = \text{query}(Q)\}$ is *weak* if it has no $\mathcal{R}_i \mathcal{D}[\![-]\!]$ function symbols and if Q is a sequence of meta-goals. An SLG-resolution is *weak* if each of its states is weak. $\square$

The correctness, i.e., the soundness and "relative completeness" follows from the theorems for general SLG-resolution. It should be noted that a meta-goal's program part does not need to be ground until it becomes the selected meta-goal, so some of the dynamic properties of the language are preserved. In case a non-ground program part is reached, an implementation could issue an error message or delay the particular meta-goal until its program part perhaps becomes ground. Finally, we observe that our normalization algorithm in this special case reduces to the algorithm given by Martelli and Montanari (1982) for plain, definite clause programs.

References

Abramson, H., and Rogers, M.H., eds., *Meta-programming in Logic Programming.* MIT Press, 1989.

Bowen, K.A., Meta-level programming and knowledge representation. *New Generation Computing* 3, pp. 359–383, 1985.

Bacha, H., Meta-level programming: A compiled approach. *Logic Programming, Proceedings of the fourth international conference*, ed. Lassez, J.-L., MIT Press, pp. 394–410, 1987.

Bacha, H., MetaProlog design and implementation. *Logic Programming, Proceedings of the fifth international conference*, ed. Kowalski, R.A., Bowen, K.A., MIT Press, pp. 1371–1387, 1988.

Bowen, K.A. and Kowalski, R.A., Amalgamating language and meta-language in logic programming. *Logic Programming*, Clark, K.L. and Tärnlund, S.Å., eds., pp. 153–172, Academic Press, 1982.

Brogi, A., Mancarella, P., Pedreschi, D., and Turini, F., Composition operators for logic theories. *Computational Logic*, ed. Lloyd, J., pp. 117–134, Springer-Verlag, 1990.

Bruynooghe, M., ed., *Proceedings of the Second Workshop on Meta-programming in Logic.* April 4–6, 1990, Leuven, Belgium.

Christiansen, H., Declarative semantics of a meta-programming language. *In:* Bruy-nooghe, 1990, pp. 159–168.

Christiansen, H., Models and resolution principles for logical meta-programming languages. *INRIA Rapport de Recherche* no. 1594, 1992.

Dovier, A., Omodeo, E.G., Pontelli, E., and Rossi, G., {log}: A logic programming language with finite sets. *Logic Programming, Proceedings of the eighth international conference*, ed. Furukawa, K., MIT Press, pp. 111–124, 1991.

van Emden, M.H. and Kowalski, R.A., The semantics of predicate logic as a programming language. *Journal of the ACM*. Vol. 23, pp. 733–742, 1976.

Giordano, L. and Martelli, A., A modal reconstruction of blocks and modules in logic programming. *Proceedings of International Logic Programming Symposium*, San Diego, pp. 239–253, 1991.

Giordano, L., Martelli, A., and Rossi, G., Extending Horn clause logic with implication goals. *Theoretical Computer Science* 95, pp. 43-74, 1991

Hill, P.M. and Lloyd, J.W., Analysis of meta-programs. *In:* Abramson, Rogers, 1989, pp. 23–51.

Kfoury, A.J., Tiuryn, J., and Urcyczyn, P., The undecidability of the semi-unification problem. *Proc. 22nd Annual ACM Symposium on Theory of Computing*, pp. 468–476, 1990.

Kowalski, R., *Logic for problem solving*. North-Holland, 1979.

Leiß, H., Semi-unification and type inference for polymorphic recursion. Technical Report INF2-ASE-5-89, *Siemens AG*, München, 1989.

Leiß, H. and Henglein, F., A decidable case of the semi-unification problem. *Lecture Notes in Computer Science* 520, pp. 318–327, Springer-Verlag, 1991

Lloyd, J.W., *Foundations of logic programming*, Second, extended edition. Springer-Verlag, 1987.

Martelli, A. and Montanari, U., An efficient unification algorithm. *ACM Transaction on Programming Languages and Systems* 4, pp. 285–282, 1982.

Miller, D., Lexical scoping as universal quantification, *Proc. of Sixth International Conference on Logic Programming*, eds. Levi, G. and Martelli, M., MIT Press, pp. 268–283, 1989.

Nait Abdallah, M.A., Ions and local definitions in logic programming, *Lecture Notes in Computer Science* 210, pp. 60–72, Springer-Verlag, 1986.

Nait Abdallah, M.A., A logico-algebraic approach to the model theory of knowledge., *Theoretical Computer Science* 66, pp. 205–232, 1989.

Subrahmanian, V.S., A simple formulation of the theory of metalogic programming. *In:* Abramson, Rogers, 1989, pp. 65–101.

Model theoretic semantics for Demo

Piero A. Bonatti

Dipartimento di Informatica, Università di Pisa
Corso Italia 40, 56125 Pisa, Italy

Abstract. A useful 3-valued provability predicate $Demo_5$ is defined, that can provably simulate non-classical logics and connectives (like autoepistemic logic and negation-as-failure) and that, despite its 3-valued semantics, is able to capture thoroughly classical notions of provability and unprovability. $Demo_5$ is largely tolerant to self-reference, and can be axiomatised consistently. Its semantics provides a guide to the development of effective axiomatisations and meta-interpreters that do not fall under the scope of the well known negative results on incompleteness and inconsistency.

1 Motivations

The idea of metaprogramming through a provability predicate, amalgamated within the object-level language, dates back to a famous work by Bowen and Kowalski [5]. The representation adopted there, has been directly inspired by Gödel's coding of arithmetic and provability.

Several authors (like, for example, Turner [22], Perlis [16] [17], Davies [7], Attardi and Simi [1]) prefer a "flattened" representation of this kind, despite the hard technical difficulties that it involves. It seems that an amalgamated, non-indexed form of meta-knowledge can represent notions like belief and knowledge more naturally than the non-amalgamated approaches (like Weyhrauch's FOL [25]) and than the indexed approaches derived from the work of Turing [21] and Feferman [8] (like the extension of Prolog proposed by Costantini and Lanzarone [6]).

The applications of the provability predicates suggested by Bowen and Kowalski include: definition and application of axiom schemas and inference rules, higher order reasoning and, last but not least, non-monotonic reasoning. In particular, non-monotonic reasoning can be accomplished through rule schemata like

$$innocent(x) \leftarrow person(x), \neg Demo(\lceil guilty(x) \rceil)$$
$$can\_fly(x) \leftarrow bird(x), \neg Demo(\lceil \neg can\_fly(x) \rceil)$$

that exploit a formalisation of non-theorems, expressed by facts of the form $\neg Demo()$. Independently of their practical use, it seems natural to require that a logic of provability can deal with non-theorems. After all, their "logical" nature is not sensibly different from that of theorems: as Varzi showed in [24], the non-theorems of propositional logic can be generated by an axiomatic system.

However, the "classical" amalgamated formalisations of provability (see [20], for example) cannot reason about non-theorems. The reason of this weakness is intimately related to *Löb's theorem*, which states that, for every formula ϕ,

$$\vdash Pr(\lceil \phi \rceil) \to \phi \text{ iff } \vdash \phi$$

where $\lceil \phi \rceil$ is the *name* of ϕ and Pr is a *provability predicate*, i.e. a formula (expressing provability) that satisfies *Löb's derivability conditions*:

D1. $\vdash \phi \Rightarrow \vdash Pr(\lceil \phi \rceil)$

D2. $\vdash Pr(\lceil \phi \rceil) \wedge Pr(\lceil \phi \to \psi \rceil) \to Pr(\lceil \psi \rceil)$

D3. $\vdash Pr(\lceil \phi \rceil) \to Pr(\lceil Pr(\lceil \phi \rceil) \rceil)$.

Löb's theorem is related to non-theorems as follows: suppose that, for some ϕ, $\vdash \neg Pr(\lceil \phi \rceil)$ holds. This implies that $\vdash Pr(\lceil \phi \rceil) \to \phi$, so, by Löb's theorem, $\vdash \phi$ must hold, and hence, by $D1$, $\vdash Pr(\lceil \phi \rceil)$ must hold, too. It follows that $\vdash \neg Pr(\lceil \phi \rceil) \wedge Pr(\lceil \phi \rceil)$. As a consequence, if the axiomatisation $\vdash$ is consistent and satisfies $D1 - D3$, then for *every* formula ϕ, $\vdash \neg Pr(\lceil \phi \rceil)$ cannot hold. In other words, this class of formalisations cannot say of any formula that it is *not* a theorem.

The formalisations proposed by Bowen and Kowalski can prove some facts about unprovability. In general, Löb's theorem is not applicable to their provability predicate *Demo*, because the derivability conditions are not universally satisfied. For example, if *Demo* is defined like the well-known *vanilla* meta-interpreter, then $D1$ is satisfied only when ϕ is a goal.

The main weaknesses of this class of axiomatisations are caused by the fact that *Demo* is defined through Clark's completion. First, many non-stratified meaningful programs are given an inconsistent axiomatisation. Several alternative semantics have been proposed to overcome the drawbacks of Clark's completion, but which are the "right" semantics for reasoning about provability is still an open question.

Secondly, SLDNF resolution (that is, the intended proof procedure for interpreting programs) is not complete for the class of programs with consistent Clark's completion. For example, if Ax is the completion of an axiomatisation of *Demo* such that, for some atom A,

$$Ax \not\models Demo(\lceil A \rceil) \text{ and}$$
$$Ax \not\models \neg Demo(\lceil A \rceil)$$

and if the definition of p in program P is

$$\{(p \leftarrow Demo(\lceil A \rceil)), (p \leftarrow \neg Demo(\lceil A \rceil))\}. \tag{1}$$

Then p is a logical consequence of $P \cup Ax$ but SLDNF resolution cannot prove p, because subgoals $\leftarrow Demo(\lceil A \rceil)$ and $\leftarrow \neg Demo(\lceil A \rceil)$ fail. In order to avoid this problem, *Demo* should be given a *hierarchical* definition, which seems too restrictive.

The applications of *Demo* call for general *positive* results. As the complexity of meta-interpreters increases, the risk of falling under the scope of incompleteness and inconsistency results increases too, so we need a "guide" to the axiomatisation of *Demo*. The main results in this area are negative, like Löb's theorem, hence they do not provide hints on how provability should be formalised, but rather an upper bound to its capabilities. So there is need for results that characterise general classes of consistent *and* expressive axiomatisations of provability. Such classes should be described in a perspicuous and concise way, in order to make the consistency theorems easily applicable.

In this paper we propose a solution to the deficiencies listed above. It is based on a semantic approach, inspired by the works on autoepistemic logics ([15], [2]), that provides clear and concise "functional specifications" for *Demo*.

In section 2 two different forms of self-reference are illustrated, that cause complete axiomatisations of *Demo* to be inconsistent. In section 3 the 3-valued semantics of *Demo* is defined. In sections 4, 5 and 6, our formalisation will be related to the classical provability predicates, auto-epistemic logics and negation-as-failure. Finally, the main results of the paper will be summarised and some of their possible applications will be discussed.

2 Completeness and Self-Reference

Let $\mathcal{L}$ be a first-order language such that, for every formula ϕ, there is a term $\lceil\phi\rceil$ whose interpretation is ϕ. We assume that there is a distinguished predicate $Demo_S$ in $\mathcal{L}$, whose intended meaning is provability from a set of $\mathcal{L}$-formulae S. Let $\mathcal{L}_O \subseteq \mathcal{L}$ be the sublanguage where $Demo_S$ does not occur. Finally, let $\vdash$ be the classical provability relation for $\mathcal{L}$, and consider the problem of defining an *extension* of $\vdash$, denoted by $\vdash_X$, that is *conservative* with respect to $\mathcal{L}_O$ and embodies a *complete* axiomatisation of $Demo_S$.

Formally, $\vdash_X$ should satisfy the following conditions, for all $S \subseteq \mathcal{L}$, for all $\phi \in \mathcal{L}$ and for all $\phi_O \in \mathcal{L}_O$:

$$
\begin{aligned}
& S \vdash \phi \Rightarrow S \vdash_X \phi && \textit{(extension)} \\
& \nvdash \phi_O \Rightarrow \nvdash_X \phi_O && \textit{(conservativity)} \\
& \nvdash_X Demo_S(\lceil\phi\rceil) \Rightarrow \vdash_X \neg Demo_S(\lceil\phi\rceil) && \textit{(completeness)}
\end{aligned}
$$

For the sake of simplicity, we will restrict our attention to the relations $\vdash_D$ defined below. For all $D \subseteq \mathcal{L}$, let

$$ Ax(D) \stackrel{def}{=} \{Demo_S(\lceil\phi\rceil) \mid \phi \in D\} \cup \{\neg Demo_S(\lceil\phi\rceil) \mid \phi \notin D\} $$

and define

$$ S \vdash_D \phi \stackrel{def}{=} S \cup Ax(D) \vdash \phi $$

Remark. The main point of this section is not affected by this restriction. In particular, examples 1 and 2 below apply to every $\vdash_X$ that satisfies the above three properties and the following condition, that is not really restrictive, being satisfied by most monotonic logics:

$$(S_1 \cup S_2 \vdash_X \phi \text{ and } \vdash_X S_2) \Rightarrow S_1 \vdash_X \phi$$

In order to capture the intended meaning of $Demos$, $\vdash_D$ should satisfy the following condition:

$$\vdash_D Demos(\lceil \phi \rceil) \Leftrightarrow S \vdash_D \phi$$

By definition of $\vdash_D$, $\vdash_D Demos(\lceil \phi \rceil)$ holds iff $\phi \in D$, therefore D should be a solution of the following fixpoint equations:

$$
\begin{aligned}
D &= \{\phi \mid S \vdash_D \phi\} \\
&= \{\phi \mid S \cup Ax(D) \vdash \phi\} \qquad\qquad\qquad (2) \\
&= \{\phi \mid S \cup \{Demos(\lceil \psi \rceil) \mid \psi \in D\} \cup \{\neg Demos(\lceil \psi \rceil) \mid \psi \notin D\} \vdash \phi\}
\end{aligned}
$$

Remark. This equation is essentially similar to the definition of the stable expansions of an autoepistemic theory:

$$
\begin{aligned}
T &= \{\phi \mid S \vdash_T \phi\} \\
&= \{\phi \mid S \cup \{L\psi \mid \psi \in T\} \cup \{\neg L\psi \mid \psi \notin T\} \vdash \phi\}
\end{aligned}
$$

Equation 2 has no consistent solutions in far too many cases.

Example 1. If there is a self-referential statement ϕ such that

$$S \vdash \phi \leftrightarrow \neg Demos(\lceil \phi \rceil) \qquad\qquad\qquad (3)$$

then it is easy to show that no solution of equation (2) can be consistent (this is nothing but Gödel's theorem: if (3) holds, then $S \cup Ax(D)$ must be inconsistent, because it contains a complete formalisation of provability; therefore, by (2), D must be inconsistent.) Self-referential statements like ϕ can be built whenever $\mathcal{L}$ contains a function (usually called $Subs$) that computes substitutions over names of formulae (see, for example, [20] for more details). The existence of $Subs$ is considered as a minimal requirement on both the computational capabilities and the expressive power of the theory (see [20] and [17]).

Example 2. Self-reference may arise as the meta-level counterpart of (3). For example, assume $S = \{p \leftarrow \neg Demos(\lceil p \rceil)\}$. Obviously, (3) does not hold for $\phi = p$. However, it is easy to see that $S \vdash_D p$ iff $\vdash_D \neg Demos(\lceil p \rceil)$. So we obtain the meta-level counterpart of (3): for all $D \subseteq \mathcal{L}$

$$S \vdash_D p \text{ iff } S \vdash_D \neg Demos(\lceil p \rceil).$$

Therefore, if D is a consistent solution of equation (2), then $p \in D$ if and only if $p \notin D$ (in fact, by (2), $p \in D$ iff $S \vdash_D p$, and $S \vdash_D p$ iff $S \vdash_D \neg Demos(\lceil p \rceil)$ iff $S \not\vdash_D Demos(\lceil p \rceil)$ iff $p \notin D$). It follows that, under this definition of S, equation (2) has no consistent solutions. Note that this type of self-reference may arise independently of the existence of functions like $Subs$, and depends on how S uses $Demos$.

We call the type of self-reference illustrated by Example 1 *intrinsic*, because it derives from minimal requirements on the expressive power of the theory (namely, the existence of *Subs*). The type of self-reference illustrated by Example 2 will be called *contingent* since it depends on the choice of S. Intrinsic self-reference has been the core of the works on truth and paradoxes (see [14] for an overview), and of the works on provability from Gödel on, while contingent self-reference has been extensively studied in the area of non-monotonic reasoning, since it arises in default logic [19], non-monotonic logics [12], [13] and autoepistemic logics [15].

The problems illustrated by examples 1 and 2 suggest that the conditions posed on $\vdash_X$ should be relaxed in some way. In general, the works on amalgamated and non-indexed provability are centered around an incomplete axiomatisation of a complete semantics of $Demo_S$, so that

$$Demo_S(\lceil \phi \rceil) \text{ is } true \iff S \vdash_X \phi \tag{4}$$

$$Demo_S(\lceil \phi \rceil) \text{ is } false \iff S \not\vdash_X \phi \tag{5}$$

hold, but at least one of the following facts does not hold:

$$\vdash_X Demo_S(\lceil \phi \rceil) \iff Demo_S(\lceil \phi \rceil) \text{ is } true \tag{6}$$

$$\vdash_X \neg Demo_S(\lceil \phi \rceil) \iff Demo_S(\lceil \phi \rceil) \text{ is } false \tag{7}$$

In these works, the facts about provability and unprovability that are *actually* captured by the axiomatisation are usually hidden in its closure. On the contrary, in this paper, we preserve (6) and (7), and relax (4) and (5), in order to specify the provability/unprovability notions that are captured by $Demo_S$ *at the same level of abstraction* provided by (4) and (5) – which are appealing for their simplicity, as opposed to the level of detail provided by a set of axiom schemata.

For this purpose, we introduce a 3-valued semantics for $Demo_S$. This is natural and intuitive if we regard $Demo_S$ as a formalisation of a proof procedure that – like most inference engines, theorem provers and intepreters of logic programming languages – may answer "yes" or "no" to a given query, but may also be unable to answer. Furthermore, the 3-valued semantics can solve both intrinsic and contingent self-reference at the same time.

3 Three-valued semantics

In order to give $Demo_S$ a 3-valued semantics, and preserve the classical meaning of the "object level" sublanguage $\mathcal{L}_O$ at the same time, we define a model theory for $\mathcal{L}$ by extending the classic 2-valued interpretations of $\mathcal{L}_O$ with a 3-valued truth assignment for $Demo_S$.

In the following let $Mod(\mathcal{L}_O)$ be the set of classical (2-valued) interpretations of $\mathcal{L}_O$. For all $\mathcal{M} \in Mod(\mathcal{L}_O)$, denote by $|\mathcal{M}|$ the domain of $\mathcal{M}$. Let ρ be any variables-assignment ($\rho : Var(\mathcal{L}) \longrightarrow |\mathcal{M}|$). For all $\mathcal{L}_O$-terms τ and all $\mathcal{L}_O$-formulae ϕ, we denote by $\mathcal{M}[\tau]\rho$ the value of τ in $\mathcal{M}$ under ρ, and write $\mathcal{M}, \rho \models \phi$ iff $\mathcal{M}$ satisfies ϕ under ρ. Sometimes, when there are no free variables in τ and ϕ, we will omit ρ.

Definition 1. A $\mathcal{D}$-interpretation is a structure $(\mathcal{M}, \mathcal{D})$, where

1. $\mathcal{M} \in Mod(\mathcal{L}_O)$
2. $\mathcal{L} \subseteq |\mathcal{M}|$
3. for all $\phi \in \mathcal{L}$, $\mathcal{M}[\lceil \phi \rceil] = \phi$
4. $\mathcal{D}$ is a pair of *disjoint* sets of formulae: $\mathcal{D} = \langle \mathcal{D}^T, \mathcal{D}^F \rangle$; intuitively, $\mathcal{D}^T$ is the set of *formally provable formulae*, while $\mathcal{D}^F$ is the set of *formally unprovable formulae*.

The satisfaction relation of $\mathcal{M}$ is extended to $(\mathcal{M}, \mathcal{D})$ so that, for all $\mathcal{L}$-predicates $p \neq Demos$ and for all $\mathcal{L}$-terms $\tau, \tau_1, \dots, \tau_n$:

$$(\mathcal{M}, \mathcal{D}), \rho \models_T p(\tau_1, \dots, \tau_n) \text{ if } \mathcal{M}, \rho \models p(\tau_1, \dots, \tau_n)$$
$$(\mathcal{M}, \mathcal{D}), \rho \models_F p(\tau_1, \dots, \tau_n) \text{ if } \mathcal{M}, \rho \not\models p(\tau_1, \dots, \tau_n)$$
$$(\mathcal{M}, \mathcal{D}), \rho \models_T Demos(\tau) \quad \text{if } \mathcal{M}[\tau]\rho \in \mathcal{D}^T$$
$$(\mathcal{M}, \mathcal{D}), \rho \models_F Demos(\tau) \quad \text{if } \mathcal{M}[\tau]\rho \in \mathcal{D}^F$$
$$(\mathcal{M}, \mathcal{D}), \rho \models_U Demos(\tau) \quad \text{if } \mathcal{M}[\tau]\rho \in \mathcal{L} \setminus (\mathcal{D}^T \cup \mathcal{D}^F)$$
$$(\mathcal{M}, \mathcal{D}), \rho \models_F Demos(\tau) \quad \text{if } \mathcal{M}[\tau]\rho \notin \mathcal{L}$$

We also write $(\mathcal{M}, \mathcal{D}) \models_V \phi$ (where $V = T, U, F$) iff, for all ρ, $(\mathcal{M}, \mathcal{D}), \rho \models_V \phi$.

We give a specific meaning to the negation symbol, namely:

$$(\mathcal{M}, \mathcal{D}), \rho \models_T \neg\phi \text{ if } (\mathcal{M}, \mathcal{D}), \rho \models_F \phi$$
$$(\mathcal{M}, \mathcal{D}), \rho \models_F \neg\phi \text{ if } (\mathcal{M}, \mathcal{D}), \rho \models_T \phi$$
$$(\mathcal{M}, \mathcal{D}), \rho \models_U \neg\phi \text{ if } (\mathcal{M}, \mathcal{D}), \rho \models_U \phi$$

At this stage we do not want to make any other assumption about the valuation scheme, because several schemes will be examined in the following. So, in the rest of this section, we simply assume that $(\mathcal{M}, \mathcal{D}) \models_T \phi$, $(\mathcal{M}, \mathcal{D}) \models_F \phi$ and $(\mathcal{M}, \mathcal{D}) \models_U \phi$ are defined for all $\mathcal{L}$-formulae ϕ, and that, for all $\mathcal{D}$, there is a provability relation $\vdash_{\mathcal{D}}$ that is sound and complete with respect to $\mathcal{D}$-interpretations, so that

$$S \vdash_{\mathcal{D}} \phi \text{ iff, for all } \mathcal{M}, (\mathcal{M}, \mathcal{D}) \models_T S \Rightarrow (\mathcal{M}, \mathcal{D}) \models_T \phi. \tag{8}$$

Definition 2. A $\mathcal{D}$-model of a set of formulae S is a $\mathcal{D}$-interpretation that satisfies S. A set of formulae S is $\mathcal{D}$-consistent iff S has a $\mathcal{D}$-model.

Next we specify the intended meaning of $Demos$. The truth condition, (4), will be preserved, while the falsity condition, (5), will be relaxed according to the following rather conservative principle:

(**Strong Unprovability Principle**) the statement $Demos(\lceil \phi \rceil)$ is *false* iff ϕ would not follow from S even if we extended the definition of $Demos$.

Thus we get the following definition:

$$Demos(\lceil \phi \rceil) \text{ is } true \Leftrightarrow S \vdash_{\mathcal{D}} \phi \tag{9}$$
$$Demos(\lceil \phi \rceil) \text{ is } false \Leftrightarrow \text{ for all } \mathcal{D}' \supseteq \mathcal{D}, S \not\vdash_{\mathcal{D}'} \phi \tag{10}$$

where $\supseteq$ is extended to pairs of sets in the natural way: $\langle D_1^T, D_1^F \rangle \supseteq \langle D_2^T, D_2^F \rangle$ iff $D_1^T \supseteq D_2^T$ and $D_1^F \supseteq D_2^F$.

By (8), $\vdash_{\mathcal{D}} Demos(\lceil \phi \rceil)$ holds iff $\phi \in \mathcal{D}^T$ and $\vdash_{\mathcal{D}} \neg Demos(\lceil \phi \rceil)$ holds iff $\phi \in \mathcal{D}^F$. Therefore, $\vdash_{\mathcal{D}}$ is a complete axiomatisation of $Demos$ (in the sense that (6) and (7) are satisfied) if and only if $\mathcal{D}$ is a solution of the following equations, that are inherently similar to the fixpoint conditions for the *generalised stable expansions* of 3-valued autoepistemic logics (see [2] or [4]):

$$\mathcal{D}^T = \{\phi \mid S \vdash_{\mathcal{D}} \phi\} \tag{11}$$

$$\mathcal{D}^F = \{\phi \mid \text{ for all } \mathcal{D}' \supseteq \mathcal{D}, S \not\vdash_{\mathcal{D}'} \phi\} \tag{12}$$

4 Relations with Provability Predicates

The right ground for the comparison with the classical provability predicates is the case where S is insensitive to the definition of $Demos$, hence contingent self-reference cannot arise. Accordingly, in this section we assume that $S \subseteq \mathcal{L}_O$. Then classical logic can be preserved even if a 3-valued semantics is adopted. Of course, a suitable valuation scheme is needed.

Definition 3. We say that $(\mathcal{M}, \mathcal{D})$ extends $(\mathcal{M}', \mathcal{D}')$, denoted by

$$(\mathcal{M}, \mathcal{D}) \supseteq (\mathcal{M}', \mathcal{D}')$$

if and only if $\mathcal{M} = \mathcal{M}'$ and $\mathcal{D} \supseteq \mathcal{D}'$.
An interpretation $(\mathcal{M}, \mathcal{D})$ is **complete** iff $\mathcal{D}^T \cup \mathcal{D}^F = \mathcal{L}$.
A **completion** of $(\mathcal{M}, \mathcal{D})$ is any complete interpretation $(\mathcal{M}', \mathcal{D}') \supseteq (\mathcal{M}, \mathcal{D})$.

Note that the complete interpretations of $\mathcal{L}$ coincide with its classical models, and that the completions of $(\mathcal{M}, \mathcal{D})$ are the classical models of $\mathcal{L}$ that extend $(\mathcal{M}, \mathcal{D})$.

Definition 4. Supervaluation [23]:
$(\mathcal{M}, \mathcal{D}) \models_T \phi$ iff ϕ is (classically) true in all the completions of $(\mathcal{M}, \mathcal{D})$.
$(\mathcal{M}, \mathcal{D}) \models_F \phi$ iff ϕ is (classically) false in all the completions of $(\mathcal{M}, \mathcal{D})$.
$(\mathcal{M}, \mathcal{D}) \models_U \phi$ otherwise.

With supervaluation, $\vdash_{\mathcal{D}}$ is strictly related to classical derivability:

Lemma 5. *Let $\vdash_{\mathcal{D}}$ be any proof procedure satisfying the completeness condition (8). Then, for all $S_O \subseteq \mathcal{L}_O$ and all $\phi \in \mathcal{L}$,*

$$S_O \vdash_{\mathcal{D}} \phi \text{ iff } S_O \cup Ax(\mathcal{D}) \vdash \phi \tag{13}$$

where $Ax(\mathcal{D}) \stackrel{def}{=} \{Demos(\lceil \phi \rceil) \mid \phi \in \mathcal{D}^T\} \cup \{\neg Demos(\lceil \phi \rceil) \mid \phi \in \mathcal{D}^F\}$.

Note that we might take (13) as a definition of $\vdash_\mathcal{D}$. Note also that, if $\mathcal{D}$ is a solution of the fixpoint conditions (11) and (12), then

$$Ax(\mathcal{D}) \vdash Demo_S(\lceil \phi \rceil) \quad \text{iff} \quad S \cup Ax(\mathcal{D}) \vdash \phi$$
$$Ax(\mathcal{D}) \vdash \neg Demo_S(\lceil \phi \rceil) \quad \text{iff} \quad \text{for all } \mathcal{D}' \supseteq \mathcal{D},\ S \cup Ax(\mathcal{D}') \not\vdash \phi$$

therefore $Ax(\mathcal{D})$ provides an extension of $\vdash$, conservative on $\mathcal{L}_O$, containing an axiomatisation of $Demo_S$ that completely captures our semantics.

The notions of provability and unprovability formalised by $Ax(\mathcal{D})$ are especially intuitive for an interesting class of non-self-referential formulae, namely the formulae that can be expressed by a modal syntax. More formally, let $\mathcal{L}_O(L)$ be the modal language obtained by extending $\mathcal{L}_O$ with the modal operator L. Every $\mathcal{L}_O(L)$-formula ϕ can be translated into an $\mathcal{L}$-formula ϕ^*, by recursively replacing $L\psi$ by $Demo_S(\lceil \psi^* \rceil)$. For example,

$$(\neg L(p \vee Lq))^* = \neg Demo_S(\lceil p \vee Demo_S(\lceil q \rceil) \rceil)$$

Lemma 6. *Let $S \subseteq \mathcal{L}_O$ and let $\mathcal{D}$ be a solution of the fixpoint conditions (11) and (12). If ϕ is an $\mathcal{L}_O(L)$-formula, then $\phi^* \in \mathcal{D}^T \cup \mathcal{D}^F$.*

Since $\phi^* \in \mathcal{D}^T \cup \mathcal{D}^F$ holds, $\not\vdash_\mathcal{D} Demo_S(\lceil \phi^* \rceil)$ implies $\vdash_\mathcal{D} \neg Demo_S(\lceil \phi^* \rceil)$. Consequently, the meaning of $Demo_S(\lceil \phi^* \rceil)$ is thoroughly traditional:

$$Ax(\mathcal{D}) \vdash Demo_S(\lceil \phi^* \rceil) \quad \text{iff} \quad S \cup Ax(\mathcal{D}) \vdash \phi^* \tag{14}$$
$$Ax(\mathcal{D}) \vdash \neg Demo_S(\lceil \phi^* \rceil) \quad \text{iff} \quad S \cup Ax(\mathcal{D}) \not\vdash \phi^* \tag{15}$$

To what extent are Löb's derivability conditions satisfied by these formalisations of provability?

Lemma 7. *If $\mathcal{D}$ is a solution of (11) and (12) then*

$$S \vdash_\mathcal{D} \phi \Rightarrow S \vdash_\mathcal{D} Demo_S(\lceil \phi \rceil).$$

Moreover, if $\phi \in \mathcal{D}^T \cup \mathcal{D}^F$, then the following facts hold:

$$S \vdash_\mathcal{D} Demo_S(\lceil \phi \rceil) \rightarrow \phi$$
$$S \vdash_\mathcal{D} Demo_S(\lceil \phi \rceil) \rightarrow Demo_S(\lceil Demo_S(\lceil \phi \rceil) \rceil).$$

In particular, if ϕ follows classically from $S \cup Ax(\mathcal{D})$, or if $\phi = \psi^*$ for some $\psi \in \mathcal{L}_O(L)$, then Löb's conditions hold for ϕ.

But we still have to show that (11) and (12) have a solution, and that S is consistent with it. The next results show that there is a *least* such solution, that can be taken as the canonical definition of $Demo_S$. The proof exploits a standard fixpoint construction, based on the following operator:

Definition 8. Operator $\Phi_S : \mathcal{L}^2 \longrightarrow \mathcal{L}^2$ is defined by:

$$\Phi_S(\mathcal{D}) = \langle \Phi_S(\mathcal{D})^T, \Phi_S(\mathcal{D})^F \rangle$$
$$\Phi_S(\mathcal{D})^T = \{\phi \mid S \vdash_\mathcal{D} \phi\}$$
$$\Phi_S(\mathcal{D})^F = \{\phi \mid \text{for all } \mathcal{D}' \supseteq \mathcal{D},\ S \not\vdash_{\mathcal{D}'} \phi\}$$

Note that the fixpoint conditions are satisfied iff $\mathcal{D}$ is a fixpoint of $\varPhi_S$.

Lemma 9. *If* $S \subseteq \mathcal{L}_O$ *then* $\varPhi_S$ *is monotonic, that is* $\mathcal{D} \supseteq \mathcal{D}' \Rightarrow \varPhi_S(\mathcal{D}) \supseteq \varPhi_S(\mathcal{D}')$.

Therefore, by Tarski's theorem, operator $\varPhi_S$ has a least fixed point $lfp(\varPhi_S)$ such that, for some ordinal α, $lfp(\varPhi_S) = \varPhi_S^\alpha(\langle \emptyset, \emptyset \rangle)$. From this fact, the next theorem follows easily:

Theorem 10. *If* S *is a set of* $\mathcal{L}_O$-*formulae, then* $\mathcal{D}_S = lfp(\varPhi_S)$ *is the least definition of Demos such that* $\vdash_{\mathcal{D}_s}$ *contains a complete axiomatisation of Demos, that is*

$$\vdash_{\mathcal{D}_s} Demos(\lceil \phi \rceil) \text{ iff } S \vdash_{\mathcal{D}_s} \phi$$

$$\vdash_{\mathcal{D}_s} \neg Demos(\lceil \phi \rceil) \text{ iff } \text{ for all } \mathcal{D}' \supseteq \mathcal{D}_S, \ S \nvdash_{\mathcal{D}_s} \phi$$

Moreover, if S *is (classically) consistent, then* S *is* $\mathcal{D}_S$-*consistent, or, equivalently,* $S \cup Ax(\mathcal{D}_S)$ *is classically consistent.*

A nice feature of $\mathcal{D}_S$ is that it has not only the impredicative definition based on the fixpoint conditions, but also a transfinite inductive construction – based on the ordinal sequence $\{\varPhi_S^\beta(\langle \emptyset, \emptyset \rangle)\}_\beta$.

Note that our formalisation does not suffer from intrinsic self-reference: the consistency result above holds independently of what naming mechanism is adopted.

5 Relations with Autoepistemic Logic

Non-monotonic reasoning is one of the intended applications of $Demos$. So, in this section, we show that our axiomatisations of $Demos$ can simulate autoepistemic logic. This is quite an encouraging result: in fact, autoepistemic logic is one of the three main non-monotonic formalisms, and it can simulate another such formalism, that is, default logic (see [11]).

We will compare the logic of $Demos$ with 3-valued autoepistemic logic ([2] and [4]), rather than with Moore's original logic [15], because the former logic behaves like the latter on non-problematic theories (like modal-operator-free theories and *stratified* theories [9]), but is more tolerant to contingent self-reference in the other cases, thus providing a more robust model of beliefs.

Autoepistemic logics are meant to model the beliefs of an ideally rational and introspective agent. They are based on a propositional modal language $\mathcal{L}_O(\hookleftarrow, L)$, obtained by extending a standard propositional language $\mathcal{L}_O$ with a modal operator L – to be read as *"I believe"* – and with a non-standard implication symbol $\hookleftarrow$ with the following meaning:

$(\mathcal{M}, \mathcal{D}), \rho \models_T \phi_1 \hookleftarrow \phi_2$ if $(\mathcal{M}, \mathcal{D}), \rho \models_T \phi_2$ implies $(\mathcal{M}, \mathcal{D}), \rho \models_T \phi_1$
$(\mathcal{M}, \mathcal{D}), \rho \models_F \phi_1 \hookleftarrow \phi_2$ otherwise.

Note that $\hookleftarrow$ is equivalent to classical implication when its arguments are defined (e.g. when $\phi_1, \phi_2 \in \mathcal{L}_O$).

The possible closures of a set of axioms S, according to the 3-valued autoepistemic semantics, are called *Generalised Stable Expansions* of S (GSE, for short). They can be represented as pairs $\langle B^+, B^- \rangle$, where B^+ is the set of beliefs of the agent, and B^- is his set of disbeliefs.

Definition 11. An $\mathcal{L}_O(\hookleftarrow, L)$-formula ϕ is *implicative* iff $\phi \in \mathcal{L}_O$ or

$$\phi = \forall(\phi_O \hookleftarrow \phi_E)$$

where $\phi_O \in \mathcal{L}_O$ and ϕ_E is a *purely epistemic* $\mathcal{L}_O(L)$-formula, that is, propositional symbols occur only within the scope of L.
A set of $\mathcal{L}_O(\hookleftarrow, L)$-formulae S is *implicative* iff all of its members are implicative.

An implicative set of formulae S has a GSE iff it is coherent with respect to classical deduction and necessitation. In this case, S has a *least* GSE, denoted by $GSE(S)$, that provides the canonical autoepistemic consequences of S.

Remark. Implicative theories are quite expressive. Implicative formulae allow to derive arbitrarily complex non-modal formulae from an arbitrarily complex modal formula. Having only non-modal conclusions, implicative formulae cannot force unmotivated beliefs, that are incompatible with the assumptions on the agent's rationality.

After this very quick summary, we are ready to state the correspondence between the logic of $Demo_S$ and autoepistemic logic. For this purpose, we need a suitable valuation scheme:

Definition 12. Strong Kleene's valuation: Let $\rho[d/v]$ be the assignment that binds variable v to d, and agrees with ρ on all the other variables. Define

$$(\mathcal{M}, \mathcal{D}), \rho \models_T \phi_1 \vee \phi_2 \text{ iff } (\mathcal{M}, \mathcal{D}), \rho \models_T \phi_1 \text{ or } (\mathcal{M}, \mathcal{D}), \rho \models_T \phi_2$$
$$(\mathcal{M}, \mathcal{D}), \rho \models_F \phi_1 \vee \phi_2 \text{ iff } (\mathcal{M}, \mathcal{D}), \rho \models_F \phi_1 \text{ and } (\mathcal{M}, \mathcal{D}), \rho \models_F \phi_2$$
$$(\mathcal{M}, \mathcal{D}), \rho \models_T \exists v\, \phi(v) \text{ iff } (\mathcal{M}, \mathcal{D}), \rho[d/v] \models_T \phi(v) \text{ for some } d \in |\mathcal{M}|$$
$$(\mathcal{M}, \mathcal{D}), \rho \models_F \exists v\, \phi(v) \text{ iff } (\mathcal{M}, \mathcal{D}), \rho[d/v] \models_F \phi(v) \text{ for all } d \in |\mathcal{M}|$$

($\rightarrow$, $\wedge$ and $\forall$ are defined as usual in terms of $\neg$, $\vee$ and $\exists$)

Now Φ_S is not monotonic, due to connective $\hookleftarrow$. But it is still monotonic with respect to $\mathcal{L}_O(L)$-formulae, and has a fixpoint that can be constructed through a semi-inductive process (see Herzberger [10]):

Lemma 13. *If $S = T^* = \{\phi^* \mid \phi \in T\}$, for some implicative T, then Φ_S has a least fixed point $lfp(\Phi_S)$. Moreover, $lfp(\Phi_S) = \Phi_S^\alpha$, for some ordinal α, where $(\Phi_S^\alpha)_\alpha$ is the semi inductive process defined below:*

$$\Phi_S^0 = \langle \emptyset, \emptyset \rangle$$
$$\Phi_S^{\alpha+1} = \Phi_S(\Phi_S^\alpha)$$
$$\Phi_S^\lambda = \bigcup_{(\alpha < \lambda)} \bigcap_{(\alpha \leq \beta < \lambda)} \Phi_S(\Phi_S^\beta)$$

where λ is a limit ordinal.

The proof is based on the fact that $\mathcal{L}_O(L)$-formulae monotonically increase, until they become stable at some α; after that, the formulae involving $\hookleftarrow$ become progressively stable – according to the nesting level of $\hookleftarrow$ – so that $\Phi_S^{\alpha+\omega}$ is a fixpoint of Φ_S. The definition of semi inductive processes allows the limit elements to be consistent even if Φ_S is not monotonic.

Theorem 14. *Let T be an implicative set of $\mathcal{L}_O(\hookleftarrow, L)$-formulae, let $S = T^*$ and let $\mathcal{D}_S = lfp(\Phi_S)$.*
T has a GSE iff $\mathcal{D}_S^T$ is consistent. Equivalently, $\mathcal{D}_S^T$ is consistent iff T is coherent with respect to classical inference and necessitation. Moreover, for all $\phi \in \mathcal{L}_O(L)$,

$$\phi \in GSE(S)^+ \text{ iff } \phi^* \in \mathcal{D}^T$$
$$\phi \in GSE(S)^- \text{ iff } \phi^* \in \mathcal{D}^F.$$

It follows that we can use our axiomatisation of $Demo_S$ to compute the autoepistemic consequences of S: for all $\phi \in \mathcal{L}_O(L)$,

$$\phi \in GSE(S)^+ \text{ iff } \vdash_{\mathcal{D}_S} Demo_S(\lceil \phi^* \rceil)$$
$$\phi \in GSE(S)^- \text{ iff } \vdash_{\mathcal{D}_S} \neg Demo_S(\lceil \phi^* \rceil).$$

This formalisation is perfectly tolerant to both forms of self-reference: from Theorem 14 it follows easily that, if S is coherent, then $\vdash_{\mathcal{D}_S}$ is consistent with S (i.e. $S \not\vdash_{\mathcal{D}_S} false$), independently of what naming mechanism is adopted and of what use is made of $Demo_S$.

Example 3. Let $S = \{p \hookleftarrow \neg Demo_S(\lceil p \rceil)\}$. As in Example 2, the following fact holds:

$$S \vdash_D p \text{ iff } S \vdash_D \neg Demo_S(\lceil p \rceil).$$

However, S has a consistent axiomatisation – complete with respect to our semantics – such that: $\not\vdash_{\mathcal{D}_S} Demo_S(\lceil p \rceil)$ and $\not\vdash_{\mathcal{D}_S} \neg Demo_S(\lceil p \rceil)$. The axiomatisation can still capture all tautologies (e.g. $\vdash_{\mathcal{D}_S} Demo_S(\lceil p \vee \neg p \rceil)$) and all non-tautologies, but the ones implied by p (e.g. $\vdash_{\mathcal{D}_S} \neg Demo_S(\lceil \neg p \rceil)$ and $\vdash_{\mathcal{D}_S} \neg Demo_S(\lceil q \rceil)$).

The provability relation $\vdash_{\mathcal{D}_S}$ can still be characterised in terms of classical provability.

Lemma 15. *Let S be an implicative set of formulae and define*

$$S(\mathcal{D}) = \{\phi \mid \phi \in (S \cap \mathcal{L}_O) \text{ or } ((\phi \hookleftarrow \phi_M) \in S \text{ and } \vdash_{\mathcal{D}} \phi_M)\}.$$

Then, for all $\mathcal{L}_O$-formulae ϕ, $S \vdash_{\mathcal{D}} \phi$ iff $S(\mathcal{D}) \vdash \phi$.

Note that checking whether $\vdash_\mathcal{D} \phi_M$ is very simple: it suffices to evaluate ϕ_M in $\mathcal{D}$ according to Strong Kleene's valuation.

This property and Lemma 15 suggest how to devise an algorithm to compute $\Phi_S(\mathcal{D})^T$, given $\mathcal{D}$. $\Phi_S(\mathcal{D})^F$ can be computed (or approximated) by using proof-theoretic and/or model-theoretic methods for generating non-theorems. Note that *any* proof procedure for first-order logic can be integrated with a method for generating unprovable facts.

6 Relations with Negation as Failure

In this section we show how different forms of negation-as-failure can be captured by $Demo_S$. We restrict our attention to (possibly infinite) ground programs. This hypothesis is not really restrictive, because every logic program is equivalent to its ground instantiation, in the sense that the two programs have the same Herbrand models.

Definition 16. For all ground clauses $C = A \leftarrow B_1, \ldots, B_n, \neg B_{n+1}, \ldots, \neg B_m$

$$TR_1(C) \stackrel{def}{=}$$
$$A \leftarrow Demo_S(\lceil Demo_S(\lceil B_1 \rceil) \rceil) \wedge \ldots \wedge Demo_S(\lceil Demo_S(\lceil B_n \rceil) \rceil) \wedge$$
$$\wedge Demo_S(\lceil \neg Demo_S(\lceil B_{n+1} \rceil) \rceil) \wedge \ldots \wedge Demo_S(\lceil \neg Demo_S(\lceil B_m \rceil) \rceil)$$

For all ground programs P, $TR_1(P) \stackrel{def}{=} \{TR_1(C) \mid C \in P\}$.

We still assume that Strong Kleene's valuation is used. It follows easily from Lemma 13 that, if $S = TR_1(P)$ for some P, then there is a fixpoint $lfp(\Phi_S) = \Phi_S^\alpha$.

Denote by $P \vdash_{SLDNF} A$ the fact that the goal $\leftarrow A$ has a successful SLDNF-derivation in P.

Theorem 17. *For all finite ground programs* P, *let* $S = TR_1(P)$ *and let* $\mathcal{D}_S = lfp(\Phi_S)$. *Then, for all ground atoms* A,

$$P \vdash_{SLDNF} A \quad \text{iff} \quad \vdash_{\mathcal{D}_S} Demo_S(\lceil A \rceil)$$
$$P \vdash_{SLDNF} \neg A \quad \text{iff} \quad \vdash_{\mathcal{D}_S} \neg Demo_S(\lceil A \rceil).$$

In other words, the axiomatisation of $Demo_S$ perfectly captures SLDNF resolution (by the way, this proves that SLDNF resolution is a complete proof procedure for the atomic consequences of this kind of $\mathcal{L}(\leftrightarrow)$-theories.)

Definition 18. For all ground clauses $C = A \leftarrow B_1, \ldots, B_n, \neg B_{n+1}, \ldots, \neg B_m$

$$TR_2(C) \stackrel{def}{=} A \leftrightarrow B_1 \wedge \ldots \wedge B_n \wedge \neg Demo_S(\lceil B_{n+1} \rceil) \wedge \ldots \wedge \neg Demo_S(\lceil B_m \rceil).$$

For all ground programs P, $TR_2(P) \stackrel{def}{=} \{TR_2(C) \mid C \in P\}$.

Also in this case, by Lemma 13, there is a fixpoint $lfp(\Phi_S) = \Phi_S^\alpha$.

Theorem 19. *Let P be a (possibly infinite) propositional logic program and let M_P be its well-founded model. Let $S = TR_2(P)$ and let $\mathcal{D}_S = lfp(\Phi_S)$. Then, for all ground atoms A,*

$$M_P \models_T A \text{ iff } \vdash_{\mathcal{D}_s} Demo_S(\lceil A \rceil)$$
$$M_P \models_F A \text{ iff } \vdash_{\mathcal{D}_s} \neg Demo_S(\lceil A \rceil).$$

Therefore, with a simple change in the translation, we can capture the well-founded semantics of negation as failure through the axiomatisation of $Demo_S$. TR_1 and TR_2 clarify the relationships between the two forms of negation-as-failure in terms of provability.

Note that our formalisation does not suffer from the problems affecting Clark's completion. In fact, it gives a consistent semantics to every logic program and, due to Strong Kleene's valuation, the definition of p illustrated in (1) does not entail p, hence the constructive character of SLD resolution is preserved.

7 Conclusions and Perspectives

We have introduced a useful 3-valued semantics for $Demo_S$: the corresponding logic is quite tolerant to both intrinsic and contingent self-reference, while the other existing approaches focus on the former and seem to disregard the latter; furthermore, our logic has interesting relations with classical provability predicates, can simulate autoepistemic reasoning (hence default reasoning), and captures two different forms of negation-by-failure. These results support the claim that non-classical forms of reasoning can be simulated by adding only one non-classical feature to a first-order language. This can be relevant for the development of flexible logic programming languages.

As we argued in Section 2, a 3-valued semantics is natural if we regard $Demo_S$ as the formalisation of a partial proof-procedure.

Moreover, the approach based on complete axiomatisations of a partial semantics is elegant: through conditions like (9)-(10) and (14)-(15), the facts about provability and unprovability that are actually captured by the axiomatisation are described at a good level of abstraction. This improves the understanding of the capabilities of the axiomatisation and provides a concise characterisation of a large class of facts about provability and unprovability, that can be axiomatised without running into inconsistency (under the reasonable assumption that S is consistent, as specified by Theorems 10 and 14).

In this respect, the complete axiomatisation $Ax(\mathcal{D}_S)$ is useful, even if it may not be recursively enumerable (which is not surprising, since it captures a large part of the complement of a first order theory), for it suggests what an effective amalgamated meta-predicate can try to capture without falling into the well-known pathologies of self-reference. The inductive (or semi-inductive) construction of D_S provides a more detailed guide to the definition of an effective provability predicate. For example, for all computable approximations of Φ_S – that is, for all computable Φ such that $\Phi(\mathcal{D}) \subseteq \Phi_S(\mathcal{D})$ for all $\mathcal{D}$ – $Ax(\Phi^\omega)$ provides a consistent axiomatisation of $Demo_S$. The definition of Φ may rely on the similarities between $\vdash_{\mathcal{D}}$ and $\vdash$, and on an increasing number of techniques for computing (part of) the non-theorems of a theory, ranging from purely syntactical methods (like in [24]) to model-checking methods.

Extending the framework presented in this paper to the 2-place *Demo* predicate of Bowen and Kowalski is an easy exercise – especially if we let the names of theories range over well-behaved theories like the translations of implicative theories (this does not prevent *Demo* to formalise inconsistent theories, anyway.) Every theorem of this paper has a direct counterpart in the extended framework, so, for example, it is possible to model a multi-agent autoepistemic logic (by exploiting the relations with autoepistemic logics) as well as modular logic programming (thanks to the relations with negation-as-failure). Further work should try to extend the theory domain, either by characterising larger classes of well-behaving theories or by devising new frameworks, even more tolerant to self-reference.

References

1. G. Attardi, M. Simi. *Reflections about reflection*. Proc. of KR91, Principles of Knowledge Representation and Reasoning, 2nd Int. Conf. (1991).
2. P. A. Bonatti. *A more general solution to the multiple expansion problem*. In Proc. Workshop on Non-Monotonic Reasoning and Logic Programming, NACLP'90.
3. P. A. Bonatti. *Beliefs as stable conjectures*. In Proc. of the First Int. Workshop on Logic Programming and Non-Monotonic Reasoning, Washington D.C., 1991, MIT Press.
4. P. A. Bonatti. *A family of three valued autoepistemic logics*. To appear in the Proc. of the II Congresso dell'Associazione Italiana per l'Intelligenza Artificiale (AI*IA), 1991.
5. K. A. Bowen, R. A. Kowalski. *Amalgamating language and metalanguage in logic programming*. In K. L. Clark, S. Tarnlund (eds.) Logic Programming, Academic Press (1982).
6. S. Costantini, G. A. Lanzarone. *A metalogic programming language*. In G. Levi, M. Martelli (eds), Logic Programming: Proc. of the Sixth Int. Conference, MIT Press, (1989).
7. N. Davies. *Toward a first-order theory of reasoning agents*. Proc. of 9th European Conference on Artificial Intelligence, Stockholm (1990).
8. S. Feferman. *Transfinite recursive progressions of axiomatic theories*. J. of Symbolic Logic 27 (1962), 259-316.
9. M. Gelfond. *On Stratified Autoepistemic Theories*. In Proceedings AAAI-87, 207-211, 1987.
10. H. G. Herzberger. *Notes on Naive Semantics*. In the Journal of Phil. Logic 11 (1982), 61-102. Reprinted in [14].
11. K. Konolige. *On the relation between default theories and autoepistemic logic*. In Proc. of IJCAI'87, Int. Joint Conf. on Artificial Intelligence, Milano, 1987, pp. 394-401. An extended version appears as *On the relation between default logic and autoepistemic theories* in Artificial Intelligence 35, pp. 343-382.
12. D. McDermott, J. Doyle. *Non-monotonic logic I.* In Artificial Intelligence, 13:41-72, 1980.

13. D. McDermott. *Non-monotonic logic II: nonmonotonic modal theories.* In the Journal of the Association for Computing machinery, 29:33-57, 1982.
14. R. L. Martin (ed.). *Recent essays on truth and the liar paradox.* Clarendon Press, Oxford, 1984.
15. R. Moore. *Semantical considerations on nonmonotonic logics.* In Artificial Intelligence 25, (1985), pp. 75-94.
16. D. Perlis. *Languages with self-reference I: Foundations.* Artificial Intelligence 25, (1985), 301-322.
17. D. Perlis. *Languages with self-reference II: Knowledge, Belief and Modality.* Artificial Intelligence 34, (1988), 179-212.
18. T. Przymusinski. *On the declarative and procedural semantics of logic programs.* Journal of Automated Reasoning, 4, 1988.
19. R. Reiter. *A logic for default reasoning.* In Artificial Intelligence 13 (1980) 81-132.
20. C. Smoryński. *Self-reference and modal logic.* Springer-Verlag 1985.
21. A. M. Turing. *Systems of logic based on ordinals.* Proc. of the London Mathematical Society, ser. 2, vol. 45 (1939), 161-228.
22. R. Turner. *Truth and Modality for Knowledge Representation.* MIT Press.
23. B. Van Fraassen. *Singular terms, truth value gaps and free logic.* In The Journal of Philosophy 63 (1966), pp. 481-95.
24. A. Varzi. *Complementary sentential logics.* Bulletin of the Sector of Logic 19, (4), 1990, pp 112-116.
25. R. Weyhrauch. *Prolegomena to a theory of mechanized formal reasoning.* In Artificial Intelligence 13 (1980), pp 133-170.

Hierarchical Meta-Logics:
Intuitions, Proof Theory and Semantics*

Fausto Giunchiglia[1,2] Luciano Serafini[1] Alex Simpson[3]

[1] IRST, Povo, 38050 Trento, Italy
[2] DIST, University of Genoa, Via Opera Pia 11A, Genova, Italy
[3] LFCS – Dept. of CS, University of Edinburgh, EH1 1HN, Scotland

Abstract. The goal of this paper is to provide a possible foundation for meta-reasoning in the fields of artificial intelligence and computer science. We first investigate the relationship that we want to hold between meta-theory and object-theory. We then outline a methodology in which reflection rules serve to deductively generate a meta-theory from its object theory. Finally, we apply this methodology and define a hierarchical meta-logic, namely a formal system generating an entire meta-hierarchy, which is sound and complete with respect to a semantics formalising the desired meta/object relationship.

1 Introduction

Meta-reasoning, as well as being an interesting field of research in its own right, is also an area of substantial applicative potential. This has been illustrated by much recent work in formal reasoning and automated deduction.

Historically, meta-reasoning was first formalised and investigated by logicians, the pioneering work being that of Gödel [10] and Tarski [17]. Gödel and Tarski concentrated on issues such as incompleteness, consistency and the formalisation of truth; later work has followed this direction, investigating the ability of different meta-theories to *represent* object-level properties.

More recently, meta-reasoning has been taken up by researchers in artificial intelligence (AI) and computer science (CS). The goal, however, is quite different: meta-reasoning is studied as a tool for *controlling* object-level inference in automated deduction systems. The important issue here is how meta-reasoning can be used to drive object-level deduction in a correct, effective and efficient manner. Some examples in proof checking are [11, 18, 9], some examples in theorem proving are [4, 3, 2].

This paper is a first step towards providing a foundation for meta-reasoning in the fields of AI and CS. From this perspective we understand and formalise *meta-reasoning as deduction in a logical (declarative) meta-theory*. Our interest is in studying how object-level deduction and meta-level deduction can be related to each other, *both proof-theoretically and semantically*. Towards this end, we outline a general methodology which we illustrate by defining a *hierarchical*

---

* We thank Frank van Harmelen for his comments on an early draft of the paper.

meta-logic in which a number of important issues are successfully dealt with. Note that our requirement of a logical meta-theory contrasts with much of the current practice in theorem proving and automated deduction in which the so called "meta-language" is nothing more than a programming language (*eg.* in LCF the "meta-language" is ML [11]). Proof plans [3] are a first step towards real meta-theoretic reasoning but the meta-theory is still partially procedural and the reasoning is performed by planning rather than by theorem proving.

Despite our logical approach, our motivations lead us to have working hypotheses quite different from those of logicians. In particular, we require the object and meta theory to be two *distinct* logical theories, each theory with its own language, axioms and deductive machinery. This approach was first suggested in [18] in which the possibility of having a hierarchy of logical meta-theories, deductively linked to each other by *reflection rules*, was first outlined. These ideas have been pushed further and put into place in [9]. However a theoretical foundation to this work has never previously been given.

The next three sections of the paper describe in detail the methodology we will follow. In section 2 we motivate our choice of keeping meta-theory and object-theory distinct. In section 3 we investigate the relationship between meta-theory and object-theory highlighting the issues of soundness and completeness. In section 4 we describe how the meta-theory can be deductively generated from the object-theory by the use of reflection rules. The above methodology is applied in section 5 in which we define a formal system, MK, a hierarchical meta-logic. The section contains two subsections. In subsection 5.1, following the methodology of section 4, we give a proof-theory spanning the whole hierarchy of meta-theories in which meta-theory and object-theory are connected by reflection rules In subsection 5.2 we provide a semantics, motivated by the considerations outlined in section 3, in which the models of the meta-theories are related to the models of their object-theories.

2   The Meta-Theory and Object-Theory are Distinct

The role of a meta-theory is to describe various properties of a given logical theory (the *object-theory*). A meta-theory will thus, itself, be a logical theory[4] whose language is rich enough to express whichever properties of the object theory we are interested in. A meta-theory of a first-order object-theory, for example, might contain such relations as "x is a free-variable in formula y" and "the result of substituting the term w for all free-occurences of the variable x in the formula y is the formula z" as well as "names" for all the formulae, terms and variables in the object-language. A particularly important meta-theoretic predicate is the "theoremhood" predicate expressing the relation "the formula x is a theorem of the object-theory".

In much existing work the meta-theory and the object-theory are the same. This is necessary when the meta-theory is encoded in the object-theory following the approach of Gödel [10] (see [13] for a recent example). Even when the

---

[4] We restrict our attention to meta-theories which are classical theories.

encoding approach is not adopted, it is still possible to amalgamate the meta and object theories (as in *eg.* [1]). However we work with distinct object and meta-theories. There are in fact many advantages in separating the two:

- we do not confuse deduction in the object-theory with deduction in the meta-theory.
- We are thus able to study the relation between the proof-theory of the object-theory and that of the meta-theory, and
- provide a semantics for the meta-theory in which the models of the meta-theory are related to the models of the object-theory.
- From an applicational perspective, the proof-theory and semantics provide a foundation for the development of automated deduction systems using meta-theoretic reasoning. The intuitions, proof theory and semantics presented in this paper, even if more general, can be seen as formalizing some of the ideas behind the GETFOL system [9], (an extension /re-implementation of the FOL system [18]).
- Finally, from the point of view of theorem proving, keeping the object-theory and meta-theory distinct gives more structure to the search space and reduces the branching rate.[5]

A commitment to separate object and meta theories presents us with an additional difficulty. If the object-theory and meta-theory are not separated (*eg.* meta-statements are added in a conservative extension of the object-theory) then the meta-theory can be used with the same level of safety as the object-theory, for the meta-theory is consistent whenever the object-theory is. In separating the two theories we have the additional burden of proving the meta-theory consistent. An important gain, however, is that we may even have a consistent meta-theory to an inconsistent object-theory. This gives a strong intuitive motivation, very important in AI (see [7]), for separating the theories as we would like to be able to reason consistently about inconsistency.

3 The Object-Theory/Meta-Theory Relationship

Let us assume that we have an object-theory with provability relation $\vdash_O$ and a meta-theory of this object-theory with provability relation $\vdash_M$. We must determine what relationship we require between the two theories.

We make some minimal expressivity assumptions about the meta-theory. First, we require naming as a primitive operation: for every formula, ϕ, of the object-language there must be a constant, "ϕ", in the meta-language, the *name*[6] of ϕ. Second, we require a unary predicate, T, in the meta-language intended to express the property of theoremhood in the object-theory.

---

[5] This contrasts with the usual approach of constraining the search by modifying the search algorithm, an approach which has the drawback of adding extra-complexity to the control.

[6] The only requirement on naming is that if "ϕ" and "ψ" are identical constants then ϕ and ψ are identical formulae.

For the meta-theory to be sound, properties of the object-theory expressed by theorems of the meta-theory must, indeed, hold. Soundness represents a minimal correctness requirement on a meta-theory and fortunately, in general, it is easy to provide sound meta-theories. In addition one may hope for completeness: *ie.* that all formulae of the meta-language which express true properties of the object-theory are theorems of the meta-theory[7]. In the case of "theoremhood" the requirements are:

$$\vdash_O \phi \Longleftrightarrow \vdash_M T(``\phi") \tag{1}$$
$$\nvdash_O \phi \Longleftrightarrow \vdash_M \neg T(``\phi") \tag{2}$$

The two $\Longleftarrow$ implications correspond to soundness, the two $\Longrightarrow$ implications correspond to completeness. (1) is easily satisfied; the $\Longrightarrow$ implication of (2), though, cannot hold whenever the meta-theory is recursively enumerable and theoremhood of the object-theory is not recursive — which will often be the case.

The unachievability (in general) of completeness raises the question of how close to the unobtainable ideal the meta-theory should be expected to get. There seem to be two very different approaches to this problem: one is to approximate the complete meta-theory to within (say) a particular proof-theoretical bound (we have much of the work in formalised proof-theory in mind see *eg.* [15]); the other is to weaken the notion of correspondence between the meta-theory and the object-theory and give a meta-theory which is complete with respect the new relationship.

In this paper we adopt the second approach and require a meta-theory which satisfies (1). This gives us also the $\Longleftarrow$ direction of (2) which follows from (1) as long as the meta-theory is consistent (and so we have soundness). However, this is not enough! From the point of view of automated deduction we would like formulae expressing valid inferences in the object-theory to be theorems of the meta-theory. Thus, for instance, we must require that the formalisations of principles such as modus-ponens are theorems of the meta-theory, *ie.* that:

$$\vdash_M T(``\phi \supset \psi") \supset (T(``\phi") \supset T(``\psi")) \tag{3}$$

What is required is a new (non arbitrary) relationship between meta-theory and object-theory which is strong enough to ensure that principles such as (1) and (3) hold, but which rejects the unachievable $\Longrightarrow$ implication of (2). In section 5.2 such a relation will be given semantically, by relating models of the meta-theory to models of the object-theory.

---

[7] It has been argued [3] that for theorem proving, the completeness and, even, the correctness of the meta-theory may not be desirable properties. This is a valid position, however, in most applications, the correctness and, if feasible, the completeness of the meta-theory are desirable properties.

4 Using Reflection to Generate the Meta-Theory

That the meta-theory must stand in a certain relationship to the object-theory is a necessary correctness criterion for the meta-theory. However, from the point of view of formal reasoning it is important that such properties be exploitable proof theoretically so that results in the meta-theory can be used in object-level deduction and results in the object-theory can be used in meta-level deduction. With such deductive connections, results can be obtained in the object-theory not only by reasoning within the confines of the theory but also by reasoning about the reasoning which can be so performed. Vice-versa, certain meta-level statements can be proved by "reflecting up" object-level properties into their meta-level representation.

To achieve this deductive link we require rules of inference which provide the desired connection by passing the locus of inference from one theory to the other. We therefore investigate providing *reflection rules* between the meta-theory and object-theory[8]. This notion of inference between different theories is a special case of the more general notion of a *bridge rule* in a *multilanguage system* (see [6]). One example of reflection rules for the meta-theoretic predicate T are (respectively "reflection up" and "reflection down"):

$$\frac{\vdash_O \phi}{\vdash_M T(``\phi")} \qquad \frac{\vdash_M T(``\phi")}{\vdash_O \phi}$$

The usual approach is to consider such reflection rules between an object-theory and a *pre-existing* (axiomatised) meta-theory standing in an appropriate relationship to the object-theory. So, *eg.*, if (1) were to hold, then reflection up and reflection down would be sound reflection rules between the two theories. On the other hand, under this approach, all the correct inter-theory inferences are *a priori* possible. The reflection rules are just "admissible" inferences between the two theories and thus, in principle, dispensible. In particular, it is hard to make sense of any notion of a *complete set of rules, ie.* a set of rules which support *all* the deductions between object-theory and meta-theory which are sound with respect to the meta/object relationship. From the point of view of formal reasoning one would like such a concept to be meaningful, *eg.* one would like to say whether the reflection up and reflection down rules given above provide such a complete set for the T predicate.

We propose an alternative and compelling approach: we suggest that *reflection rules be used in a creative role* in which, rather than being admissible rules of inference between an object-theory and a pre-existing meta-theory, they serve to generate the meta-theory from the object-theory via inter-theory deduction. Instead of having an axiomatised meta-theory we envisage a meta-theory whose properties emerge deductively from the object-theory via the application of the reflection rules.

---

[8] Such rules between two distinct theories were first suggested in [18] where they were called "reflection principles" after [5]. However "reflection principle" as in [5] is a substantially different idea (see [8] for discussion) so we prefer the more suggestive "reflection rules".

A first advantage of the approach is that the notion of a complete set of rules can now be defined: a set of rules being complete if the induced meta-theory stands in the desired relation to the object-theory. It still remains to be shown that the approach is feasible, *ie.* that it is possible to provide a complete set of reflection rules with respect to an intuitive meta/object relationship. In section 5.1 we provide generalisations of reflection up and reflection down which turn out to be complete with respect to the semantic meta/object relationship defined in section 5.2.

A second advantage of the approach is that the reflection rules are independent of the particular object-theory that they are applied to. The reflection rules should produce a sound and complete meta-theory for any axiomatised object-theory. Thus the methodology we follow is one of providing a *meta-logic*, a formal system which produces theorems both in the object-theory and in the meta-theory. From an AI or CS perspective this is important as (in contrast with mathematical logic in which interest is concentrated, in each case, on a single theory, *eg.* Peano Arithmetic) in these fields the object-theory varies and depends upon the application.

5 A Hierarchical Meta-Logic: MK

So far we have restricted our attention to the case in which there is just one object-theory and its meta-theory; however, as the meta-theory is itself a logical theory, it is possible to generalise to a hierarchy, indexed by the natural numbers, where theory $i + 1$ is the meta-theory of theory i. The usual assumptions are that the properties of the object-theory expressed in the meta-language and the relationship between meta-theory and object-theory are the same at each level in the hierarchy. This generalisation, technically natural and easy to perform, is motivated, from a theorem proving and AI perspective, by the fact that reasoning happens at many meta-levels and in a manner which is independent of the level [18, 9].

The assumption of uniformity throughout the hierarchy leads to the possibility of capturing the whole infinite tower of theories within a single all-encompassing formal system schematic on the levels. Thus, each meta-language can be defined as partially generated from its object-language by including names for all the represented linguistic elements of the object-language and closing under the application of the meta-linguistic primitives to these names[9]. Moreover, each theory, closed under the rules of classical logic, can be linked to its adjacent object-theory by the same reflection rules. We call the resulting formal-systems, containing a whole hierarchy of meta-theories, *hierarchical meta-logics*.

---

[9] The meta-language will, in general, only be partially determined in this fashion because it may well be desirable for a meta-language to contain other linguistic elements, for example one may require the meta-language to extend the object-language (as in [17]) in which case the meta-language would contain not only the names of the elements of the object language but also the very elements themselves.

In the following we develop an example of hierarchical meta-logic, called MK (first defined in [7]) covering a simple case: the initial object theory is propositional and the only property we are interested in expressing in each meta-theory is theoremhood in the theory below. It is important to point out that, even if relatively simple, MK makes full use of the methodology developed in the previous sections and it is complicated enough to illustrate its power. Moreover, as the technical results will show, MK, has properties that make it interesting in its own right (for more about MK see [7]).

5.1 Proof Theory of MK

First we define the hierarchy of languages in MK. The languages[10] L_0, L_1 ... etc. are defined from the (possibly empty) sets of propositional constants P_0, P_1 ... etc. to be the least sets satisfying:

1. $p \in P_i \implies p \in L_i$ 2. $\bot \in L_i$
3. $\phi, \psi \in L_i \implies \phi \supset \psi \in L_i$ 4. $\phi \in L_i \implies T("\phi") \in L_{i+1}$

Conditions 1., 2. and 3. just ensure that all the languages in the hierarchy contain the boolean closure of their propositions (including the logical constant for falsity, $\bot$, and all their propositional constants). That only the implication connective, $\supset$, is included is for compactness of presentation[11]; all the other connectives are defined in the usual way, in particular $\neg\phi$ will be used as an abbreviation for $\phi \supset \bot$.

Condition 4. ensures that the meta-languages express the theoremhood of any formula in the object language. The flexibility on the definition of P_{i+1} means that the meta-language L_{i+1} is only partially generated from its object-language L_i via 2. 3. and 4. Two interesting cases are: $P_1 = P_2 = ... = \emptyset$, in which case every language L_{i+1} is a "pure" meta-language; and $P_0 = P_1 = ...$, in which case every meta-language is an extension of its object-language.

MK uses a new natural-deduction framework which has the important property of allowing derivations that span different theories; in particular *an assumption in one theory may influence the deduction in another*, this is the crucial idea which enables (3) to be derivable in the system. To keep track of the level at which inference takes place the inference rules are defined over the set of indexed-formulae:

$$L = \{\langle\phi, i\rangle \mid \phi \in L_i\}$$

where i is the index of ϕ in $\langle\phi, i\rangle$.

The "local" inference rules for each theory are just the usual rules for classical natural-deduction (for explanation of the notation and terminology see [14]):

---

[10] A language is defined as the set of its well-formed formulae.
[11] Implication and falsity have been chosen as primitive because of the eventual natural-deduction definition of the formal system.

$$\bot \quad \frac{\begin{array}{c}[\,\langle\neg\phi,\,i\rangle\,]\\ \langle\bot,\,i\rangle\end{array}}{\langle\phi,\,i\rangle}$$

$$\supset\!I \quad \frac{\begin{array}{c}[\,\langle\phi,\,i\rangle\,]\\ \langle\psi,\,i\rangle\end{array}}{\langle\phi\supset\psi,\,i\rangle} \qquad \supset\!E \quad \frac{\langle\phi\supset\psi,\,i\rangle \quad \langle\phi,\,i\rangle}{\langle\psi,\,i\rangle}$$

The reflection rules, TI and TE, are seemingly just the "reflection up" and "reflection down" already presented:

$$TI \quad \frac{\langle\phi,\,i\rangle}{\langle T(``\phi"),\,i+1\rangle} \qquad TE \quad \frac{\langle T(``\phi"),\,i+1\rangle}{\langle\phi,\,i\rangle}$$

However they are more general than "reflection up" and "reflection down" in that their premises and consequences do not need to be theorems in their theory but may depend on some undischarged assumptions. The TI rule is restricted to be applicable only when all undischarged assumptions upon which the premiss $\langle\phi,\,i\rangle$ depends have index $\geq i+1$; intuitively this prevents the case in which a consequence of an assumption in a theory is treated, by the meta-theory, as a theorem of that theory.

Figure 5.1 gives a graphical representation of the structure of MK.

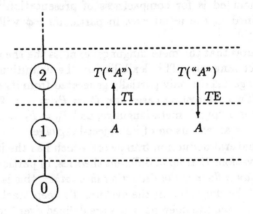

Fig. 1. The structure of MK

Derivations can span multiple languages. The derivability relation $\Gamma \vdash \langle\phi,\,i\rangle$ is defined to hold if there is a derivation in MK of $\langle\phi,\,i\rangle$ in which all the undischarged assumptions are contained in the set $\Gamma \subseteq L$. However, because of the restriction on TI, we can assume that the conclusion of a derivation never depends on an assumption with a lower index. Accordingly we define the derivability relation only for sets of assumptions Γ which contain no indexed-formula with index $< i$. Applying the usual principle of natural deduction, the theorems of the logic are the indexed formulae which have a derivation containing no undischarged assumption, *ie.* $\langle \phi, i \rangle \in L$ is a theorem if $\vdash \langle \phi, i \rangle$.

MK clearly satisfies the analogue of (1):

$$\vdash \langle \phi, i \rangle \iff \vdash \langle T(\text{``}\phi\text{''}), i+1 \rangle$$

by the applicability of *TI* in one direction and *TE* in the other.

Moreover, as desired, the analogue of (3) also holds, that is:

$$\vdash \langle T(\text{``}\phi \supset \psi\text{''}) \supset (T(\text{``}\phi\text{''}) \supset T(\text{``}\psi\text{''})), i+1 \rangle$$

This is provable in the following manner:

$$
\cfrac{
  \cfrac{
    \cfrac{
      \cfrac{\langle T(\text{``}\phi \supset \psi\text{''}), i+1 \rangle}{\langle \phi \supset \psi, i \rangle} \qquad \cfrac{\langle T(\text{``}\phi\text{''}), i+1 \rangle}{\langle \phi, i \rangle}
    }{\langle \psi, i \rangle}
  }{\langle T(\text{``}\psi\text{''}), i+1 \rangle}
  }{\cfrac{\langle T(\text{``}\phi\text{''}) \supset T(\text{``}\psi\text{''}), i+1 \rangle}{\langle T(\text{``}\phi \supset \psi\text{''}) \supset (T(\text{``}\phi\text{''}) \supset T(\text{``}\psi\text{''})), i+1 \rangle}}
\tag{4}
$$

In fact it can be shown, that under the obvious translation, $(\cdot)^+$, of formulae in $\bigcup_{i \in \mathbb{N}} L_i$ to propositional modal formulae (preserving the logical structure and mapping $T(\text{``}\phi\text{''})$ to $\square\phi^+$), the set of theorems of MK is embedded in that of the modal logic K. The theorems of the logic are characterised by the following equivalence (proved in [7]):

$$\vdash \langle \phi, i \rangle \iff \vdash_K \phi^+$$

One important corollary is that all the theories are consistent. A second corollary is that the theorems with index 0 are just the propositional tautologies and so the meta-hierarchy is (correctly) conservative over the object theory.

The logic presented so far is axiom-free and thus conforms to our approach of having the relationship between object-theory and meta-theory emerge deductively rather than via an axiomatisation. On the other hand, for MK to be of any practical use, we must show that the relationship between meta-theory and object-theory is maintained even if we add theoretic axioms. The addition of axioms is not in contradiction with our approach. The relationship between object and meta-theory is part of the logic of the meta-hierarchy and it is this relationship which is accounted for by the rules of inference (this is why we call MK a meta-logic); however, different theories in the hierarchy must be specified by the addition of non-logical axioms at the appropriate levels. The situation is analogous to that of classical natural-deduction in which Hilbert-style logical axioms are avoided, but non-logical axioms are still required to axiomatise a theory. With MK we have provided a natural-deduction system for the the whole meta-hierarchy.

The sets of axioms $\Omega_i \subseteq L_i$ of each theory can be collected together in one set $\Omega \subseteq L$ defined by $\Omega = \{\langle \phi, i \rangle \mid \phi \in \Omega_i\}$. The indexed-axioms (henceforth just axioms) in Ω are used in deductions in the standard natural-deduction fashion: in a deduction tree an axiom can be used as a leaf which depends on no assumptions. We write $\Gamma \vdash_\Omega \langle \phi, i \rangle$ if there is a derivation of $\langle \phi, i \rangle$ from the set of axioms $\Omega \subseteq L$ in which all the undischarged assumptions are contained in the set $\Gamma \subseteq L^{12}$.

The rules of inference of the logic ensure that the relationship between meta-theory and object-theory is maintained for arbitrary sets of axioms. The analogues of (1) and (3) still hold:

$$\vdash_\Omega \langle \phi, i \rangle \iff \vdash_\Omega \langle T(\text{``}\phi\text{''}), i+1 \rangle$$

$$\vdash_\Omega \langle T(\text{``}\phi \supset \psi\text{''}) \supset (T(\text{``}\phi\text{''}) \supset T(\text{``}\psi\text{''})), i+1 \rangle$$

The consistency of every theory is clearly no longer guaranteed as there may be an axiom of the form $\langle \perp, i \rangle$; however, an important property is that inconsistency is, in general, localised: given a set of axioms Ω such that $\vdash_\Omega \langle \perp, i \rangle$ it is not necessarily the case that $\vdash_\Omega \langle \perp, i+1 \rangle$. A precise result is that for any set of axioms Ω if there exists a maximum index i in Ω then $\not\vdash_\Omega \langle \perp, i+1 \rangle$, so in particular no finite set of axioms can render all theories inconsistent (proved in [7]).

We have thus fulfilled the obligation (see end of section 2) to prove the meta-theory consistent. Given an axiomatisation of theory i, by a set of axioms of index i, its meta-theory (ie. theory $i+1$) is consistent. In particular theory $i+1$ is consistent even when theory i is inconsistent, a property which we have argued to be desirable (see end of section 2).

5.2 Semantics of MK

In section 3 we have informally discussed about the relationship between meta-theory and object-theory. In this section, we make the notion precise by providing a semantic formulation of this relationship with respect to which MK is both sound and complete.

Taking any one object-theory we want $T(\text{``}\phi\text{''})$ to be provable in its meta-theory if and only if ϕ is true in all classical models of the object-theory (by soundness and completeness this corresponds to ϕ being provable in the object-theory). However, as we are providing a meta-logic, we want this relationship to hold for any object-theory, ie. for any subset of the models of the object-logic. We thus take subsets of models of the object-logic as models of the meta-theory. In any such model of the meta-theory the interpretation of "ϕ", that is ϕ, belongs to the extension of T if and only ϕ holds in every object-theory model in the subset.

Generalising to the hierarchy, any meta-theory is also object-theory of the theory one level above. This forces us to consider, at each level, sets of models where each of the models of each meta-theory is linked to a subset of the models of its object-theory. To formalise this notion we introduce a *visibility relation R*.

---

[12] The same restriction is placed on Γ, namely that it contains no index $< i$; no restriction is placed on Ω (note that TI can be applied to an axiom).

We thus construct the models of MK over structures, frames, which consist of "points" (that are to be interpreted as classical models of the theories) together with a the visibility relation R which relates points associated to a meta-theory to points associated to its object-theory. Technically, the approach developed is close to the "possible worlds" semantics of modal logic. However we have a different frame structure: the sets of points are separated to model the hierarchy of independent theories; and the visibility relation is defined to mimic the behaviour of the reflection rules, TI and TE.

Definition 1 frame. A frame is a structure of the form $\langle\langle W_0, W_1, \dots \rangle, R\rangle$ where the W_i are mutually disjoint (possibly empty) sets of *points* and R is a relation satisfying:

1. $R \subseteq \bigcup_{i \in \mathbb{N}} (W_{i+1} \times W_i)$
2. For each $w_i \in W_i$ there must exist a $w_{i+1} \in W_{i+1}$ such that $w_{i+1} R w_i$

Even before the semantics is given formally, the restrictions on the relation R can be motivated intuitively given that R is to determine the relationship between meta-theory and object-theory. The relation acts to "connect" each meta-theory with its object-theory and so is only defined between points in theories adjacent in the hierarchy. Moreover it is the meta-theories which depend on their object-theories so the points of the meta-theory see the points of the object-theory and not vice-versa. This captures the intuition that we want the metatheory emerge deductively from the object theory via the application of the reflection rules. The restriction that every point must be visible is to ensure that the meta-theory does not prove properties which do not hold in the object-theory, in other words to ensure the soundness of the meta-theory. The dual restriction that all points in W_{i+1} must see a point in W_i is not enforced. This condition would prevent the case in which there is an inconsistent object-theory with a consistent meta-theory.

The models of the hierarchy are defined over the frames by interpreting each point of W_i as a classical model of the ith theory. W_0, W_1, ... etc. are thus interpreted as the sets of models for the each level of the hierarchy and R is the visibility relation. The interpretation of each point of W_i as a classical model of the ith theory is given by a valuation, v, of the propositional constants in P_i at that point.

Definition 2 model. A model is a structure of the form $\langle\langle W_0, W_1, \dots \rangle, R, v\rangle$, where $\langle\langle W_0, W_1, \dots \rangle, R\rangle$ is a frame and v is a valuation of propositional constants on the points, that is a relation of type $v \subseteq \bigcup_{i \in \mathbb{N}} (W_i \times P_i)$.

Figure 2 gives a graphical representation of the semantics of MK. The satisfaction of a formula in L_i at a point W_i is defined inductively on the structure of the formula. This definition proceeds in the usual way for the boolean connectives; the meta-theoretic relation $T(``\phi")$ is satisfied at a point in W_{i+1} if and only if ϕ is satisfied in all points of W_i visible to it — the intuitive justification for this definition is, as discussed above, the equivalence of theoremhood and validity for classical logics.

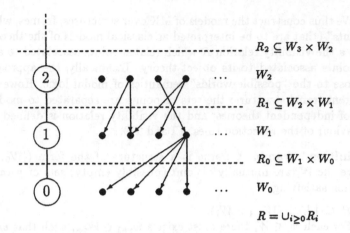

Fig. 2. The structure of the models of MK

Definition 3 satisfaction of a formula at a point. The satisfaction relation $\models_{\mathcal{M}}$ for a model $\mathcal{M} = \langle\langle W_0, W_1, \dots \rangle, R, v\rangle$, of type $\models_{\mathcal{M}} \subseteq \bigcup_{i\in\mathbb{N}} (W_i \times L_i)$, is defined inductively on the complexity of the formula by:

- $w_i \not\models_{\mathcal{M}} \bot$;
- $w_i \models_{\mathcal{M}} p$ iff $\langle w_i, p\rangle \in v$ (where $p \in P_i$);
- $w_i \models_{\mathcal{M}} \phi \supset \psi$ iff $w_i \not\models_{\mathcal{M}} \phi$ or $w_i \models_{\mathcal{M}} \psi$;
- $w_{i+1} \models_{\mathcal{M}} T(\text{``}\phi\text{''})$ iff $w_i \models_{\mathcal{M}} \phi$ holds for all $w_i \in W_i$ satisfying $w_{i+1} R w_i$.

The validity of an indexed-formula $\langle\phi, i\rangle$ in a model corresponds to the satisfaction in the model of the formula at all points of W_i.

Definition 4 validity in a model. In a model $\mathcal{M} = \langle\langle W_0, W_1, \dots \rangle, R, v\rangle$ the validity of an indexed formula $\langle\phi, i\rangle \in L$ is defined by:

$$\mathcal{M} \models \langle\phi, i\rangle \iff \forall w_i \in W_i.\ w_i \models_{\mathcal{M}} \phi$$

The validity of a set of indexed formulae $\Gamma \subseteq L$ is defined by:

$$\mathcal{M} \models \Gamma \iff \forall \langle\phi, i\rangle \in \Gamma.\ \mathcal{M} \models \langle\phi, i\rangle$$

It is now possible to state the completeness theorem for the logic (proved in [16]).

Theorem 5 semantic characterisation of theoremhood. *The two statements below are equivalent:*

1. $\vdash_{\Omega} \langle\phi, i\rangle$
2. For all models $\mathcal{M}$, $\mathcal{M} \models \Omega \Longrightarrow \mathcal{M} \models \langle\phi, i\rangle$

The above completeness result characterises theoremhood in the logic. Importantly, only the points at level i in all models are required to characterise theoremhood in theory i. This fact is interesting as the derivation of a theorem may span many levels (see *eg.* derivation (4)) — however, the interaction between the different levels is taken care of by the structure of the frames.

The semantics can also be used to characterise the full derivability relation. This is important as it allows us to understand the interaction among assumptions in different theories. For this purpose the concept of a *k-chain* in a frame will be introduced. A k-chain is an infinite sequence of points in which the first point is to be interpreted as a model of the kth theory and in which each successive point "sees" its predecessor via the accessibility relation. The idea is that the satisfaction of the derivability relation $\Gamma \vdash_\Omega \langle A, k \rangle$ in a model validating the axioms in Ω is determined by the satisfaction of the formulae (at the the appropriate points) in each k-chain of the model's frame. This gives a semantic reason for not considering indexed formulae with index $< k$ in the derivability relation, for in a k-chain there is no point at which such a formula could be interpreted.

Formally a k-chain in a model $\mathcal{M} = \langle\langle W_0, W_1, \dots \rangle, R, v\rangle$ is an infinite sequence of points $\langle w_k, w_{k+1}, \dots\rangle$ satisfying $w_i \in W_i$ and $w_{i+1} R w_i$ for each $i \geq k$.

If $c = \langle w_k, w_{k+1}, \dots \rangle$ is a k-chain then the notation $c[i]$ will be used to refer to w_i (for $i \geq k$).

If $\mathcal{M}$ is a model and $\Gamma \subseteq L$ has no index $< k$ then define $\Gamma \models_\mathcal{M} \langle \phi, i \rangle$ to hold if and only if every k-chain c of $\mathcal{M}$ satisfies

$$(\forall \langle \psi, i \rangle \in \Gamma. \ c[i] \models_\mathcal{M} \psi) \implies c[k] \models_\mathcal{M} \phi$$

We can now give the semantic characterisation of derivability (proved in [16]).

Theorem 6 semantic characterisation of derivability. *The two statements below are equivalent:*

1. $\Gamma \vdash_\Omega \langle \phi, k \rangle$
2. For all models $\mathcal{M}$. $\mathcal{M} \models \Omega \implies \Gamma \models_\mathcal{M} \langle \phi, k \rangle$

6 Conclusions

The goal of this paper has been to provide a methodology for giving a foundation to the use of meta-reasoning in AI and CS. As a consequence, we have focused on how meta-reasoning, (*ie.* meta-level deduction), and reasoning, (*ie.* as object level deduction) can be effectively related. This is one of the primary requirements for meta-reasoning to be usable in automated deduction systems.

We have first highlighted the properties that a pair meta-theory/ object theory should have, given the restriction that the meta-theory must be mechanizable on a machine (leading us to a relationship rejecting property (2)). We have then concentrated on how to give a proof-theory capturing the required relationship. Given our requirement of mechanizability, that such a proof-theory can be defined is of primary interest. We have also shown that our approach produces a meta-logic which is independent of the particular axiomatised object-theory.

In the last part of the paper, the technical development has naturally followed from the line of thought drawn in the initial sections. The formal system developed, MK, has a proof theory which is indeed sound and complete with respect to a semantics capturing the relationship between meta-theory and object-theory.

Even if the technical development has been essentially restricted to propositional reasoning, the intuitions and the formal framework (both proof-theoretical and semantical) should be extendible at least to the first-order case. Such an extension is currently under investigation. Our goal is to provide foundations to work of the kind described in [9] and implemented in the GETFOL system.

References

1. K.A. Bowen and R.A. Kowalski. Amalgamating language and meta-language in logic programming. In S. Tarlund, editor, *Logic Programming*, pages 153–173, New York, 1982. Academic Press.
2. R.S. Boyer and J.S. Moore. *A Computational Logic*. Academic Press, 1979. ACM monograph series.
3. A. Bundy. The Use of Explicit Plans to Guide Inductive Proofs. In R. Luck and R. Overbeek, editors, *Proc. of the 9th Conference on Automated Deduction*. Springer-Verlag, 1988. Longer version available as DAI Research Paper No. 349, Dept. of Artificial Intelligence, Edinburgh.
4. A. Bundy and B. Welham. Using meta-level inference for selective application of multiple rewrite rules in algebraic manipulation. *Artificial Intelligence*, 16(2):189–212, 1981. Also available as DAI Research Paper 121, Dept. Artificial Intelligence, Edinburgh.
5. S. Feferman. Transfinite Recursive Progressions of Axiomatic Theories. *Journal of Symbolic Logic*, 27:259–316, 1962.
6. F. Giunchiglia and L. Serafini. Multilanguage first order theories of propositional attitudes. In *Proceedings 3rd Scandinavian Conference on Artificial Intelligence*, Roskilde University, Denmark, 1991. IRST-Technical Report 9001-02, IRST, Trento, Italy.
7. F. Giunchiglia and L. Serafini. Multilanguage hierarchical logics (or: how we can do without modal logics). Technical Report 9110-07, IRST, Trento, Italy, 1991. Submitted to Artificial Intelligence Journal. Extended abstract in Proc. META-92, Uppsala, Sweden, 1992.
8. F. Giunchiglia and A. Smaill. Reflection in constructive and non-constructive automated reasoning. In J. Lloyd, editor, *Proc. Workshop on Meta-Programming in Logic Programming*. MIT Press, 1989. IRST Technical Report 8902-04. Also available as DAI Research Paper 375, Dept. of Artificial Intelligence, Edinburgh.
9. F. Giunchiglia and P. Traverso. Plan formation and execution in a uniform architecture of declarative metatheories. In M. Bruynooghe, editor, *Proc. Workshop on Meta-Programming in Logic*, 1990. IRST Technical Report 9003-12.
10. K. Goedel. Über formal unentscheidbare Sätze der Principia Mathematica und verwandter Systeme I. *Monatsh. Math. Phys.*, 38:173–98, 1931. English translation in [12].
11. M.J. Gordon, A.J. Milner, and C.P. Wadsworth. *Edinburgh LCF - A mechanised logic of computation*, volume 78 of *Lecture Notes in Computer Science*. Springer Verlag, 1979.

12. J. Van Heijenoort. *From Frege to Gödel: a source book in Mathematical Logic,
    1879-1931.* Harvard University Press, Cambridge, Mass, 1967.
13. T.B. Knoblock and R.L. Constable. Formalized Metatheory in Type Theory. Technical Report TR 86-742, Dept. Computer Science, Cornell University, 1986.
14. D. Prawitz. *Natural Deduction - A proof theoretical study.* Almquist and Wiksell,
    Stockholm, 1965.
15. H. Schwichtenberg. Proof theory: Some applications of cut-elimination. In
    J. Barwise, editor, *Handbook of Mathematical Logic.* North Holland Publishing
    Company, 1977.
16. A. Simpson. The Semantics of a Multilanguage System. Technical report, IRST,
    1990. Ref. no. 9009-05.
17. A. Tarski. *Logic, Semantics, Metamathematics.* Oxford University Press, 1956.
18. R.W. Weyhrauch. Prolegomena to a theory of Mechanized Formal Reasoning.
    Artificial Intelligence. Special Issue on Non-monotonic Logic, 13(1), 1980.

Negation and Control in
Automatically Generated Logic Programs

Geraint A Wiggins

DRᴇᴀM Group,
Department of Artificial Intelligence,
University of Edinburgh,
80 South Bridge, Edinburgh EH1 1HN,
Scotland.

Abstract. I discuss issues of control and floundering during execution of automatically synthesised logic programs. The process of program synthesis can be restricted, without loss of generality, so that the only negated calls appearing in a program are unifications, as in [10]. If called nonground, these predicates are *delayed* in a logic programming language with a flexible execution rule. [10] presents proposals for automatically delaying calls to predicates until they are instantiated appropriately. This can be implemented easily and safely in my synthesis approach, using meta-level knowledge about inductive proof. This technique is more general, more reliable and less laborious than the original.

1 Introduction

In this paper, I discuss an aspect of my work on the Whelk logic program synthesis system, which is further explained in [3], [13] and [12].

An important assumption made in the synthesis process is that one is synthesising predicates in the *all-ground* mode (*ie* with all arguments fully instantiated), and that a predicate synthesised in this way will be usable in other less fully instantiated modes. In general, this is not a safe assumption, because the synthesised program may contain calls which lead to unbounded recursion as a result of the presence of unbound variables in the top level conjecture.

Floundering under negation is a related problem. If programs are called in the all-ground mode, floundering is minimised. Nevertheless, the problem can still arise, and is exacerbated by use of the synthesised predicates in other modes.

This paper explains how these two problems can be solved through the use of meta-knowledge about inductive proof, which is the basis of the Whelk system. Section 2 outlines the Whelk system. Section 3 covers floundering under negation and the realisation of a solution originally proposed by [10], as a user-independent side effect of the synthesis technique. Section 4 shows that automatic generation of delay declarations can be more general and more reliable in the proof-based synthesis paradigm than when working with *a priori* existing programs. Section 5 summarises, draws conclusions and outlines future research.

Note that, while the examples here demonstrate the technique at work in Whelk synthesising Gödel programs, its operation is general, and is dependent on neither system.

2 Program Synthesis and Transformation in Whelk

2.1 Introduction

Whelk is a Gentzen Sequent Calculus proof development system for a first order typed logic with equality. An adaptation of the *proofs as programs* paradigm [9, 6] to synthesise functional programs allows us in certain circumstances automatically to derive programs from the proofs elaborated in Whelk; the necessary changes to the technique are detailed in [3, 13]. We call the adapted version *proofs as relational programs*.

The synthesised relational programs stand in a close structural relationship to the corresponding proofs. In the longer term, this will allow us to plan the construction of proofs which will lead to particular kinds of program (*eg* ones which use efficient algorithms), using adaptations of *proof planning* techniques described in [5, 2, 4, 8].

The key idea is to view the execution of our desired logic program *with no uninstantiated arguments* as evaluation of a boolean valued function. Then, we prove a *specification conjecture* which postulates that for all possible arguments of the correct type, there exists some truth value logically equivalent to the truth of the specification of that program. This is equivalent to proving that the specification is decidable, which we write thus, reading ∂ as "It is decidable whether..." (I have omitted the types here to avoid clutter):

$$\vdash \forall \overline{a}.\partial S(\overline{a})$$

where $S(\overline{a})$ is the specification of the program we wish to synthesise, and

$$\vdash \forall \overline{a}.\partial S(\overline{a}) \quad \textit{iff} \quad \vdash \exists P.\forall \overline{a}.S(\overline{a}) \leftrightarrow P(\overline{a}) \text{ and } P, \text{ a relation, is decidable}$$

Various other ways of expressing the conjecture are discussed in [13]. A fuller description of the system is given in [12].

2.2 Example: zero/1

For example, a specification conjecture which, when proven in Whelk, would lead to the construction of the zero/1 predicate, which is true iff its argument is zero, would be:

$$\vdash \forall n.\partial(n = 0)$$

While this example is sufficiently simple to make the synthesis procedure unnecessary, it still serves as a useful example of the general framework in which synthesis is performed. See [13] for the full elaboration of the synthesis proof.

The resulting *pure logic program* [3] is

```
zero(n) ↔ n=0 ∧ true ∨ ~n=0 ∧ false.
```

and (automatic) conversion to Gödel [7] yields:

```
MODULE Zero.
IMPORT Naturals.
PREDICATE Zero : Natural.
Zero(n) <- n=0.
```

2.3 Induction and Recursion

The proofs as programs technique relies on an intimate relationship between proofs by induction and recursive programs. The Whelk logic is arranged so that the identification of a subconjecture with an induction hypothesis leads to the inclusion of a recursive call in the synthesised program.

In the current Whelk prototype, we are limited to primitive and two-step induction on natural numbers and parametric lists. In principle, however, there is no reason for such a restriction — later, we will extend the system to include more powerful induction schemes. This will allow us to generate various algorithms for a given specification; for example, while bubble sort corresponds with primitive inductive proof of the sorting specification, quicksort corresponds with course-of-values induction. Thus, correct selection of induction schemes is crucial to the successful operation of the technique, and work is proceeding on its automation.

2.4 Non-ground use of synthesised programs

The introduction of recursive calls raises a problem. In synthesising programs this way, we suppose that we are using the all-ground mode, and *assume* that calls to the synthesised predicates which are not fully instantiated will work.

Given a language with a fixed computation rule, like Prolog, it is easy to give a counterexample to this supposition. Consider the following (logically correct) program and call.

```
append([H|T1],L,[H|T2]) :- append(T1,L,T2).
append([],L,L).

?- append(X,[],Y).
```

This is clearly a logically correct specification of the append/3 relation, but the program never terminates, because the base case appears after the step case.

A safer specification and call might be given in Gödel:

```
MODULE Append.

IMPORT Lists.

PREDICATE Append: List(Integer) * List(Integer) * List(Integer).

DELAY Append(X,Y,Z) UNTIL NONVAR(X) \/ NONVAR(Z).

Append([h|t1],l,[h|t2]) :- Append(t1,l,t2).
Append([],l,l).

[] <- Append(x,[],y).
```

in which case the program terminates with the message "Floundered". We would like the delay declaration which causes this (desirable) behaviour to be generated automatically.

I will address this problem in Section 4. First, let us consider a related problem to do with the use of negation as failure in logic programming languages.

3 Prevention of Floundering under Negation

3.1 The Problem

In his PhD thesis, [10], Lee Naish discusses the problems with the idea of negation in logic programming languages. It is unnecessary here to repeat that detail; I merely summarise by reiterating that there is a serious problem in languages involving negation but no explicit quantifiers. To borrow and slightly adapt Naish's example, consider the goal

$$?- \text{not}(X=a), X=b.$$

in Prolog. If not is called first, as one would normally expect, the goal fails – incorrectly – whereas otherwise it succeeds with an instantiation of b for X. In Gödel, on the other hand, the negated conjunct is *delayed* until X is ground, so the rightmost conjunct is executed first, giving the correct result.

The problem is that, by not(X=a), we mean $\exists x.x \neq a$. What actually happens, because of negation as failure, is that we evaluate $\neg \exists x.x = a$ (*ie* $\forall x.x \neq a$). This is only a problem if we bind variables in the goal; for example, a call of not(X=X) evaluates to the negation of $\forall x.x = x$, *ie* $\neg \exists x.x \neq x$.

This problem generalises to the floundering of arbitrary negated goals, if any variable in the scope of a negation is bound during evaluation. The worst case is where a variable thus bound is later used outside the negation (where it remains unbound). Some languages partly defuse this problem by use of the unnamed variable "_" to mean a variable which is universally quantified, and therefore unimportant in terms of floundering due to any binding of that variable under negation. I find this solution unsatisfying, especially in a language like Gödel whose syntax already contains the necessary quantifiers.

3.2 The Solution

In [10, p19ff], Naish proposes an elegant means of removing floundering under negation. The solution is simply partially to evaluate and replace the negated predicates with new positive ones which compute the negated original. The $\neq$ predicate is then used to ensure failure where the old (unnegated) version would succeed and *vice versa*.

By removing negated goals other than =, we restrict the problem of floundering under negation to those goals, thus improving things greatly. However, it was shown in Section 3.1 that floundering can still be a problem even with such a restriction. Fortunately, a complete solution to this problem was also proposed in Section 3.1: we simply delay the negated goals (viewing them, if we wish, as constraints) until they can be executed without resulting instantiation of the variables in them. A less refined, but substantially easier to implement, approach is to require that the negated goals are delayed until either they are fully ground or there are no other goals to execute. In this case, we would wish the program to terminate with an error message warning of floundering.

Naish has already shown this approach to be correct. However, its implementation in Whelk is of interest as it neatly demonstrates the system at work.

3.3 The Implementation

Negation is Whelk is as in other constructive logic systems. The Law of the Excluded Middle does not in general hold, and ¬P means that, if P is true, a contradiction can be derived. The two sequent calculus rules governing negation may be written thus:

$$\neg \text{ introduction } \frac{\Gamma, A \vdash \{\}}{\Gamma \vdash \neg A} \qquad \neg \text{ elimination } \frac{\Gamma, \neg A, \Delta \vdash A}{\Gamma, \neg A, \Delta \vdash \{\}}$$

where Γ, Δ are sequences of formulæ, A is a formula, and $\{\}$ is contradiction.

The point about these rules is that they operate only on conjectures and hypotheses whose negation is the outermost operator. The result of this is that the negated formulæ (A in the rules above) must be evaluated by rewriting or in some other way *before* the operations involving the negation itself. Further, the negation is not reflected into the synthesised program by explicit introduction of ¬, but by switching the boolean value associated with the formula from true to false or *vice versa*. An example will help here.

3.4 Example: notmember/2 (base case)

[10] gives the example of negating the member/2 predicate. This example serves nicely both here and in Section 4 so I use it too. We start off with the specification conjecture:

$$\vdash \forall n.\forall l.\partial \neg \text{member}(n, l)$$

We also need the definition of member/2 (which is logically equivalent to the completion of the more familiar Horn clause member/2 definition):

$$\forall x.\neg \text{member}(x, [\,]) \tag{1}$$

$$\forall x.\forall h.\forall t.\text{member}(x, [h|t]) \leftrightarrow x = h \lor \text{member}(x, t) \tag{2}$$

The proof is by induction on l. I give the rules of the calculus as I use them. Here, I explain only the base case of the induction; the step case is left for Section 4. I present the full detail of the proof here so that the reader may form an intuition for how the technique works; the reader not needing such an intuition may skip to Section 3.5. Again, I omit types for legibility. The proof is presented in refinement style (*ie* effectively backwards). Note that I use $A(x/y)$ to mean "A with x replaced by y" because the more usual notation is ambiguous with Prolog's and Gödel's list notation.

Proof (Base Case):

$$\vdash \forall n.\forall l.\partial \neg \text{member}(n, l)$$

Apply $\forall$ introduction $\dfrac{\Gamma, v \vdash A}{\Gamma \vdash \forall v.A}$ on n and l, synthesising the program fragment

$$\text{notmember}(n, l \ldots) \leftrightarrow \ldots$$

and leaving us with the subconjecture

$$n, l \vdash \partial\neg member(n, l)$$

Apply list induction $\dfrac{\Gamma, v, \Delta \vdash A\langle[]/v\rangle \quad \Gamma, v, \Delta, v_0, v_1, A\langle v_1/v\rangle \vdash A\langle[v_0|v_1]/v\rangle}{\Gamma, v, \Delta \vdash A}$ on l,

to give two cases: base case (3); and step case (4), which I defer to Section 4.4:

$$n, l \vdash \partial\neg member(n, []) \tag{3}$$

$$n, l, v_0, v_1, \partial\neg member(n, v_1) \vdash \partial\neg member(n, [v_0|v_1]) \tag{4}$$

and with the following constructed program fragment:

$$notmember(n, l, \ldots) \leftrightarrow notmember_l(n, l, \ldots)$$
$$notmember_l(n, l, \ldots) \leftrightarrow n = [] \wedge \ldots \vee \exists v_1.\exists v_0.y = [v_0|v_1] \wedge \ldots$$

The auxiliary predicate, $notmember_l$, is introduced as a result of the induction rule application; auxiliaries are required in the synthesis of recursive predicates, because, in the absence of explicit higher-order induction terms (such as are used in type theory), we need names by which to refer to them when we come to the recursive call.

Next, on subconjecture (3), above, rewrite using definition (1) to give the subconjecture:

$$n, l, \forall x.\neg member(x, []) \vdash \partial\neg member(n, [])$$

Apply $\forall$ elimination $\dfrac{\Gamma, v_0, \forall v_1.A, \Delta, A\langle v_0/v_1\rangle \vdash B}{\Gamma, v_0, \forall v_1.A, \Delta \vdash B}$ with n on the lemma to give:

$$n, l, \forall x.\neg member(x, []), \neg member(n, []) \vdash \partial\neg member(n, [])$$

Apply ∂_{true} introduction $\dfrac{\Gamma \vdash A}{\Gamma \vdash \partial A}$ giving subconjecture

$$n, l, \forall x.\neg member(x, []), \neg member(n, []) \vdash \neg member(n, [])$$

and constructed program

$$notmember(n, l) \leftrightarrow notmember_l(n, l)$$
$$notmember_l(n, l) \leftrightarrow n = [] \wedge true \vee \exists v_1.\exists v_0.y = [v_0|v_1] \wedge \ldots$$

Apply axiom $\dfrac{}{\Gamma, A, \Delta \vdash A}$ which completes this branch of the proof.

3.5 How does the method work?

The proof above may be divided into two distinct sections: before the application of ∂ introduction, and after it. These are called the *synthesis* and *verification* parts of the proof. In the synthesis part, many of the rules contribute to the construction of the new program; in the verification part, they show that synthesised program is correct.

Now, recall from Section 3.3 that the rules for negation are defined thus:

$$\neg \text{ introduction } \dfrac{\Gamma, A \vdash \{\}}{\Gamma \vdash \neg A} \qquad\qquad \neg \text{ elimination } \dfrac{\Gamma, \neg A, \Delta \vdash A}{\Gamma, \neg A, \Delta \vdash \{\}}$$

and that I have chosen to give no rule which will allow us to introduce ¬ under ∂. Therefore negation must be left until the last step in the synthesis part of the proof; this is the point at which the "witness" for the decidability of our specification is supplied: either true or false. Appropriate rewrite rules (*eg* de Morgan's Laws) are allowed, so that a conjecture may be rewritten into a suitable form for this to be possible. The choice of true or false determines the polarity of the part of the synthesised predicate corresponding with the current branch of the proof, using the ∂ introduction rules (one of which was used above)

$$\partial_{true} \text{ introduction } \frac{\Gamma \vdash A}{\Gamma \vdash \partial A} \qquad \partial_{false} \text{ introduction } \frac{\Gamma \vdash \neg A}{\Gamma \vdash \partial A}$$

The rule will introduce a ¬ if appropriate, and proof proceeds with the verification that the "witness" was the correct one. The "witness", and not the ¬, will appear in in the synthesised program, controlling its success or failure in the way we want — this is why we need two rules to introduce one operator.

Because of the restrictions on proof rules for negation explained above, it is impossible to introduce a negation into a synthesised program by application of a proof rule. However, it might still be possible to do so by introduction of lemmas, or by cutting in generalisations, and so on, if this were to involve insertion of part of a synthesised program from "outside" a given proof.

To plug this loophole, we require that any formula introduced into the system and used in program construction, unless it is a formula decidable by first order unification without reference to external definitions (*eg* an equality between canonical terms), must be elaborated in the above style, with synthesis and verification proof. This approach is enforced by the proof system and cannot be circumvented. Since the proof system is restricted to handling negation in the way demonstrated above, it is impossible for negations other than ≠ (between canonical terms and/or variables) to appear. These goals are acceptable, as we are able to deal with them by delaying (and/or by viewing them as constraints) in any reasonable logic programming language.

Finally, note that the initial conjecture of the above example need not be the top level conjecture of a theorem, but may be produced by prior application of rules to a more complicated conjecture. Similarly, it is possible in many circumstances partially to evaluate such conjectures as parts of larger formulæ; this generally produces interleaving as in [1], which I discuss in [11].

4   Automatic Generation of Delay Declarations

4.1   More General Delay Declarations

Having suggested how we may go about using the built-in delay capability of (*eg*) Gödel, it is now appropriate to ask how we might go about extending the idea to generating our own delay declarations.

The idea is that we want to prevent unbounded recursion in our synthesised programs by preventing predicates from being called before the arguments which control their recursion are sufficiently instantiated so to do.

The approach I will propose in this section is closely related to that suggested in [10, p26ff], in that it uses meta-knowledge about the recursive structure of programs and data-types to infer which arguments to a predicate are significant in controlling recursion. In my approach, however, because of the extra meta-knowledge contained in a Whelk proof, we can be definite about which the significant arguments are, and we can reach this conclusion much more easily and reliably, as a side-effect of the proof process, requiring no *post hoc* analysis.

4.2 Detecting Recursive Arguments

Naish's approach to detecting recursive arguments in a predicate is based on *post hoc* analysis of an existing program. This carries with it all the implications of any technique founded on the same precepts: even if the recursive structure is obvious to the informed human reader, an automated analysis may be laborious and difficult. In arbitrary programs, the significant arguments may be very hard to spot, even for experienced programmers. Naish presents a compact algorithm to carry out the process, but acknowledges that its worst case complexity is exponential with the size of the input (with the comment that this seems in practice not to be significant, and that linear behaviour is the norm; other authors have more efficient algorithms based on the same idea).

Naish's summary of the algorithm runs thus ([10, p34]):

```
for-each pair L, of unifiable clause heads and recursive calls do
    if the head is as general as the call then
        terminate with failure
    else
        for-each argument I, less general in the head do
            add a wait declaration to wait group L,
              with 0 in argument I and 1 in all other arguments
        end-for
    end-if
end-for
Allwaits = { W | W is the intersection of one wait from each group }
Waits = { W | W ∈ Allwaits ∧ ∀V. V ∈ Allwaits → W ⊄ V }
```

Waits is the value we want here. It is a set of *wait declarations*. A *wait declaration* is a specification of which arguments must be non-variable for a given predicate to be called. For example:

$$?- \text{wait } p(0,1,1).$$

states that the predicate p/3 must only be executed when its first argument is at least partly instantiated. The algorithm generates a number of wait declarations for each predicate in the program over which it works - some of these can usually be discarded because they are subsumed by others. When applied to an appropriate logic programming language, the wait declarations change the order of execution, delaying the analysed predicates until their significant arguments are non-variable.

Generation of these declarations is based on the structurally recursive datatypes in the arguments to the program being analysed: this can be seen in the inner **for-each** loop – the instantiation of the clause head and the recursive call are compared and it is required that the head argument be *less general* – that is, *more* instantiated. Thus, if the predicate is called with this argument instantiated, the data is broken down by unification; given a well-founded recursive datatype in the argument position, this gives well-founded recursion in the predicate. For example, list destruction involves an argument in the head of the form [H|T], and a corresponding recursive argument of the form T. Note, incidentally, that Naish's algorithm does not require that the recursive call be on the same variable as named in the head variable – *ie* in this example, the variable need not be T; any variable will do. This causes over-generality in the loop checking and thence over-caution in the wait declarations — sometimes perfectly executable programs are delayed so much that they flounder.

Naish's algorithm cannot deal with recursive datatypes whose constructors are not explicit in a program (*eg* integers); nor is mutual recursion apparently covered (unless this is included in the "recursive calls" in the first **for-each** of the algorithm, which would be begging hard questions).

Naish demonstrates his algorithm working with two quite hard examples: n-queens, and a term-ordering predicate. It works well in these restricted cases.

4.3 Delay Generation from Inductive Proof Structure

While the technique I propose is very similar to Naish's in its theoretical basis, in terms of execution it is fundamentally different. As I mentioned in Section 2, inductive proof and the construction of recursive programs in Whelk correspond one-to-one — if we prove a synthesis conjecture by induction, we necessarily get a recursive program.

Now, "recursive" datatypes are defined inductively. Thus, they are known *a priori* to be well-founded, and so can be used to control recursion in the same way as they provide a basis for induction. What is more, when we choose a particular variable as the induction variable (as with l in Section 3), we always construct a corresponding argument position in the synthesised program — as a direct result of applying the induction rule. Therefore, it is trivial to generate a wait declaration (or, better, a more general *delay* declaration as in Gödel), for that argument.

4.4 Example: notmember/2 (step case)

I now return to the notmember/2 example, starting from where we left off: conjecture (4) in Section 3.4. I will use the definition of member/2 given in Section 3 and an axiom about the decidability of equality:

$$\vdash \forall x. \forall y. x = y \lor \neg x = y \tag{5}$$

Recall that the proof of the base case (from Section 3.4) has given this constructed program fragment:

$$\text{notmember}(n, l) \leftrightarrow \text{notmember}_l(n, l)$$
$$\text{notmember}_l(n, l) \leftrightarrow n = [] \wedge \text{true} \vee \exists v_1.\exists v_0.y = [v_0|v_1] \wedge \ldots$$

We reached this stage by application of induction on l, and proof of the resulting base case. Because our synthesised program arises from a proof by induction on l, its recursion is necessarily controlled by any argument(s) corresponding with l. Therefore, we can generate a simple delay declaration (here in Gödel):

DELAY Notmember(n,l) UNTIL NONVAR(l)

The proof now proceeds as follows. Again, I present the full detail of the proof; the reader not interested in that detail should skip to Section 4.5.

Proof (step case):

$$n, l, v_0, v_1, \partial\neg\text{member}(n, v_1) \vdash \partial\neg\text{member}(n, [v_0|v_1])$$

Rewrite according to definition (2):

$$n, l, v_0, v_1, \partial\neg\text{member}(n, v_1) \vdash \partial(\neg(n = v_0 \vee \text{member}(n, v_1)))$$

Rewrite under de Morgan law:

$$n, l, v_0, v_1, \partial\neg\text{member}(n, v_1) \vdash \partial(\neg n = v_0 \wedge \neg\text{member}(n, v_1))$$

Apply $\wedge$ introduction under ∂ $\dfrac{\Gamma \vdash \partial C \qquad \Gamma \vdash \partial D}{\Gamma \vdash \partial(C \wedge D)}$ giving subconjectures

$$n, l, v_0, v_1, \partial\neg\text{member}(n, v_1) \vdash \partial\neg n = v_0 \tag{6}$$
$$n, l, v_0, v_1, \partial\neg\text{member}(n, v_1) \vdash \partial\neg\text{member}(n, v_1) \tag{7}$$

and constructed program fragment

$$\text{notmember}(n, l) \leftrightarrow \text{notmember}_l(n, l)$$
$$\text{notmember}_l(n, l) \leftrightarrow n = [] \wedge \text{true} \vee \exists v_1.\exists v_0.y = [v_0|v_1] \wedge \ldots \wedge \ldots$$

On subconjecture (6) (omitting the induction hypothesis, which we do not need here), introduce decidability axiom (5):

$$n, l, v_0, v_1, \ldots, \forall x.\forall y.x = y \vee \neg x = y \vdash \partial\neg n = v_0$$

Substitute values by $\forall$ elimination, as before:

$$n, l, v_0, v_1, \forall x.\forall y.x = y \vee \neg x = y, n = v_0 \vee \neg n = v_0 \vdash \partial\neg n = v_0$$

Apply $\vee$ elimination $\dfrac{\Gamma, A \vee B, \Delta, A \vdash C \qquad \Gamma, A \vee B, \Delta, B \vdash C}{\Gamma, A \vee B, \Delta \vdash C}$ to give subconjectures

$$n, l, v_0, v_1, \forall x.\forall y.x = y \vee \neg x = y, n = v_0 \vee \neg n = v_0, n = v_0 \vdash \partial\neg n = v_0 \tag{8}$$
$$n, l, v_0, v_1, \forall x.\forall y.x = y \vee \neg x = y, n = v_0 \vee \neg n = v_0, \neg n = v_0 \vdash \partial\neg n = v_0 \tag{9}$$

and constructed program fragment

$$\text{notmember}(n, l) \leftrightarrow \text{notmember}_1(n, l)$$
$$\text{notmember}_1(n, l) \leftrightarrow n = [\,] \wedge \text{true} \vee$$
$$\exists v_1 . \exists v_0 . y = [v_0 | v_1] \wedge$$
$$(n = v_0 \wedge \ldots \vee \neg n = v_0 \wedge \ldots) \wedge \ldots$$

Apply ∂ introduction, as before, with false in (8) and true in (9):

$$n, l, v_0, v_1, \forall x . \forall y . x = y \vee \neg x = y, n = v_0 \vee \neg n = v_0, n = v_0 \vdash \neg \neg n = v_0$$
$$n, l, v_0, v_1, \forall x . \forall y . x = y \vee \neg x = y, n = v_0 \vee \neg n = v_0, \neg n = v_0 \vdash \neg n = v_0$$

The constructed program fragment is now:

$$\text{notmember}(n, l) \leftrightarrow \text{notmember}_1(n, l)$$
$$\text{notmember}_1(n, l) \leftrightarrow n = [\,] \wedge \text{true} \vee$$
$$\exists v_1 . \exists v_0 . y = [v_0 | v_1] \wedge$$
$$(n = v_0 \wedge \text{false} \vee \neg n = v_0 \wedge \text{true}) \wedge \ldots$$

This part of the program is now complete; the rest of this branch of the proof is trivial verification, using rules already demonstrated above.

Finally, to subconjecture (7), above:

$$n, l, v_0, v_1, \partial \neg \text{member}(n, v_1) \vdash \partial \neg \text{member}(n, v_1)$$

apply axiom as before. Note that in this case, the application of the induction rule causes an appropriate recursive call to be associated with the induction hypothesis. This is now inserted into the constructed program, to give the finished article:

$$\text{notmember}(n, l) \leftrightarrow \text{notmember}_1(n, l)$$
$$\text{notmember}_1(n, l) \leftrightarrow n = [\,] \wedge \text{true} \vee$$
$$\exists v_1 . \exists v_0 . y = [v_0 | v_1] \wedge$$
$$(n = v_0 \wedge \text{false} \vee \neg n = v_0 \wedge \text{true}) \wedge$$
$$\text{notmember}_1(n, v_1)$$

4.5 The Notmember/2 Module

The proof gives rise, automatically, to the Gödel module shown in Figure 1. Consider the behaviour of the program called with the goal

```
[] <- Notmember(0,[0|t]).
```

We wish this to fail, and indeed it does so. Now, suppose we give the following goal, which we would like explicitly to flounder, with an error message:

```
[] <- Notmember( 0, [h|t] ).
```

```
MODULE Notmember.

IMPORT Lists.
IMPORT Naturals.

PREDICATE Notmember: Natural * List(Natural).

Notmember(n,l)    <- Notmember_1(n,l)

PREDICATE Notmember_1: Natural * List(Natural).

DELAY Notmember(n,l) UNTIL NONVAR(l).

Notmember_1(n,l)  <- l=[] & true \/
                     Some [v1] Some [v0] y=[v0|v1] & ~n=v0 &
                                                Notmember_1(n,v1)
```

Fig. 1. The Gödel Module for Notmember/2

Our delay declaration admits this, as far as the first call to Notmember_1/2, because the second argument is non-variable. However, at this point there will be a call to =/2 with one argument uninstantiated. Then, the default behaviour is to delay, so the recursive call is made to Notmember/2. This time, however, the second argument is a fully uninstantiated variable, so this call too is delayed. Therefore, the whole computation flounders, explicitly, as we would wish, and an error is reported.

4.6 More Subtle Control Generation

While this approach will work well for many cases, some similar problems are harder. Let us return now to the append/3 definition I gave in Section 2. I gave the delay declaration

DELAY Append(x,y,z) UNTIL NONVAR(x) \/ NONVAR(z).

requiring that *either* the first *or* the last argument be instantiated before the predicate is executed.

The append/3 predicate is different from member/2 in that its recursion is controlled by either of two variables – x or z in the declaration above. How can we spot this? Simply enough, it arises from the use of a more powerful induction scheme: simultaneous primitive recursion on two variables, as defined by the following rule (writing the preconditions of the rule vertically):

$$\frac{\begin{array}{c}\Gamma, u, v, \Delta \vdash A\langle[]/u\rangle \\ \Gamma, u, v, \Delta, u_0, u_1 \vdash A\langle[u_0|u_1]/u\rangle\langle[]/v\rangle \\ \Gamma, u, v, \Delta, u_0, v_0, u_1, v_1, A\langle u_1/u\rangle\langle v_1/v\rangle \vdash A\langle[u_0|u_1]/u\rangle\langle[v_0|v_1]\rangle/v\rangle \end{array}}{\Gamma, u, v, \Delta \vdash A}$$

This scheme is equivalent to primitive induction on x, followed by primitive induction on y in x's step case. Use of such a scheme begs the question: how do we know which scheme to use? This is outside the scope of this paper, but is addressed in [4].

There follows a sketch of the proof. Note that we are synthesising the program from an equivalent definition of append/3 here. This is not a failure of the technique — it is impossible to specify the simple example program in any other logical terms. We gain from the proof process because the delay declarations are generated for us, and because our specification is shown to be realisable.

4.7 Example: append/3

Lemmas:

$$\vdash \forall x. app([], x) = x \tag{10}$$

$$\vdash \forall h. \forall t. \forall y. app([h|t], y) = [h|app(t, y)] \tag{11}$$

$$\vdash \forall x. \forall y. x = y \lor \neg x = y \tag{12}$$

$$\vdash \forall h. \forall t. \neg[h|t] = [] \tag{13}$$

$$\vdash \forall h_1. \forall h_2. \forall t_1. \forall t_2. [h_1|t_1] = [h_2|t_2] \leftrightarrow h_1 = h_2 \land t_1 = t_2 \tag{14}$$

Proof:

$$\vdash \forall x. \forall y. \forall z. \partial(app(x, y) = z)$$

Introduce x, y and z. Apply simultaneous primitive induction on x and z:

$$x, y, z \vdash \partial(app([], y) = z) \tag{15}$$

$$x, y, z, x_0, x_1 \vdash \partial(app([x_0|x_1], y) = []) \tag{16}$$

$$x, y, z, x_0, x_1, z_0, z_1, \partial(app(x_1, y) = z_1) \vdash \partial(app([x_0|x_1], y) = [z_0|z_1]) \tag{17}$$

In (15), and use (12) above to case split on equality:

$$x, y, z, y = z \vdash \partial(app([], y) = z) \tag{18}$$

$$x, y, z, \neg y = z \vdash \partial(app([], y) = z) \tag{19}$$

(18) is completed by ∂_{true} introduction, (19) by ∂_{false}, verified using (10).

For (16), rewrite under = using (11); then ∂_{false} introduction can be verified by (13). In (17), the step case, rewrite the left hand side of the equation in the conjecture according to (11):

$$x, y, z, x_0, x_1, z_0, z_1, \partial(app(x_1, y) = z_1) \vdash \partial([x_0|app(x_1, y)] = [z_0|z_1])$$

Next, rewrite using (14) to give:

$$x, y, z, x_0, x_1, z_0, z_1, \partial(app(x_1, y) = z_1) \vdash \partial(x_0 = z_0 \land app(x_1, y) = z_1)$$

Finally, $\land$ introduction gives us two subgoals:

$$x, y, z, x_0, x_1, z_0, z_1, \partial(app(x_1, y) = z_1) \vdash \partial(x_0 = z_0) \tag{20}$$

$$x, y, z, x_0, x_1, z_0, z_1, \partial(app(x_1, y) = z_1) \vdash \partial(app(x_1, y) = z_1) \tag{21}$$

(20) is solved with a case split on (12), as for conjecture (15). (21) is identical to the induction hypothesis, and so we have finished the whole proof. The finished program then looks like this:

$$append(x, y, z) \leftrightarrow append_l(x, y, z)$$
$$append_l(x, y, z) \leftrightarrow x = [] \land (y = z \land true \lor \neg y = z \land false) \lor$$
$$\exists v_0.\exists v_1.x = [v_0|v_1] \land$$
$$(z = [] \land false \lor$$
$$\exists v_2.\exists v_3.z = [v_2|v_3] \land$$
$$(v_0 = v_2 \land true \lor \neg v_0 = v_2 \land false) \land$$
$$append_l(v_1, y, v_3)$$

Since we performed the proof by simultaneous induction on x and z, we can generate the delay declaration we need in the same way as before, *dis*joining the requirements that each argument be non-variable since clearly *either* induction variable, and not necessarily *both* will be enough to control the recursion. Thus, the general form of delay declarations generated by Whelk will be disjunctive. *Con*joined delays will never arise, because of the form of the programs; each argument is represented by exactly one variable, and each induction corresponds with the introduction of a new recursive predicate. Since it is not possible to apply two proof rules at once, only disjunctive delays can be generated.

Finally, it is worth mentioning that the technique will still work even if the proof is elaborated by two separate applications of ordinary primitive induction on x and then z – though the program produced will be slightly uglier. This, however, is not important, as we still have the specification to reason with.

5 Conclusion and Further Work

In this paper, I have explained how an existing technique may be applied in a new way within my program synthesis system to allow the automatic generation of delay declarations, thus allowing my synthesised programs to benefit from facilities for control within modern logic programming languages.

The new approach to the technique relies on meta-knowledge about the synthesised program encorporated in the synthesis proof, and not on *post hoc* analysis. Thus, the information it has to work with is complete and correct, and need never be guessed by analysing the syntactic form of a program. Therefore, the technique applies to datatypes which do not have explicit constructor functions, unlike earlier approaches. It is trivial to generate delay declarations for many recursive programs synthesised by the proofs as programs technique, modulo the question of choosing the right induction scheme which is addressed elsewhere.

This approach is possible only because of the initial decision to reason with specifications instead of programs. It is just one of several areas (see *eg* [8]) where this approach facilitates program construction and manipulation.

The next step in this work will be to build these ideas into the existing Whelk system. This begs a question of the level at which such reasoning should take place: either at the object level, in Whelk, or at the meta-level, in the CLAM proof planner, with which we will automate the search for synthesis proofs over the next few years. It seems likely that the latter option is best, in which case a full implementation will be some time away.

6 Acknowledgements

This work is funded by ESPRIT Basic Research Action #3012, "Computational Logic". Thanks to Alan Bundy and the DREAMers for their interest and support; and to John Lloyd and Wolfgang Bibel for raising the problems addressed here.

References

1. M. Bruynooghe, D. de Schreye, and B. Krekels. Compiling control. *Journal of Logic Programming*, pages 135–162, 1989.

2. A. Bundy, A. Smaill, and J. Hesketh. Turning eureka steps into calculations in automatic program synthesis. In S.L.H. Clarke, editor, *Proceedings of UK IT 90*, pages 221–6, 1990. Also available from Edinburgh as DAI Research Paper 448.

3. A. Bundy, A. Smaill, and G. A. Wiggins. The synthesis of logic programs from inductive proofs. In J. Lloyd, editor, *Computational Logic*, pages 135–149. Springer-Verlag, 1990. Esprit Basic Research Series. Also available from Edinburgh as DAI Research Paper 501.

4. A. Bundy, A. Stevens, F. van Harmelen, A. Ireland, and A. Smaill. Rippling: A heuristic for guiding inductive proofs. Research Paper 567, Dept. of Artificial Intelligence, Edinburgh, 1991. To appear in Artificial Intelligence.

5. A. Bundy, F. van Harmelen, J. Hesketh, and A. Smaill. Experiments with proof plans for induction. *Journal of Automated Reasoning*, 7:303–324, 1991. Earlier version available from Edinburgh as DAI Research Paper No 413.

6. R.L. Constable. Programs as proofs. Technical Report TR 82-532, Dept. of Computer Science, Cornell University, November 1982.

7. P. Hill and J. Lloyd. The Gödel Report. Technical Report TR-91-02, Department of Computer Science, University of Bristol, March 1991. Revised March 1992.

8. P. Madden. *Automated Program Transformation Through Proof Transformation*. PhD thesis, University of Edinburgh, 1991.

9. Per Martin-Löf. *Intuitionistic Type Theory*. Bibliopolis, Naples, 1984. Notes by Giovanni Sambin of a series of lectures given in Padua, June 1980.

10. L. Naish. *Negation and control in Prolog*, volume 238 of *Lecture notes in Computer Science*. Springer Verlag, 1986.

11. G. A. Wiggins. The improvement of prolog program efficiency by compiling control: A proof-theoretic view. In *Proceedings of the Second International Workshop on Meta-programming in Logic*, Leuven, Belgium, April 1990. Also available from Edinburgh as DAI Research Paper No. 455.

12. G. A. Wiggins. Synthesis and transformation of logic programs in the Whelk proof development system. In K. R. Apt, editor, *Proceedings of JICSLP-92*, 1992.

13. G. A. Wiggins, A. Bundy, H. C. Kraan, and J. Hesketh. Synthesis and transformation of logic programs through constructive, inductive proof. In K-K. Lau and T. Clement, editors, *Proceedings of LoPSTr-91*, pages 27–45. Springer Verlag, 1991. Workshops in Computing Series.

Transforming Normal Programs by Replacement *

Annalisa Bossi, Nicoletta Cocco, Sandro Etalle

Dipartimento di Matematica Pura ed Applicata,
Università di Padova,
Via Belzoni 7, 35131 Padova, Italy.
email: bossi,cocco,etalle@pdmat1.unipd.it
fax: ++39-49-8758596

Abstract. The replacement transformation operation, already defined in [28], is studied wrt normal programs. We give applicability conditions able to ensure the correctness of the operation wrt Fitting's and Kunen's semantics. We show how replacement can mimic other transformation operations such as thinning, fattening and folding, thus producing applicability conditions for them too. Furthermore we characterize a transformation sequence for which the preservation of Fitting's and Kunen's semantics is ensured.

1 Introduction

Program transformation is now a widely accepted technique for the systematic development of correct and efficient programs, see [6,12,15,28,13,18,8,21,22,4] to quote just a few papers on this topic. A main concern when transforming a program is the preservation of its meaning. In order to express the meaning of a program we need to choose a *semantics*. Unfortunately, as regards logic programs, on one hand there is no general agreement on which semantics is the best one, on the other hand a transformation can be correct with respect to one semantics and incorrect with respect to another one. For instance, in the program

$$\{ p(X) \leftarrow q(X), q(X). \qquad q([a, Y]). \qquad q([Z, b]). \}$$

the duplicated atom $q(X)$ in the first clause is superfluous when considering the least Herbrand model semantics and then it can be safely deleted from the body of the clause. The same operation is not safe when the computed answers semantics is considered [3]: in fact the answer substitution $X = [a, b]$ would be missed in the transformed program. The first papers on logic programs transformation considered definite programs and the least Herbrand model semantics. *Normal programs* have been taken into consideration only recently [19,11,25,24,23] together with suitable applicability conditions for guaranteeing the preservation of the meaning of the program. For normal programs the problem of choosing a sensible semantics is in fact more serious, given the logical and computational problems related to the introduction of negation and the amount of semantic proposals (see [26,27] for an almost complete panorama).

In this paper we concentrate on one transformation operation for normal programs: *the replacement*. This operation has been introduced for definite programs

* This work has been partially supported by "Progetto Finalizzato Sistemi Informatici e Calcolo Parallelo" of CNR under grant n. 89.00026.69

by Tamaki and Sato in [28] and after that it has been rather neglected by people working on program transformation apart from Sato himself [23], Maher [19] and Gardner and Shepherdson [11]. It consists in substituting part of a clause body with an equivalent conjunction of literals. It is a very general transformation able to mimic other operations, such as thinning, fattening [3] and folding, which can be seen as particular instances of replacement. We have defined *applicability conditions* able to guarantee the correct application of replacement with respect to *Fitting's and Kunen's semantics for normal programs*. We choose these semantics for several reasons. Three-valued logic is, in our opinion, very reasonable for dealing with normal programs, since it can represent the fact that some query produce a successful answer, some fail, some other produce no answer because of circularity in its derivation. Both the semantics are defined by means of a monotonous immediate consequence operator and this is exploited in our applicability conditions. Moreover, Kunen's semantics corresponds to the top-down evaluation procedure given by SLDNF-resolution when this is complete (for allowed programs and queries [26,2]), while Fitting's semantics corresponds to a bottom-up evaluation procedure, when the program is stratified [2]. Our applicability conditions for replacement are undecidable in general, but other decidable conditions can be derived for special cases. In the paper we consider two such cases when replacement mimics folding.

Structure of the paper: the next section briefly recalls the main definitions related to Fitting's and Kunen's semantics. The definitions of *dependency level of a literal wrt a clause* is also given. In section 3 the definitions of *equivalence* and *semantic delay of a conjunction of literals wrt another one* are given and these concepts are used to define the applicability conditions for replacement in normal programs. Section 4 shows how thinning and fattening can be interpreted as special cases of replacement, thus yielding, as a consequence, conditions for a safe application of these operations to normal programs. Two different definitions of folding are also considered and the corresponding applicability conditions are derived from the ones of replacement. These conditions are easily checkable either syntactically or by considering the transformation history of the program. A short concluding section follows.

2 Preliminaries

We assume that the reader is familiar with the basic concepts of logic programming; throughout the paper we use the standard terminology of [17] and [1]. We consider *normal programs*, that is finite collections of *normal rules*, $A \leftarrow L_1, \ldots, L_m$. where A is an atom and $L_1, \ldots, L_m$ are literals. Let P be a normal program, B_P denotes the Herbrand base and *ground(P)* the set of ground instances of rules of P.

2.1 Fitting's and Kunen's semantics for normal programs

We briefly recall here the definitions of Fitting's and Kunen's semantics, for more details see [10], [16] and [27]. Both semantics are based on Kleene's three-valued logic [14] where the truth values are *true, false* and *undefined*. The usual logical connectives have value *true* (or *false*) when they have that value in ordinary two-valued logic for all possible replacements of *undefined* by *true* or *false*, otherwise they have the value *undefined*.

The usual Clark's completion definition, $Comp(P)$, [7] is extended to three-valued logic by replacing $\leftrightarrow$, in the completed definitions of the predicates, with $\cong_3$, Lukasiewicz's operator of "having the same truth value". This saves $Comp(P)$ from the inconsistency that it can have in two-valued logic. For example the program $P = \{p \leftarrow \neg p.\}$ has $Comp(P) = \{p \cong_3 \neg p\}$ which has a model with p *undefined*.

Definition 1 (three-valued interpretation and model). *A three-valued (or partial) interpretation I is an ordered couple, (T, F), of disjoint sets of ground atoms. The atoms in T (resp. F) are considered to be true (resp. false) in I.*
T is the positive part of I and is referred as I^+; equivalently F is denoted by I^-. Atoms which do not appear in either set are considered to be *undefined*.
A three-valued model is a three-valued interpretation which is also a model of $Comp(P)$.

Let I and J be two partial interpretations. $I \subseteq J$ iff $I^+ \subseteq J^+$ and $I^- \subseteq J^-$.
$I \models_{HU} \Psi$ indicates that the first order formula Ψ is *true* in the interpretation I, when the underlying universe is the Herbrand Universe.

Definition 2 (Fitting's operator and Fitting's semantics). Let P be a normal program and let (T, F) be a partial interpretation. *Fitting's three-valued immediate consequence operator, Φ_P, is defined as* $\Phi_P(T, F) = (T_1, F_1)$ *where*
$T_1 = \{A \mid$ there exists a clause $A \leftarrow L_1, \ldots, L_m.$ in $ground(P)$ and
$\qquad L_1, \ldots, L_m$ are *true* in $(T, F)\}$;
$F_1 = \{A \mid$ for all clauses $A \leftarrow L_1, \ldots, L_m.$ in $ground(P)$,
$\qquad$ the conjunction of literals $L_1, \ldots, L_m$ is *false* in $(T, F)\}$.
Fitting's three valued model of P, $Fit(P)$, is the least fixed point of the associated operator Φ_P.

We adopt the notation:
$\Phi_P^0(T, F) = (T, F)$;
$\Phi_P^{\alpha+1}(T, F) = \Phi_P(\Phi_P^\alpha(T, F))$;
$\Phi_P^\alpha(T, F) = \cup_{\delta < \alpha} \Phi_P^\delta(T, F)$, when α is a limit ordinal.
When the argument is omitted, we assume it to be $(\emptyset, \emptyset)$: $\Phi_P^\alpha = \Phi_P^\alpha(\emptyset, \emptyset)$. It follows directly from the definition that Φ_P is a monotonic operator, hence it converges to its least fixed point, which is given by Φ_P^α, for some ordinal α; then $Fit(P) = \Phi_P^\alpha$. Φ_P being monotone but not continuous, α could be greater than ω. From definition 2 we have the following.

Remark. If a ground atom A is *true* (resp. *false*) in Φ_P^α, where α is a limit ordinal, then there exists a successor ordinal $\beta < \alpha$ such that A is *true* (resp. *false*) in Φ_P^β.

$Fit(P)$ is the minimal three-valued Herbrand model and it is equal to the intersection of all three-valued Herbrand models of $Comp(P)$. Kunen instead proposes to consider as normal programs' semantics the intersection of all (not just Herbrand ones) three-valued models of $Comp(P)$.

Definition 3 (Kunen's semantics). Let P be a normal program. Kunen's semantics of P, $Kun(P)$, is the three-valued interpretation resulting from the intersection of all the three-valued models of $Comp(P)$.

Remark.
- $Kun(P) = \Phi_P^\omega$, hence $Kun(P) \subseteq Fit(P)$.
- $Kun(P)$ coincides with the computable part of the Fitting's model. If $Kun(P) \neq Fit(P)$, then $Kun(P)$ is not a model of P.
- $Kun(P)$ is the set of all logical consequences of $Comp(P)$.

2.2 Dependency degree

Let us consider the following normal program:

$$P = \{ \begin{array}{l} c1 : p \leftarrow \neg q, s. \\ c2 : q \leftarrow r. \\ c3 : r. \\ c4 : s \leftarrow q. \end{array} \}$$

The definitions of the atoms p, q, s and r, all depend from clause $c3$. Informally we could say that *the dependency degree of the predicate p over clause c3 is two*, as the shortest derivation path from a clause having head p to $c3$ contains two arcs: the first from $c1$ to $c2$, through the negative literal $\neg q$; the second from $c2$, to $c3$, through the atom r. Similarly, the dependency degree of q and s on $c3$ are respectively one and two and the dependency degree of r on $c3$ is zero. The next definition formalises this intuitive notion. The atom A and the clause cl are assumed to be standardized apart.

Definition 4 (dependency degree). Let P be a program, cl a clause of P and A an atom. *The dependency degree of A (and $\neg A$) on cl, $depen_P(A, cl)$, is*

0 if A unifies with the head of cl;

n+1 if A does not unify with the head of cl and n is the least integer such that there exists a clause $C \leftarrow C_1, \ldots, C_k$. in P, whose head unifies with A via mgu, say, θ, and, for some i, $depen_P(C_i\theta, cl) = n$.

A *is independent from cl when no such n exists.*

The definition can be extended to conjunctions of literals. *The conjunction of literals* $(L_1 \wedge \ldots \wedge L_n)$ *is independent from cl iff all its components are independent from cl; otherwise the dependency degree of* $(L_1 \wedge \ldots \wedge L_n)$ *on cl is equal to the least dependency degree of one of its elements on cl, $depen_P((L_1 \wedge \ldots \wedge L_n), cl) = inf\{depen_P(L_i, cl)$, where $1 \leq i \leq n\}$.

3 Applicability conditions for the replacement operation

The replacement operation has been introduced by Tamaki and Sato in [28] for definite programs. Syntactically it consists in substituting a conjunction, $\tilde{C}$, of literals with another one, $\tilde{D}$, in the body of a clause.

Definition 5 (replacement). Let $c : A \leftarrow \tilde{J}, \tilde{C}, \tilde{H}$. be a clause in a normal program P and $\tilde{D}$ be a conjunction of literals. Replacing $\tilde{C}$ with $\tilde{D}$ in c consists of substituting c' for c, where $c' : A \leftarrow \tilde{J}, \tilde{D}, \tilde{H}$. $replace(P, c, \tilde{C}, \tilde{D}) \stackrel{\text{def}}{=} P\backslash\{c\} \cup \{c'\}$.

Some *applicability conditions* are necessary in order to ensure the preservation of the semantics through the transformation. Such conditions depend on the semantics we associate to the program. In [28] definite programs are considered; the applicability condition requires that $\tilde{C}$ and $\tilde{D}$ are logically equivalent in P and that the size of the smallest proof tree for $\tilde{C}$ is greater or equal to the size of the smallest proof tree for $\tilde{D}$. Gardner and Shepherdson, in [11], give different conditions for preserving procedural (SLDNF) semantics and the declarative one. Such conditions are based on Clark's (two valued) completion of the program. Also Maher, in [19,20], studies replacement wrt Success set, Finite Failure Set, Ground Finite Failure Set and Perfect Model semantics. Sato, in [23], considers also replacement of tautologically equivalent formulas in first order programs. We consider the replacement operation for normal programs and state some applicability conditions which ensure that Fitting's and Kunen's semantics are preserved by the transformation.

We say that the replacement operation is *acceptable* only if it does not change the Herbrand base of the program. From now on, we shall consider only acceptable replacements.

3.1 Replacement wrt Fitting's semantics

We now introduce some new definitions for expressing relations between first order formulas, such as conjunctions of literals, in terms of their semantic properties. They are used for defining general applicability conditions for transformation operations, that is conditions based only on the semantics of the program to be transformed.

Definition 6 (equivalence wrt Fitting's semantics). Let E, F be first order formulas and P be a normal program. We can define an order relation based on Fitting's semantics of P in this way: E *is less defined or equal to F wrt* $Fit(P)$, $E \preceq_{Fit(P)} F$, iff for each ground substitution θ, if $E\theta$ is *true* (resp. *false*) in $Fit(P)$, then $F\theta$ is *true (false)* in $Fit(P)$ as well. F *is equivalent to E in* $Fit(P)$, $F \cong_{Fit(P)} E$, iff $E \preceq_{Fit(P)} F$ and $F \preceq_{Fit(P)} E$.

Note that $F \cong_{Fit(P)} E$ iff $Fit(P) \models_{HU} \forall (F \cong_3 E)$.

Consider now the following definite program.

$$P = \{ \ m(X) \quad \leftarrow \ n(s(X)).$$
$$n(0).$$
$$n(s(X)) \ \leftarrow \ n(X). \quad \}$$

The predicates m and n have exactly the same meaning, but in order to refute the goal $\leftarrow m(s(0))$. we need four resolution steps, while for refuting $\leftarrow n(s(0))$. two steps are sufficient. Each time $\leftarrow n(t)$. has a refutation (or finitely fails) with j resolution steps, $\leftarrow m(t)$. has a refutation (or fails) with k resolution steps, where $k \leq j + 2$. We can formalise this intuitive idea by saying that *the semantic delay of m wrt n is 2*. By transposing this idea into Fitting's semantics, we have that each time $n(t)$ is *true* (or *false*) in Φ_P^j, $m(t)$ is *true* (resp. *false*) in Φ_P^{j+2}.

Definition 7 (delay in Fitting's semantics). Let P be a normal program, E and F be first order formulas such that F is equivalent to E in $Fit(P)$.

The delay of F wrt E in Fitting's semantics is the least ordinal β such that, for each ordinal α and each ground substitution θ:

if $E\theta$ is *true* (resp. *false*) in Φ_P^α, then $F\theta$ is *true* (resp. *false*) in $\Phi_P^{\alpha+\beta}$.

Intuitively, E is *true* in Φ_P^α iff its truth has been proved from scratch in at most α steps. The semantic delay of F wrt E shows how many steps later than E, we determine the truth value of F (at worse).

Example 1. Let P be the following program:

$$P = \{\; p(0). \qquad\qquad\qquad q(0).$$
$$\quad p(s(0)). \qquad\qquad\quad q(s(X)) \leftarrow p(X).$$
$$\quad p(s(s(X))) \leftarrow p(X). \qquad\qquad\qquad\qquad \}$$

p and q both compute natural numbers and $p(X) \cong_{Fit(P)} q(X)$, but while $q(s^k(0))$ is *true* starting from Φ_P^{k+1}, $p(s^k(0))$ is *true* starting from $\Phi_P^{(k/2)+1}$. The delay of $p(X)$ wrt $q(X)$ in $Fit(P)$ is zero, in fact if for some ground term t and ordinal α, $q(t)$ is *true* (resp. *false*) in Φ_P^α, then $p(t)$ is also *true* (resp. *false*) in Φ_P^α. Vice versa, the delay of $q(X)$ wrt $p(X)$ in $Fit(P)$ is ω, in fact there exists no integer $m < \omega$ such that if, for some ground term t and ordinal α, $p(t)$ is *true* (resp. *false*) in Φ_P^α, then $q(t)$ is *true* (resp. *false*) in $\Phi_P^{\alpha+m}$.

When considering Fitting's semantics, our first requirement is the equivalence of $\tilde{C}$ and $\tilde{D}$ wrt $Fit(P)$. It would make no sense to replace $\tilde{C}$ with something which has a different meaning. Unfortunately this is not enough, in fact we need the equivalence to hold also after the transformation. The equivalence can be destroyed when $\tilde{D}$ depends on the modified clause. This is shown by the next example.

Example 2. Let P be the following definite program:

$$P = \{\quad p \leftarrow q.$$
$$cl : q \leftarrow r.$$
$$r. \qquad\qquad \}$$

$Fit(P) = (\{p, q, r\}, \emptyset)$.
p, q and r are all *equivalent* in $Fit(P)$, but if we replace r with p in the body of cl we obtain

$$P' = \{\quad p \leftarrow q.$$
$$cl' : q \leftarrow p.$$
$$r. \qquad\qquad \}$$

which is by no means equivalent to the previous program. In fact $Fit(P') = (\{r\}, \emptyset)$. We have introduced a loop and p and q are no more *true*.
Consider now the following normal program:

$$P = \{\; d : p(X) \leftarrow \neg q(X).$$
$$\quad cl : r \quad\leftarrow \ldots, \neg q(t), \ldots$$
$$\qquad\qquad \ldots \qquad\qquad\qquad \}$$

where d is the only clause defining the predicate symbol p. $p(X)$ and $\neg q(X)$ are *equivalent* in $Fit(P)$. Now, if we replace $\neg q(t)$ with $p(t)$ in cl, we obtain the following program:

$$P' = \{\; d: p(X) \leftarrow \neg q(X).$$
$$cl: r \quad\;\; \leftarrow \ldots, p(t), \ldots$$
$$\ldots \qquad\qquad\qquad\qquad \}$$

which has the same Fitting's semantics as the previous one, that is $Fit(P) = Fit(P')$. This holds even if the definition of p is dependent from cl. The point is that "there is no room for introducing a loop". We can try to clarify further the previous statement: replacing $\neg q(t)$ by $p(t)$ in cl preserves Fitting's semantics of the initial program if

- either p *does not depend on* cl or
- the *dependency level of* p *on* cl (this is how big the loop would be) *is greater or equal to the semantic delay of* $p(X)$ *wrt* $\neg q(X)$ *in* $Fit(P)$ (this is the space where the loop would be introduced).

In our example the *delay* of $p(X)$ wrt $\neg q(X)$ in $Fit(P)$ is one:
$$\Phi_P^\alpha \models_{HU} \neg q(X)\tau \quad \text{iff} \quad \Phi_P^{\alpha+1} \models_{HU} p(X)\tau \text{ and}$$
$$\Phi_P^\alpha \models_{HU} q(X)\tau \quad \text{iff} \quad \Phi_P^{\alpha+1} \models_{HU} \neg p(X)\tau.$$
d is the only clause defining predicate p and $d \neq cl$, then $depen_P(p(X), cl) > 0$, thus satisfying the above conditions.

Theorem 8 (applicability conditions wrt Fitting's semantics). *Let P be a normal program, $cl : A \leftarrow \tilde{J}, \tilde{C}, \tilde{H}.$ be a clause of P where $\tilde{J}, \tilde{C}, \tilde{H}$ are conjunctions of literals. Let $\tilde{D}$ be another conjunction of literals and P' be the program resulting from the replacement of $\tilde{C}$ with $\tilde{D}$ in cl, $P' = P\backslash\{cl\} \cup \{cl' : A \leftarrow \tilde{J}, \tilde{D}, \tilde{H}.\}$. Let X be the set of variables of $\tilde{C}$ local wrt cl and not in $\tilde{D}$,*
$$X = var(\tilde{C})\backslash(var(\tilde{D}) \cup var(A) \cup var(\tilde{H}) \cup var(\tilde{J}));$$
Y be the set of variables of $\tilde{D}$ local wrt cl' and not in $\tilde{C}$,
$$Y = var(\tilde{D})\backslash(var(\tilde{C}) \cup var(A) \cup var(\tilde{H}) \cup var(\tilde{J})).$$
If $\exists X \tilde{C}$ is equivalent to $\exists Y \tilde{D}$ in $Fit(P)$ and one of the following two conditions holds:

1. *$\tilde{D}$ is independent from cl;*
2. *the dependency degree of $\tilde{D}$ on cl is greater or equal to the delay in $Fit(P)$ of $(\exists Y \tilde{D})$ wrt $(\exists X \tilde{C})$;*

then $Fit(P) = Fit(P')$.

The theorem is a direct consequence of the following two lemmas.

Lemma 9. *If $\exists Y \tilde{D} \preceq_{Fit(P)} \exists X \tilde{C}$, then $Fit(P) \supseteq Fit(P')$.*

Lemma 10. *Let $Z = (var(\tilde{C}) \cup var(\tilde{D})) \setminus (X \cup Y)$ be the set of non-local variables in $\tilde{C}$ and $\tilde{D}$. If*

(i) *$\exists X \tilde{C} \preceq_{Fit(P)} \exists Y \tilde{D}$;*
(ii) *for each ground substitution σ having Z as domain, if $\tilde{D}\sigma$ is dependent on cl, then the dependency degree of $\tilde{D}\sigma$ on cl is greater or equal to the delay in $Fit(P)$ of $(\exists Y \tilde{D})\sigma$ wrt $(\exists X \tilde{C})\sigma$,*

then $Fit(P) \subseteq Fit(P')$.

The proofs of the lemmata, in the simpler case in which $\tilde{C}$ and $\tilde{D}$ are ground literals are given in [5]. The general case is proved in [9]. The proof strictly depends on the fact that the semantics is defined by means of an immediate consequence operator.

3.2 Replacement wrt Kunen's semantics

When considering Kunen's semantics we have to give a slightly different notion of equivalence between conjunctions of literals. This is illustrated by the following example.

Example 3. Let us consider the following program:

$$P = \{ \quad p(s(X)) \leftarrow p(X).$$
$$cl : r(b) \quad \leftarrow f.$$
$$q \quad \leftarrow f. \quad \}$$

In $Fit(P)$, both f and each ground instance of $p(X)$ are *false*. Hence, from definition 6, $p(X) \cong_{Fit(P)} f$. Since p is independent from cl, we can replace f with $p(X)$ in the body of cl and, by theorem 8, the resulting program P' has the same Fitting's semantics of the original one.

$$P' = \{ \quad p(s(X)) \leftarrow p(X).$$
$$cl' : r(b) \quad \leftarrow p(X).$$
$$q \quad \leftarrow f. \quad \}$$

But such replacement operation is not safe wrt Kunen's semantics: $r(b)$ is *false* in $Kun(P)$, while it is *undefined* in $Kun(P')$. In fact even if all ground instances of $p(X)$ are *false* in $Kun(P)$, just like (all ground instances of) f, $p(X)$ and f cannot be considered *equivalent* wrt Kunen's semantics.

In Kunen's semantics we consider only equivalence below ω and define the semantic delay consequently.

Definition 11 (equivalence wrt Kunen's semantics). Let E, F be first order formulas and P be a normal program. We can define an order relation based on Kunen's semantics of P in this way: E *is less defined or equal to F wrt $Kun(P)$*, $E \preceq_{Kun(P)} F$, iff for each integer j there exists an integer k such that for each ground substitution θ,
if $E\theta$ is *true* (resp. *false*) in Φ_P^j, then $F\theta$ is *true* (*false*) in Φ_P^k.
F *is equivalent to E in in $Kun(P)$*, $F \cong_{Kun(P)} E$, iff $E \preceq_{Kun(P)} F$ and $F \preceq_{Kun(P)} E$.

Definition 12 (delay in Kunen's semantics). Let P be a normal program, E and F be first order formulas such that F is equivalent to E in $Kun(P)$.
The delay of F wrt E in Kunen's semantics is the least integer n such that, for each natural m: if $E\theta$ is *true* (resp. *false*) in Φ_P^m, then $F\theta$ is *true* (resp. *false*) in Φ_P^{m+n}.

With the above definitions theorem 8 can be transposed in a natural way for Kunen's semantics.

Theorem 13 (applicability conditions wrt Kunen's semantics). *In the hypothesis of theorem 8, if $\exists X \tilde{C}$ is equivalent to $\exists Y \tilde{D}$ in $Kun(P)$ and one of the following two conditions holds:*

1. *$\tilde{D}$ is independent from cl;*
2. *the dependency degree of $\tilde{D}$ on cl is greater or equal to the delay in $Kun(P)$ of $(\exists Y \tilde{D})$ wrt $(\exists X \tilde{C})$;*

then $Kun(P) = Kun(P')$.

The proof is like the one for Fitting's semantics; it is sufficient to consider only ordinals below ω.

4 Replacement vs other operations

The replacement operation is a very general one. In this section we show how some other transformation operations can be interpreted as special cases of replacement. In this way we indirectly obtain applicability conditions for these other operations on normal programs which ensure that Fitting's and Kunen's semantics are preserved. *A transformation operation is correct wrt Fitting's (or Kunen's) semantics if it is sound and complete wrt that semantics, that is if $Fit(P) = Fit(P')$ (resp. $Kun(P) = Kun(P')$),* where P is the initial program and P' is the result of the operation.

Thinning and Fattening

The *thin* operation allows one to eliminate superfluous literals from the body of a clause.

Definition 14 (thin). Let $c : A \leftarrow \tilde{K}, \tilde{L}.$ be a clause in a program P. *Thinning c of the literals $\tilde{L}$ in P consists of substituting c' for c, where $c' : A \leftarrow \tilde{K}.$ $thin(P, c, \tilde{L}) \stackrel{\text{def}}{=} P \backslash \{c\} \cup \{c'\}$.*

The applicability condition must guarantee that the literals are actually superfluous. Conditions have been given for the preservation of the least Herbrand model semantics [28,4] and the computed answers semantics for definite programs [18,3] and of the Well Founded semantics for normal programs [24].

Thinning can be seen as a particular case of replacement. From theorems 8 and 13, we get the new applicability conditions.

Theorem 15 (correctness of thinning). *Let P' be the result of thinning the literals $\tilde{L}$ in the body of c, and let X be the set of local variables of $\tilde{L}$. If $\exists X (\tilde{K} \wedge \tilde{L})$ is equivalent to $\tilde{K}$ in $Fit(P)$ and one of the following two conditions holds:*
  1. *$\tilde{K}$ is independent from c;*
  2. *$depen_P(\tilde{K}, c)$ is greater or equal to the delay of $\tilde{K}$ wrt $\exists X (\tilde{K} \wedge \tilde{L})$ in $Fit(P)$;*
then $Fit(P)=Fit(P')$.

The same result holds using Kunen's instead of Fitting's semantics.
The conditions given above are more restrictive than the ones given in [4,3]. In fact theorem 8 considers also the "negative" information that can be inferred from the completion of the program and distinguishes it from non-termination.
Let us consider the following program:

$$P = \{ \quad r \leftarrow q. $$
$$c : p \leftarrow p, q. \}$$

Being both *false*, p and q are *equivalent* in *Fit(P)*, but we cannot eliminate q from the body of c, as the delay of p wrt q is one (q is *false* in Φ_P^1, p is *false* in Φ_P^2), while $depen_P(p, cl) = 0$.

The *fatten* operation is the inverse of thinning. It consists in introducing redundant literals in the body of a clause. It is generally used in order to make possible some other transformations such as folding.

Definition 16 (fatten). Let $c : A \leftarrow \tilde{K}$. be a clause in a program P and $\tilde{L}$ a conjunction of literals. *Fattening c with $\tilde{L}$ in P consists of substituting c' for c, where $c' : A \leftarrow \tilde{K}, \tilde{L}$. fatten $(P, c, B) \overset{\text{def}}{=} P \backslash \{c\} \cup \{c'\}$.*

The literals added to the body of the clause must be "superfluous". Conditions have been given for the preservation of the least Herbrand model semantics [28,4] and the computed answers semantics [3] of definite programs.

Theorems 8 and 13 supply the applicability conditions also for the *fatten* operation when applied to normal programs.

Theorem 17 (correctness of fattening). *Let P' be the result of fattening the body of c with $\tilde{L}$ and let X be the set of local variables of $\tilde{L}$, $X = var(\tilde{L}) \backslash var(c)$. If $\exists X (\tilde{K} \wedge \tilde{L})$ is equivalent to $\tilde{K}$ in Fit(P) and one of the following two conditions holds:*
1. $\tilde{K} \wedge \tilde{L}$ is independent from c;
2. $depen_P(\tilde{K} \wedge \tilde{L}, c)$ is greater or equal to the delay of $\exists X (\tilde{K} \wedge \tilde{L})$ wrt $\tilde{K}$ in Fit(P);
then Fit(P)=Fit(P').

The same result holds using Kunen's instead of Fitting's semantics.

Folding

The *fold* operation consists in substituting an atom for an equivalent conjunction of literals, in the body of a clause. This operation is generally used in all the transformation systems in order to pack back unfolded clauses and to detect implicit recursive definitions. In the literature we find different definitions for this operation. This is due to the fact that it is not generally safe even for definite programs and declarative semantics and its application must be restricted by some conditions which depend on the semantics we choose. We show here two ways of using the replacement operation in order to perform folding in normal programs. The first folding depends only on the program to be transformed, while the second one depends on a transformation sequence. Both seem to be useful in program transformations.

Definition 18 (reversible folding). Let P be a normal program, $cl : R \leftarrow \tilde{J}, \tilde{K}', \tilde{L}$. and $d : Q \leftarrow \tilde{K}$ be distinct clauses of P, Y be the set of local variables of $\tilde{K}$, $Y = var(\tilde{K}) \backslash var(Q)$. If there exists a substitution θ, $dom(\theta) = var(\tilde{K}) \backslash Y$, such that $\tilde{K}' = \tilde{K}\theta$ and d is the only clause of P whose head unifies with $Q\theta$; then the result of folding $\tilde{K}'$ in cl by using d as folding clause, is the program: $P' = P \backslash \{cl\} \cup \{cl' : R \leftarrow \tilde{J}, Q\theta, \tilde{L}.\}$.

This operation corresponds to the one considered in [19,11]. To prove its correctness wrt Fitting's and Kunen's semantics we need the following lemma. The proof is omitted since it is straightforward.

Lemma 19. *Let P be a normal program, $cl : Q \leftarrow \tilde{K}$. be a clause of P, X be the set of local variables of Q, $X = var(Q) \backslash var(\tilde{K})$ and Y be the set of local variables of $\tilde{K}$, $Y = var(\tilde{K}) \backslash var(Q)$. If θ is a substitution such that $dom(\theta) = var(\tilde{K}) \backslash Y$ and cl is the only clause of P whose head unifies with $Q\theta$, then*

a) $\exists X Q\theta$ is equivalent to $\exists Y \tilde{K}\theta$ in $\mathrm{Fit}(P)$ and $\mathrm{Kun}(P)$;

b) the delay of $\exists X Q\theta$ wrt $\exists Y \tilde{K}\theta$ in $\mathrm{Fit}(P)$ and $\mathrm{Kun}(P)$ is one.

Theorem 20 (correctness of reversible folding). *The reversible fold operation is correct wrt Fitting's and Kunen's semantics.*

Proof. From lemma 19 we have that $\exists X Q\theta$ is equivalent to $\exists Y \tilde{K}\theta$ in $Fit(P)$ and the delay of $\exists X Q\theta$ wrt $\exists Y \tilde{K}\theta$ in $Fit(P)$ is one.

Since d is the only clause that unifies with $Q\theta$ and $d \neq cl$, $depen_P(Q\theta, cl) > 0$.

We can then *replace* $\tilde{K}' = \tilde{K}\theta$ with $Q\theta$ in the body of cl, thus obtaining P'. This operation coincides with the folding defined in 18. By theorem 8, we have that $Fit(P) = Fit(P')$.

A similar reasoning holds also for Kunen's semantics. □

Example 4. Let us consider the following program:

$$P = \{ \, cl : p(X) \quad \leftarrow q(X, b), \neg s(X), r(a, X). \qquad q(X, a).$$
$$d : r(Z, Y) \leftarrow q(Y, Z), \neg s(Y). \qquad\qquad q(X, b).$$
$$r(a, Y) \leftarrow p(Y). \qquad\qquad\qquad\qquad\qquad \}$$

With $\theta = \{Z = b\}$, we have that $body(d)\theta$ is a variant of $(q(X, b), \neg s(X))$ and that d is the only clause of P whose head unifies with $r(Z, Y)\theta$. Hence we can fold clause cl, thus obtaining the program:

$$P = \{ \, cl : p(X) \quad \leftarrow r(b, X), r(a, X). \qquad q(X, a).$$
$$d : r(Z, Y) \leftarrow q(Y, Z), \neg s(Y). \qquad q(X, b).$$
$$r(a, Y) \leftarrow p(Y). \qquad\qquad\qquad\qquad \}$$

We consider now a fold operation similar to the ones defined in [28,25] since it depends on the transformation history. The operation and the transformation sequence are defined in terms of each other.

Definition 21 (recursive folding). Let $P_0, \ldots, P_k$ be a transformation sequence, $cl : R \leftarrow \tilde{J}, \tilde{K}', \tilde{L}$. a clause of P_k, $d : Q \leftarrow \tilde{K}$. a non-recursive, definite clause of P_j,

where $j < k$ and $\tilde{K}$ has no local variables, that is, $var(\tilde{K}) \subseteq var(Q)$.
If there exists a substitution θ, $dom(\theta) = var(\tilde{K})$, such that $\tilde{K}' = \tilde{K}\theta$ and d is the
only clause of P_j whose head unifies with $Q\theta$ and if one of the following conditions
holds:

1. Q is independent from cl in P_k, or
2. P_k was obtained from P_j by applying the unfolding operation to all the atoms
of $\tilde{K}$;

then the result of recursive folding $\tilde{K}'$ in cl, by using d as folding clause, is the
program: $P_{k+1} = P_k\backslash\{cl\} \cup \{cl' : R \leftarrow \tilde{J}, Q\theta, \tilde{L}.\}$

It is worth noting that we require the clause d to be non-recursive but indirect
recursion is not excluded.

Definition 22 (transformation sequence). A transformation sequence is a sequence $P_0, \ldots, P_n$ where for each i, $0 < i \leq n$, P_i is obtained from P_{i-1} by applying
either an unfolding, or a thinning, fattening, reversible folding or recursive folding.

Example 5. A simple example of recursive folding is the following:

$P = \{ d : thereiszero(L) \qquad \leftarrow member(0, L).$
$\qquad member(X, [X \mid T]).$
$\qquad member(X, [H \mid T]) \leftarrow member(X, T). \qquad \}$

Predicate $thereiszero(L)$ is *true* in $Fit(P)$ when L is a list containing a zero. By
unfolding the body of d we obtain the following:

$P_2 = P\backslash\{d\} \cup \{ d_1 : thereiszero([0 \mid T]).$
$\qquad\qquad d_2 : thereiszero([H \mid T]) \leftarrow member(0, T). \qquad \}$

We can now apply recursive fold to $member(0, T)$ in the body of d_2, by using d as
folding clause; the result is:

$P_3 = P\backslash\{d\} \cup \{ d_1 : thereiszero([0 \mid T]).$
$\qquad\qquad d_3 : thereiszero([H \mid T]) \leftarrow thereiszero(T). \qquad \}$

Now predicate $thereiszero$ is recursive and independent from other definitions.

The *unfold* operation is correct wrt Fitting's and Kunen's semantics. The proof
for Fitting's model can be found in [9], while Kunen's case is an easy corollary of
the same proof.

To prove the correctness of recursive folding, we need the following results.

Lemma 23. *Let P be a normal program, $d : Q \leftarrow \tilde{K}$. a definite, nonrecursive,
clause of P and X the set of local variables of Q. Let θ be a substitution such that
$dom(\theta) = var(\tilde{K})$, and P' the program obtained by unfolding all the atoms in $\tilde{K}$.
If $\tilde{K}$ has no local variables, $var(\tilde{K}) \subseteq var(Q)$, and d is the only clause of P whose
head unifies with $Q\theta$, then*
a) $\exists XQ\theta$ is equivalent to $\tilde{K}\theta$ in $Fit(P')$ and $Kun(P')$;
b) the delay of $\exists XQ\theta$ wrt $\tilde{K}\theta$ in $Fit(P')$ and $Kun(P')$ is zero.

As regards Fitting's semantics, the proof can be found in [5]. For Kunen's semantics it can be obtained in a similar way.

Theorem 24 (correctness of recursive folding). *The recursive folding operation is correct wrt Fitting's and Kunen's semantics.*

Proof. If P_k was obtained from P_j by unfolding all the atoms in $\tilde{K}$, from lemma 23 $\exists X Q \theta$ is equivalent to $\tilde{K}\theta$ and the delay of $\exists X Q \theta$ wrt $\tilde{K}\theta$ is zero in $Fit(P_k)$ and $Kun(P_k)$. This implies that the delay of $\exists X Q \theta$ wrt $\tilde{K}\theta$ cannot be greater than $depen_P(Q\theta, cl)$ and theorems 8 and 13 are applicable. If Q is independent from cl in P_k, by theorems 8 and 13 we can as well replace $\tilde{K}' = \tilde{K}\theta$ with $Q\theta$ in cl. $\square$

Corollary 25 (correctness of a transformation sequence). *All the programs in a transformation sequence have the same Fitting's and Kunen's semantics.*

We give now an example to show that in the definition 21, it is not possible to drop the condition that $\tilde{K}$ must not have local variables.

Example 6. Let P be the program:

$$P = \{\ cl : q(a) \quad\leftarrow p(X).$$
$$p(s(X)) \leftarrow p(X). \quad \}$$

$q(a)$ is equivalent to $\exists X p(X)$ in $Fit(P)$ and the delay of $q(a)$ wrt $\exists X p(X)$ in $Fit(P)$ is one. Since the body of cl contains local variables, the unfold operation cannot reduce the semantic delay between $q(a)$ and $\exists X p(X)$. By unfolding $p(X)$ in cl we obtain:

$$P' = \{\ cl' : q(a) \quad\leftarrow p(Y).$$
$$p(s(X)) \leftarrow p(X). \quad \}$$

which is identical to P modulo renaming of variables. Thus the delay between $q(a)$ and $\exists X p(X)$ in $Fit(P)$ has not changed. This shows that lemma 23 depends on the condition on local variables.

In the last program, since $p(Y)$ in cl' is the result of an unfold operation, we could apply the *fold* operation defined in [25], using cl as the defining clause. The result would be:

$$P'' = \{\ cl'' : q(a) \quad\leftarrow q(a).$$
$$p(s(X)) \leftarrow p(X). \quad \}$$

But $q(a)$ is *undefined* in $Fit(P'')$, hence $Fit(P'') \neq Fit(P)$.
This can be used as a counterexample for proving the following corollary.

Corollary 26. *The fold operation defined in [25] does not preserve Fitting's semantics.*

5 Conclusions

In this paper we study the replacement transformation operation wrt normal programs. It consists in substituting a conjunction of literals, $\tilde{C}$, in a clause body, with an equivalent conjunction of literals, $\tilde{D}$. We propose some conditions which guarantee the preservation of Fitting's and Kunen's semantics during the transformation. The equivalence between $\tilde{C}$ and $\tilde{D}$ is obviously necessary but it is generally not sufficient. In fact we also need to preserve the equivalence after the transformation. Such equivalence can be destroyed only when $\tilde{D}$ depends on the modified clause. Hence we establish a relation between the level of dependency of $\tilde{D}$ from the clause and the difference in "semantic complexity" between $\tilde{C}$ and $\tilde{D}$. Such semantic complexity is measured by counting the number of applications of the fixed point operator which are necessary in order to determine the truth or falsity of a predicate. For Fitting's semantics this complexity can go beyond ω.

By considering replacement as a generalization of other transformation operations, such as thinning, fattening and folding, we show how replacement applicability conditions can be used also for them. A variant of the Tamaki-Sato's transformation sequence is defined which preserves Fitting's and Kunen's semantics. The applicability conditions considered for folding are rather simple since they are either syntactic or they depend on the history of the transformation.

Future work requires:

(i) to single out further cases where syntactic conditions are sufficient for a safe application of replacement and

(ii) to define applicability conditions for replacement wrt other semantics for normal programs. Actually, in [9] the Well Founded Model semantics has already been considered and similar results have been obtained. The proof however is much more complex due to the asymmetric construction of the positive and negative parts of the model.

References

1. K. Apt. Introduction to logic programming. In *Handbook of Theoretical Computer Science*, pages 493–574. Elsevier Science Publishers B.V., 1990.
2. K. R. Apt, H. A. Blair, and A. Walker. Towards a theory of declarative knowledge. In e. J. Minker, editor, *Foundation of Deductive Databases and Logic Programming*, pages 89–148. Morgan Kaufmann, 1988.
3. A. Bossi and N. Cocco. Basic transformation operations for logic programs which preserve computed answer substitutions. Technical Report 16, Dip. Matematica Pura e Applicata, Università di Padova, Italy, April 1990. to appear in Special Issue on Partial Deduction of the Journal of Logic Programming.
4. A. Bossi, N. Cocco, and S. Dulli. A method for specializing logic programs. *ACM Transactions on Programming Languages and Systems*, 12(2):253–302, April 1990.
5. A. Bossi, N. Cocco, and S. Etalle. Transforming normal program by replacement. Technical Report 18, Dip. Matematica Pura e Applicata, Università di Padova, Italy, November 1991.
6. R. Burstall and J. Darlington. A transformation system for developing recursive programs. *Journal of the ACM*, 24(1):44–67, January 1977.
7. K. L. Clark. Negation as failure rule. In H. Gallaire and G. Minker, editors, *Logic and Data Bases*, pages 293–322. Plenum Press, 1978.
8. Y. Deville. *Logic Programming. Systematic Program Development.* Addison-Wesley, 1990.

9. S. Etalle. Transformazione dei programmi logici con negazione, Tesi di Laurea, Dip. Matematica Pura e Applicata, Università di Padova, Padova, Italy, July 1991.

10. M. Fitting. A Kripke-Kleene semantics for logic programs. *Journal of Logic Programming*, (4), 1985.

11. P. Gardner and J. Shepherdson. Unfold/fold transformations of logic programs. In J.-L. Lassez and e. G. Plotkin, editors, *Computational Logic: Essays in Honor of Alan Robinson*. 1991.

12. C. Hogger. Derivation of logic programs. *Journal of the ACM*, 28(2):372–392, April 1981.

13. C. Hogger. *Introduction to Logic Programming*. Academic Press, 1984.

14. S. Kleene. *Introduction to Metamathematics*. D. van Nostrand, Princeton, New Jersey, 1952.

15. H. Komorowski. Partial evaluation as a means for inferencing data structures in an applicative language: A theory and implementation in the case of Prolog. In *Ninth ACM Symposium on Principles of Programming Languages, Albuquerque, New Mexico*, pages 255–267, 1982.

16. K. Kunen. Negation in logic programming. *Journal of Logic Programming*, (4):289–308, 1985.

17. J. Lloyd. *Foundations of Logic Programming*. Springer-Verlag, 1987.

18. J. Lloyd and J. Shepherdson. Partial evaluation in logic programming. Technical Report CS-87-09, Department of Computer Science, University of Bristol, England, 1987. to appear in Journal of Logic Programming.

19. M. Maher. Correctness of a logic program transformation system. IBM Research Report RC13496, T.J. Watson Research Center, 1987.

20. M. Maher. A transformation system for deductive databases with perfect model semantics. *Theoretical Computer Science*, to appear.

21. M. Proietti and A. Pettorossi. The synthesis of eureka predicates for developing logic programs. In N. Jones, editor, *ESOP'90, (Lecture Notes in Computer Science, Vol. 432)*, pages 306–325. Springer-Verlag, 1990.

22. M. Proietti and A. Pettorossi. Unfolding, definition, folding, in this order for avoiding unnecessary variables in logic programs. In Maluszynski and M. Wirsing, editors, *PLILP 91, Passau, Germany (Lecture Notes in Computer Science, Vol.528)*, pages 347–358. Springer-Verlag, 1991.

23. T. Sato. An equivalence preserving first order unfold/fold transformation system. In *Second Int. Conference on Algebraic and Logic Programming, Nancy, France, October 1990, (Lecture Notes in Computer Science, Vol. 463)*, pages 175–188. Springer-Verlag, 1990.

24. H. Seki. A comparative study of the well-founded and stable model semantics: Transformation's viewpoint. In D. P. W. Marek, A. Nerode and V. Subrahmanian, editors, *Workshop on Logic Programming and Non-Monotonic Logic, Austin, Texas, October 1990*, pages 115–123, 1990.

25. H. Seki. Unfold/fold transformation of stratified programs. *Journal of Theoretical Computer Science*, 86:107–139, 1991.

26. J. C. Shepherdson. Negation as failure: a comparision of Clark's completed data base and Reiter's closed world assumption. *Journal of Logic Programming*, (1):1–48, 1984.

27. J. C. Shepherdson. Negation in logic programming. In e. J. Minker, editor, *Foundation of Deductive Databases and Logic Programming*, pages 19–88. Morgan Kaufmann, 1988.

28. H. Tamaki and T. Sato. Unfold/fold transformation ol logic programs. In S. Tarnlund, editor, *2nd International Logic Programming Conference, Uppsala, Sweden, July 1984*, pages 127–138, 1984.

Meta-Programming for
Reordering Literals in Deductive Databases

Jesper Larsson Träff and Steven David Prestwich

ECRC – European Computer-Industry Research Centre
Arabellastrasse 17, D-8000 Munich 81, Germany
e-mail: traff@ecrc.de, steven@ecrc.de

Abstract. Specifying efficient evaluation strategies by meta-interpreters, and then eliminating the interpretation overhead by partial evaluation with respect to given object programs, is an elegant technique for the transformation of logic programs. In this paper we show how to apply this technique to the compilation of *instantiation based* evaluation strategies for DATALOG with negation, and hence to query optimization in deductive databases. We demonstrate our approach on two well-known optimizations, dealing with the passing of information between literals (*sideways information passing strategies* or SIPS), and the early evaluation of constraining literals (the C-transformation).

1 Introduction

There are two complementary approaches to the optimization of logic programs. Either a clever evaluation strategy can be defined under which programs are executed with improved efficiency, or the programs themselves can be transformed into equivalent versions, which are more efficient when evaluated using a standard strategy. The main advantages of the first approach is its versatility and the fact that the correctness of an evaluation strategy needs be proven only once, but a price is often paid in the form of a significant interpretation overhead. This overhead is not manifest in the transformation approach, but each transformation requires a specialized proof of correctness. Furthermore, not all run-time optimizations are possible at transformation time, simply because an interpreter can take actual data into account which are not known at transformation-time.

The specification of clever evaluation strategies by meta-interpreters for subsequent partial evaluation (in the context of logic programming sometimes referred to as "partial deduction" [8]) with respect to given object-level programs provides an attractive combination of the two approaches. Meta-interpreters are more easily written than program transformers, and correctness follows from the correctness of the specification (assuming the partial evaluator to be correct). If the interpreter can be completely eliminated by partial evaluation, the result can be as efficient as that from a special purpose transformer. However, if the meta-interpreter specifies an evaluation strategy which depends upon the groundness of certain arguments of (object level) literals – what is called an *instantiation based computation rule* in [13] – then meta-logical tests for groundness and identity of object-level variables are needed. These tests cannot in general be decided at partial evaluation time, which makes it impossible to remove the meta-level completely.

We show how, in the DATALOG case, the meta-logical tests can be replaced by unproblematic tests which are decidable at partial evaluation time. This is done by first adopting a simple naming scheme for object level variables, then using these names for annotating – or, to stay within the database terminology [17], *adorning* – the object program. The meta-logical tests can then be reformulated as tests on the adornments, which, since adornments are just ground terms, can be decided at partial evaluation time. We present a generic meta-interpreter for specifying evaluation strategies for DATALOG and instantiate it to two cases: expressing sideways information passing strategies (SIPS) [1], and specifying the computation rule underlying the C-transformation [7,15] which aims at propagating constraining literals through recursive clauses. Finally, we show how to apply partial evaluation to the generic meta-interpreter, and state sufficient conditions on the meta-interpreters for the partial evaluation process to terminate and for all meta-interpreter function symbols to be removed, giving the desired DATALOG to DATALOG transformation.

The rest of the paper is organized as follows: Section 2 presents two examples of query optimization and Section 3 contains preliminary definitions. In Section 4 we introduce a generic meta-interpreter for instantiation based computation rules and show that it can be used to specify the computation rules of both examples. Section 5 indicates how to replace the meta-logical tests of the meta-interpreter by introducing the concept of adorned program and query. In Section 6 we outline how the meta-level can be eliminated by partial evaluation. Section 7 concludes the paper.

2 Motivating examples

We consider two well-known query optimization transformations from the deductive database context. We borrow some terminology from this field, but the examples should nevertheless be readily understandable. Both transformations require knowledge of when variables become bound.

Example 1 Let a relation p be defined by the following rule:

p(X,Y) :- r(X,Z), s(Y), t(X,Y,Z).

where r, s and t are already defined relations. We can assume that r, s and t all define sets of ground facts (hence so does p). Consider a complex query ?- ..., p(c1,Y), ..., p(X,c2), ... For both occurrences of predicate p in the query the order of literals in the definition of p as written is not likely to be the order in which the query is most efficiently answered.

Evaluating p(c1,Y) with the variable Y unbound leads to s(Y) being evaluated with Y unbound, which might be costly if the relation s is large. This could easily be remedied by postponing evaluation of s till after evaluation of t(X,Y,Z) when bindings for Y will be available to guide the evaluation of s(Y).

For the second occurrence of p in the query where the variable X is unbound, early evaluation of s(Y) is the better choice. Hence, a reordering of the body literals of p, creating in the process two new instances of the p-rule

```
p1(X,Y) :- r(X,Z), t(X,Y,Z), s(Y).
p2(X,Y) :- s(Y), t(X,Y,Z), r(X,Z).
```

and replacing the query by ?- p1(c1,Y), ..., p2(X,c2), ... is likely to improve the evaluation efficiency. This is a very simple example of compilation of a *sideways information passing strategy* (SIPS) [1]. Informally, a SIPS determines the order in which predicates in rule bodies should be evaluated and how bindings of variables should be passed from one predicate to the next. The reordering specified by the SIPS is in this case purely rule-local: the literals in each rule are reordered to meet some criterion, in this case based on the hypothesis that "bound-is-better". Notice that the same rule can give rise to more than one instance in the transformed program.

Example 2 Let the ancestor relation be defined by the following two rules:

```
anc(X,Y) :- parent(X,Y).
anc(X,Y) :- parent(X,Z), anc(Z,Y).
```

The query ?- anc(X,Y), c1(X), c2(X,Y) asks for the ancestors Y of some X's with the property c1(X) where furthermore c2(X,Y) holds. Assume we know that the literals c1(X) and c2(X,Y) are most efficiently evaluated if their variables are bound. Then c1(X) and c2(X,Y) will be said to be *constraining literals*, since in this context they merely serve to restrict the set of answer tuples to the complex query. Again, evaluating the query as it stands is not the most efficient way of obtaining an answer. The literal c1(X) can be evaluated as soon as X becomes bound, which happens by evaluation of parent(X,Y).

The program and query can be transformed into the following equivalent program:

```
anc1(X,Y) :- parent(X,Y), c1(X), c2(X,Y).
anc1(X,Y) :- parent(X,Z), c1(X), anc2(Z,Y,X).

anc2(X,Y,C) :- parent(X,Y), c2(C,Y).
anc2(X,Y,C) :- parent(X,Z), anc2(Z,Y,C).
```

with query ?- anc1(X,Y). The effect of the transformation is that c1(X) is taken into account before the whole ancestor relation is constructed. The potential gain should be obvious. For the literal c2(X,Y) the effect is not as clear, since also the variable Y, which appears in the recursive literal has to be bound before evaluation.

This transformation is an example of a non-local reordering of literals. It was introduced in [7] where it is termed the C-transformation. It is worth noting that the original proof of correctness of this transformation is quite complicated. In [15] it was shown that the transformation can be obtained by meta-interpreter specialization.

3 Preliminaries

In this paper we consider only DATALOG with (stratified) negation [2]. We will, following [7], use the term *constraining literal* to denote a literal that (in a given context) is required to have all its variables bound before it can be evaluated.

Negative literals are always constraining. Comparison predicates (for example X>Y) are also natural examples of constraining literals, but others might also be considered constraining. Determining which literals are constraining in a given context is an orthogonal problem not addressed in this paper. We will assume that constraining literals C are identified by a meta-predicate **constraining**(C).

Using this terminology, we can define what is meant by a safe evaluation strategy.

Definition 1 (Safe evaluation strategy) *An evaluation strategy is* safe *for a program P if constraining literals are only evaluated whenever they are ground.*

For simplicity our meta-interpreters will be required to *preserve safety* in the following sense: if the standard (left-to-right) strategy is safe for a program P, then the strategy specified by the meta-interpreter must also be safe for P. We will not be concerned with the behaviour of the meta-interpreter on programs that are not safe under the standard strategy.

We assume [2] that all relations are either *intensional* (defined by rules only) or *extensional* (defined by facts only), that rules are *range restricted*, that is all variables in the rule head must occur in some body literal, and all variables in constraining body literals must occur in a non-constraining body literal, and that facts are ground. These assumptions imply that the evaluation of a non-constraining literal results in all variables of that literal becoming bound.

4 A meta-interpreter for instantiation based reordering

We will start by developing a generic meta-interpreter for instantiation based computation rules. The key is to make explicit the set of variables that are bound (to ground terms) prior to the evaluation of a conjunction of literals. To this end the "vanilla" interpreter is given an extra argument V representing these variables with the intention that db_eval(G,V) will hold if the conjunction G can be evaluated preserving safety under the assumption that the variables represented by V have already been bound prior to evaluation of G.

Object level conjunctions of literals will be represented as lists of literals. Object level rules will be represented by meta-facts **rule**(H,B), whereas it will be assumed that facts can be evaluated by reflection. In order to limit reflection to facts only, facts have to be classified as such by a meta-predicate **fact**. Constraining literals are assumed to be either negative literals, comparison predicates or facts.

The meta-interpreter has an explicit selection function **select** which forms new conjunctions from a conjunction of literals. The selection function must handle the (representation of) bound variables correctly: **select**(P,V,Qs,VQ,Rs,VR) must hold with Qs and Rs being conjunctions formed on the basis of P, and if V represents the variables that are bound prior to evaluation of P then VQ must be the variables that will be bound prior to evaluation of Qs, and VR the variables that will be bound prior to evaluation of Rs. The meta-interpreter (figure 1) is sufficient for dealing with the examples in Section 2.

$$
\begin{aligned}
&\textbf{db\_eval}([],V). \\
&\textbf{db\_eval}([F],V) && \leftarrow \textbf{fact}(F),\ F. \\
&\textbf{db\_eval}([C],V) && \leftarrow \textbf{comparison}(C). \\
&\textbf{db\_eval}([\text{not }F],V) && \leftarrow \textbf{not db\_eval}([F],V). \\
&\textbf{db\_eval}([R],V) && \leftarrow \textbf{rule}(H,B), \\
&&& \quad \textbf{unify}(R,H,V,NewV), \\
&&& \quad \textbf{db\_eval}(B,NewV). \\
&\textbf{db\_eval}([P|Ps],V) && \leftarrow Ps\neq[], \\
&&& \quad \textbf{select}([P|Ps],V,Qs,VQ,Rs,VR), \\
&&& \quad \textbf{db\_eval}(Qs,VQ), \\
&&& \quad \textbf{db\_eval}(Rs,VR).
\end{aligned}
$$

Fig. 1. Generic meta-interpreter for instantiation based reordering of literals

The predicate **unify**(R,H,V,NewV) holds if V represents the variables bound prior to evaluation of R, if literals R and H unify, if NewV represents the variables bound prior to evaluation of the body of H. Hence **unify**(R,H,V,NewV) handles renaming of the variables V under unification of a literal R with a head of a rule H.

To formulate the selection functions one more predicate is needed. We will assume that **binds**(P,V,VP) holds if VP represents the variables that will become bound by evaluation of literal P given that V are the variables bound prior to evaluation of P.

4.1 A selection function for SIPS

A "greedy" selection function for the SIPS discussed in Example 1 is now easy to write: **select**(P,V,Q,VQ,Rs,VR) splits the conjunction P into a most preferred predicate Q plus a conjunction Rs of the remaining literals. Predicate Q is preferred to Q' if Q has more bound arguments than Q' given the bound variables V. The variables that will be bound prior to evaluation of Rs are (those represented by) V plus those which will become bound by evaluation of Q.

$$
\begin{aligned}
&\textbf{select}([P],V,[P],V,[],NewV) && \leftarrow \textbf{binds}(P,V,NewV). \\
&\textbf{select}([P,R|Ps],V,Q,VQ,[R|Rs],VR) && \leftarrow \textbf{more\_bound}(P,R,V), \\
&&& \quad \textbf{select}([P|Ps],V,Q,VQ,Rs,VR). \\
&\textbf{select}([P,R|Ps],V,Q,VQ,[P|Rs],VR) && \leftarrow \textbf{not more\_bound}(P,R,V), \\
&&& \quad \textbf{select}([R|Ps],V,Q,VQ,Rs,VR).
\end{aligned}
$$

The predicate **more_bound**(P,Q,V) holds if P has more bound arguments than Q, given V. If P is a constraining literal, **more_bound**(P,Q,V) holds only if P is ground (given V). Thus the selection function preserves safety.

The SIPS-selection function is naive, but more sophisticated strategies can be expressed. For instance the selection function can be made less "greedy" by taking into account how the selection of a particular literal will affect the free variables of the remaining literals of the conjunction, or modified to utilize other kinds of information on the relative preferability of literals. Such information can be made available through meta-predicates, the same way in which constraining literals are assumed to be marked. Heuristics for rule-local reordering of literals are frequently discussed, for example in [2,17,14].

4.2 A selection function for the C-transformation

The C-transformation aims at propagating constraining literals to the earliest safe position in the program, that is to the earliest position where they will be ground under the standard evaluation strategy. The corresponding selection function (figure 2 below) must ensure the same effect, that is select each constraining literal for evaluation as soon as it becomes ground. We will split it into two parts.

First, a literal that can be safely evaluated is selected from the conjunction [P|Ps] – we simply choose the first, P, which isn't necessarily safe, but the choice preserves safety (cfr. Section 3). Evaluation of P binds additional variables VR and all constraining literals in Ps that can be safely evaluated with the variables in VR bound are selected. This is expressed by select_c(Ps,Cs,Rs,VR), where Cs is a conjunction of constraining literals and Rs the remaining literals of Ps (some of which may be constraining, but then will not have all variables bound). If P is a fact, it is evaluated and the conjunctions formed from [P|Ps] are Cs and Rs. If P is the head of a rule, the selection function must ensure that propagation of the constraining literals Cs continues through the body of the rule, so a clause lookup is performed. The two conjunctions formed from [P|Ps] in this case are first the rule-body conjoined with Cs, second Rs (see figure 2).

The predicate **conjunction**(B,V,Cs,NewV,Bs) holds if Bs is the conjunction formed from conjunctions B and Cs (see Section 5).

4.3 Representing object level variables

So far we have not dealt with the representation of object level variables bound prior to evaluation of a conjunction P. As will be seen, the choice is critical.

An obvious choice is to represent the object level variables by themselves, that is to amalgamate object and meta-level. To determine whether all variables of a constraining literal C have been bound, expressed by **all_bound**(C,V) in the C-transformation selection function, one has to check whether all the variables of C occur in V. So for each variable of C represented by meta-variable X it has to be checked whether it is identical to some variable in V represented by

```
select([F|Ps],V,Cs,V,Rs,VR)        ← fact(F), F,
                                     binds(F,V,VR),
                                     select_c(Ps,Cs,Rs,VR).
select([P|Ps],V,Bs,NewV,Rs,VR)     ← rule(R,B),
                                     unify(P,R,V,NewV),
                                     binds(P,V,VR),
                                     select_c(Ps,Cs,Rs,VR),
                                     conjunction(B,NewV,Cs,V,Bs).

select_c([],[],[],V).
select_c([C|Qs],[C|Cs],Rs,V)       ← constraining(C),
                                     all_bound(C,V),
                                     select_c(Qs,Cs,Rs,V).
select_c([Q|Qs],Cs,[Q|Rs],V)       ← not (constraining(C), all_bound(C,V)),
                                     select_c(Qs,Cs,Rs,V).
```

Fig. 2. The selection function for the C-transformation

meta-variable Y. In Prolog notation, this is the case if X==Y. Now X and Y are syntactically different (meta) variables and at partial evaluation time it cannot be decided whether they will at run-time eventually denote the same object level variable. Therefore the partial evaluator has to leave these tests in the output program which will then no longer be a DATALOG program and also be inefficient since part of the meta-level will still remain. Part of the problem is due to the fact that the Prolog built-in == is an impure implementation of a meta-logical predicate.

A ground representation of object level variables as advocated in [9] and subsequently made a feature of the logic programming language Gödel [5] alleviates the problems with the (impure) meta-logical tests: two meta-variables represent the same object variable iff they are bound to the same ground term. Writing meta-interpreters using a ground representation is on the other hand less straightforward, and the proof of termination of the partial evaluation process may become more involved.

In [3] a solution based on classification of (meta) variables into variables ranging over object level constants and variables ranging over object level variables is sketched. By letting the meta-interpreter maintain sets of these different types of variables, the problems with the meta-logical predicates can to some extent be circumvented.

We propose another solution. We will show that it is possible to statically derive information from the object program which allows the meta-logical tests to be replaced by equivalent tests that can be decided at partial evaluation time. This information will be attached to the program as *adornments*. Therefore it is possible even in an amalgamated language like Prolog to formulate and compile away instantiation based computation rules. The use of adornments to replace certain test by others that can be decided at partial evaluation time is reminiscent of the *binding time analysis* employed by [6].

5 Representing binding information

For each object program clause, choose a unique ground term name for each of the variables of the clause, and let each literal be *adorned* with the set of names of variables of that literal. Under the assumption (valid in the DATALOG case, cf. Section 3) that all variables of a literal become bound by evaluation of the literal, the variables bound after evaluation of any set of literals of an adorned clause are the variables whose names appear in the union of the adornments of the literals. If names of variables are chosen such that the adornment of any body literal is a subset of the adornment of each unifying head literal, the adornments can be used to keep track of when variables become bound throughout programs. The following definitions make the concept of *adorned programs* precise:

Definition 2 (Adorned clause) *Let* $H : -L_1, \ldots, L_n$ *be a DATALOG rule and choose a unique ground-term naming of the variables in the rule. An* adorned clause *is a rule*

$$H^A : -L_1^{A1}, \ldots, L_n^{An}$$

where the head-adornment A *is the set of ground terms naming the variables that will be bound after evaluation of the rule, and the adornment* Ai *for each literal* L_i *is the set of ground terms naming the set of variables that will become bound by evaluation of* L_i.

Definition 3 (Adorned program and query) *Let* P *be the set of rules of a DATALOG program. An* adorned program *is a set of adorned rules, such that for any body literal* L_i *of a rule in* P *which unifies with the head* H' *of a rule in* P *there exist adorned rules*

$$H^A : -L_1^{A1}, \ldots, L_i^{Ai}, \ldots, L_n^{An}$$

and

$$H'^{A'} : -L'_1{}^{A'1}, \ldots, L'_m{}^{A'm}$$

such that $Ai \subseteq A'$.

   Let $: -G_1, \ldots, G_m$ *be a conjunctive query. An* adorned query *is a conjunction of adorned literals*

$$: -G_1^{A1}, \ldots, G_m^{Am}$$

such that for each head H' *of a rule in* P *with which* G_i *unifies, there is an adorned rule* $H'^{A'} : -L'_1{}^{A'1}, \ldots, L'_n{}^{A'n}$ *such that* $Ai \subseteq A'$.

The idea of adorning programs to be able to decide when variables become bound is similar to the bf-adornments used for many database transformations, for example the magic-sets transformation [1,17]. The traditional adornment, however, only indicates whether argument positions are bound or not, but is not informative enough to capture *when* individual variables become bound.

Notice that the definition is independent of the evaluation order of literals, and also that constraining and non-constraining literals are treated identically. Variations on the definition are possible: by treating adornments as sets information on which position in a literal binds a certain variable is lost, but few adorned clauses are needed. Keeping more precision is possible at the cost of more adorned clauses.

Let us see how the ancestor program and query (Example 2 of Section 2) can be adorned: let x, y, z be names (ground terms) of variables. Then a possible adornment is as follows:

$$\mathtt{anc(X,Y)}^{\{x,y\}} \mathtt{:-parent(X,Y)}^{\{x,y\}}.$$
$$\mathtt{anc(U,V)}^{\{x,y\}} \mathtt{:-parent(U,W)}^{\{x,z\}}, \mathtt{anc(W,V)}^{\{z,y\}}.$$

$$\mathtt{anc(X,Y)}^{\{y,z\}} \mathtt{:-parent(X,Y)}^{\{y,z\}}.$$
$$\mathtt{anc(U,V)}^{\{y,z\}} \mathtt{:-parent(U,W)}^{\{x,z\}}, \mathtt{anc(W,V)}^{\{x,y\}}.$$

$$\mathtt{?-anc(A,B)}^{\{x,y\}}, \mathtt{c1(A)}^{\{x\}}, \mathtt{c2(A,B)}^{\{x,y\}}.$$

For adorning DATALOG programs, a variation of a simple groundness analysis suffices. We have the following proposition.

Proposition 1 *For any DATALOG program there exists an adorned program with only a finite number of adorned rules. Adornment can be done automatically using only the intensional database.*

It is now possible to write out the unspecified predicates of the meta-interpreter (cf. Section 4) assuming that we are dealing with adorned object level programs:

Adorned literal R^{AR} unifies with an adorned head of a rule H^{AH} if R and H unify and if the variables represented by the adornment AR of R appear in the head-adornment AH of H (as required by definition 3). The variables bound prior to evaluation of H are the variables of AH already bound by V:

$$\mathbf{unify}(R^{AR}, H^{AH}, \text{V}, \text{NewV}) \leftarrow R = H, AR \subseteq AH, \text{NewV} = AH \cap V$$

Notice that the size of the set NewV is bounded by the size of V.

An adorned literal P^A binds the variables represented by its adornment A:

$$\mathbf{binds}(P^A, \text{V}, \text{NewV}) \leftarrow \text{NewV} = A \cup V$$

All variables of a (constraining) literal C are bound by V if the adornment A of C is fully contained in V:

$$\mathbf{all\_bound}(C^A, V) \leftarrow A \subseteq V$$

Finally, given a set of bound variables represented by V, an adorned literal P^{AP} has more variables bound than Q^{AQ} if the cardinality of the set of P's bound variables is greater than the cardinality of the set of Q's bound variables. A constraining literal C has more bound variables than any literal Q if it has all its variables bound given V:

$$\text{more\_bound}(P^{AP}, Q^{AQ}, V) \leftarrow \text{not constraining}(P),$$
$$\|AP \cap V\| \geq \|AQ \cap V\|$$
$$\text{more\_bound}(P^{AP}, Q^{AQ}, V) \leftarrow \text{constraining}(P),$$
$$\text{all\_bound}(P^{AP}, V)$$

The predicate **conjunction**(B,NewV,Cs,V,Bs) has still not been specified. We can partly do so here. To ensure that adorned programs contain only finitely many clauses it was essential that only a finite number of different names were used. The adornments are therefore not powerful enough to allow forming conjunctions of arbitrary rule bodies, while maintaining the property that adornments properly describe when variables become bound. However, when a conjunction of constraining literals are conjoined to a rule body, clashes between names which might lead to unsafety can be avoided by eliminating from the adorned constraining literals the names of all variables that have already been bound given V.

conjunction(A,VA,B,VB,C) $\leftarrow$ V=VA\VB,
$$\text{eliminate\_clashes}(V,B,Bs),$$
$$\text{conjunction}(A,Bs,C).$$
eliminate_clashes(V,[],[]).
eliminate_clashes(V,$[C^A|Cs]$,$[C^{A'}|Cs']$) $\leftarrow A' = A \setminus V,$
$$\text{eliminate\_clashes}(V, Cs, Cs').$$

The specification of **conjunction**(A,Bs,C) will be left until the next section.

6 Eliminating the meta-level

In this section we will establish some requirements a partial evaluator has to fulfil in order for it to be able to completely eliminate the interpretation overhead of the generic meta-interpreter (figure 1).

Partial deduction of the meta-interpreter with respect to a DATALOG program must result in a DATALOG program. If partial deduction is done by unfolding only, the result will contain object level predicate symbols as function symbols. These must somehow be removed. We achieve this by basing the partial deduction on an unfold/fold transformation framework. Most partial deducers are based on unfolding alone, for example [10,11], but these must be supplemented by another transformation (renaming) to avoid clause subsumption, and in our case a transformation to eliminate function symbols [4]. The unfold/fold method is sufficient as it stands, and its correctness is well established. For our purposes we choose a framework which can handle negation in a satisfactory way; for further details on this see [12].

We need just three transformation rules: **unfold, definition** and **fold**. The **unfold** rule for logic programs is simple: select a resolving atom in a clause body and replace the clause by the set of resolvents. The **definition** rule allows us to add a new clause to the program at any time during the transformation, as long as its head predicate symbol does not already appear anywhere else in the program. We call the new clause a *definition*. Although not necessarily true of the definition rule in general, in our case the arguments of the head atom of the definition will be exactly the variables of the body. The **fold** rule is a form of inverse unfold. If an atom in a clause body is an instance of a definition body, then the atom can be replaced by the definition head, with corresponding variable bindings and subject to certain safety criteria which we shall not detail here. These three rules can be applied to a program in any order, producing a sequence of equivalent programs.

Although unfold/fold transformations cannot strictly be called "partial deduction", as defined in [10] for example, we will continue to use this term because the transformation has a similar aim and results as partial deduction. The *closedness* and *singularity* conditions mentioned in that paper are automatically satisfied, because our transformation starts from a single clause which is to be transformed, and there is no specialised query.

6.1 The partial evaluation strategy

We now impose a strategy on top of the three transformation rules to obtain the required partial evaluation strategy. Our transformation will inherit the correctness properties of the underlying transformation framework, that is it will preserve the semantics of the interpreted program.

Start from the meta-interpreter and a DATALOG object program where facts are left undefined plus a query ?-db_eval(G,V). First create a new definition (**New**(G,V) ← db_eval(G,V)) then transform as follows. If the current clause (which is initially the new definition) is of the form (**Head** ← **true**) then partial deduction halts by adding the clause to the output program. Otherwise it can be put into the form (**Head** ← **First, Rest**) (where **Rest** may be **true**). There are 4 cases for this clause:

(1) **First** is a positive literal with an extensional database predicate or comparison atom. These are unspecified at partial deduction time, so create an *auxiliary definition* (**Aux** ← **Rest**) and partially evaluate it. Then fold using the auxiliary definition, and add the result (**Head** ← **First, Aux**) to the output program.

(2) The clause is not a definition (and is therefore the product of unfolding) and **First** is a positive call to db_eval which can be folded using a definition whose body is a variant of **First**, say (**New** ← **First**) – if there is no such definition then create one and recursively partially evaluate it. Fold **First** and create an auxiliary definition (**Aux** ← **Rest**) and partially evaluate it. Then fold using the auxiliary definition, and add the result (**Head** ← **New, Aux**) to the output program.

(3) **First** is a negative literal, say **not** db_eval(F,V). So make a new definition (**New** ← db_eval) and partially evaluate it. Then fold **First** and create an auxiliary definition (**Aux** ← **Rest**) and partially evaluate it. Then fold using the auxiliary definition, and add the result (**Head** ← **not New, Aux**) to the output program.

(4) The clause is a definition (that is, it has not yet been unfolded) and **First** is positive and not an extensional database predicate. So apply the unfold rule on **First**.

6.2 Some sufficient conditions

There are two requirements to the partial evaluation strategy: the transformation must terminate, and the result of the transformation must be a DATALOG program. Sufficient conditions for these to hold are (i) the size of the two arguments of **db_eval** and the number of (object and meta level) constants both remain bounded during unfolding, (ii) no call to **db_eval** during unfolding must contain a free meta-variable. The following proposition provides useful sufficient conditions for (i) and (ii).

Proposition 2 *If the select(P, V, Qs, VQ, Rs, VR) predicate in the generic meta-interpreter (figure 1) has the property that both arguments Qs and Rs are conjunctions*

1. *formed only from literals from P and both contain fewer literals than P, or*
2. *formed only from clause lookups and literals from P in such a way that no literal produced by clause lookup appears with higher multiplicity than it has in the clause body,*

then the partial deduction process terminates with no residual meta-calls.

For a proof, see [16].

The SIPS selection function clearly has property 1, since P is simply split into a single literal Q and a conjunction Rs of the remaining literals. For the C-transformation selection function matters are a bit more complicated. Property 1 does not hold, since a clause lookup may take place, and without precautions neither does property 2. Consider the clause

```
p(X,Y) :- a(Y), p(X,Y), c(X).
```

where c(X) is assumed to be a constraining literal. Now, what happens during evaluation of a query to the p-relation? Eventually, after clause lookup and evaluation of a(Y), p(X,Y) will be selected with c(X) as a safe constraining literal, so after clause lookup the conjunction

```
a(Y), p(X,Y), c(X), c(X)
```

will have to be evaluated. After yet another few steps, the conjunction to be evaluated will be

```
a(Y), p(X,Y), c(X), c(X), c(X)
```

and so on. Fortunately, all the c(X)'s but one are redundant, so property 2 can be ensured by deleting these redundant constraining literals. This is what conjunction(A,B,C) is intended to capture:

conjunction([],C,C).
conjunction([L^A|As],B,C) ← redundant(L,A,B),
          conjunction(As,B,C).
conjunction([L^A|As],B,C) ← not redundant(L,A,B),
          conjunction(As,B,[L^A|C]).

A literal L with adornment A is redundant in the conjunction B if L appears in B with the same adornment.

7 Concluding remarks

Expressing alternative evaluation strategies by meta-interpreters, then eliminating the interpretation layer by partial evaluation, is a well-known and elegant technique. It combines the expressiveness and flexibility of meta-interpreters with the efficiency of specialized program transformations. We have shown how to apply this technique to the compilation of interpreters expressing evaluation strategies for DATALOG programs with negation, which depend upon patterns of variable bindings. The main difficulty with such evaluation strategies is that they are naturally expressed using meta-logical tests on variables for determining whether particular object level variables occur in some given literal, and such tests cannot always be decided at partial evaluation time. We replaced the meta-logical tests by unproblematic tests on program adornments and gave sufficient conditions for the interpretation layer to be completely eliminated.

The work has obvious applications to query optimization in deductive databases, as we showed on two examples: compilation of a simple sideways information passing strategy (SIPS) and C-transformation. A series of experiments with the meta-interpreters described in this paper using an existing unfold/fold partial evaluation system (developed at ECRC) shows the practicality of the method (see [16]). We found that the evaluation time of the transformed query/program plus the transformation time (partial evaluation time) was typically significantly less than the evaluation time of the original program/query.

We are investigating the application of the approach outlined in this paper to extended classes of programs, in particular programs containing function symbols. For this, however, a better formal understanding of the adornment idea is needed. It would also be interesting to see whether the transformations could be obtained by standard partial deduction [10,11], that is without the use of folding (but combined with the removal of function symbols as in [4]).

7.1 Acknowledgement

This paper is an extended version of [15]. The first author is sponsored by ESPRIT Basic Research Action no. 3012, Compulog. Torben Amtoft and Danny De Schreye read and commented critically on a previous version of this paper; this input is gratefully acknowledged. Both authors would also like to thank ECRC for supporting this work.

References

1. Catriel Beeri and Raghu Ramakrishnan. On the power of magic. In *Proceedings of the 6th ACM SIGACT-SIGMOD-SIGART Symposium on Principles of Database Systems*, pages 269–283, 1987.
2. Stefano Ceri, Georg Gottlob, and Letizia Tanca. *Logic Programming and Databases. Surveys in Computer Science*, Springer, 1990.
3. John Gallagher. Transforming logic programs by specialising interpreters. In *Proceedings of the 7th European Conference on Artificial Intelligence (ECAI-86)*, pages 109–122, 1986.
4. John Gallagher and Maurice Bruynooghe. Some low-level source transformations for logic programs. In *Proceedings of the Second Workshop on Meta-Programming in Logic*, pages 229–244, 1990.
5. Patt Hill and John W. Lloyd. *The Gödel Report*. Technical Report TR-91-02, Department of Computer Science, University of Bristol, March 1991.
6. Neil D. Jones, Peter Sestoft, and Harald Søndergaard. Mix: a self-applicable partial evaluator for experiments in compiler generation. *Lisp and Symbolic Computation*, 2(1):9–50, 1989.
7. David Kemp, Katagori Ramamohanarao, Isaac Balbin, and Krishnamurthy Meenakshi. Propagating constraints in recursive deductive databases. In *Proceedings of the North American Conference on Logic Programming*, pages 981–998, 1989.
8. Jan Komorowski. *A specification of an abstract Prolog machine and its application to Partial Evaluation*. PhD thesis, Linköping University, 1981. Also technical report LSST69.
9. John W. Lloyd and Pat Hill. Analysis of meta-programs. In *Meta-Programming in Logic Programming*, pages 23–51, MIT Press, 1988.
10. John W. Lloyd and John C. Shepherdson. Partial evaluation in logic programming. *Journal of Logic Programming*, 11(3&4):217–242, 1991.
11. Bern Martens and Danny De Schreye. Sound and complete partial deduction with unfolding based on well-founded measures. In *Proceedings of the International Conference on Fifth Generation Computer Systems (FGCS)*, pages 473–480, 1992.
12. Taisuke Sato. *A First Order Unfold/Fold System*. Technical Report TR-90-17, Electrotechnical Laboratory, 1990.
13. Danny De Schreye and Maurice Bruynooghe. On the transformation of logic programs with instantiation based computation rules. *Journal of Symbolic Computation*, 7(2):125–154, 1989.
14. Günther Specht and Oliver Krone. Zur Steuerung und Optimierung der SIP-Auswahl in der Magic Set Transformation. In *GWAI-91, 15. Fachtagung für Künstliche Intelligenz*, pages 33–42, Springer, September 1991.
15. Jesper Larsson Träff and Steven David Prestwich. Meta-programming for propagating constraining literals for query optimization in deductive databases. In Laks Lakshmanan, editor, *ILPS workshop on deductive databases*, pages 91–100, October 1991.
16. Jesper Larsson Träff and Steven David Prestwich. *Meta-Programming for Reordering Literals in Deductive Databases*. Technical report ECRC-92-3, ECRC, 1992.
17. Jeffrey D. Ullman. *Principles of Database and Knowledge-Base Systems: the new technologies*. Volume 2, Computer Science Press, 1989.

Propagation : a New Operation in a Framework for Abstract Interpretation of Logic Programs

Maurice Bruynooghe and Gerda Janssens

Department of Computer Science, Katholieke Universiteit Leuven
Celestijnenlaan 200A, B-3001 Heverlee, Belgium
e-mail: {maurice,gerda}@cs.kuleuven.ac.be

Abstract. Most frameworks for abstract interpretation of logic programs provide abstract operations which safely approximate their concrete counterpart. That means, given a concrete state S described by an abstract state AS and a concrete operation O producing $O(S)$, the corresponding abstract operation AO produces $AO(AS)$ which must describe $O(S)$. We sketch a framework which relaxes this condition and requires that the safe approximation condition is only reached after a propagation step which reexecutes – at the abstract level – all operations leading to $O(S)$ until a stable abstract state is reached. We illustrate the novel framework with a mode analysis which, notwithstanding a very simple abstract domain, reaches on several examples similar precision as mode analysis systems using much complexer abstract domains.

1 Introduction

Almost all frameworks for abstract interpretation of logic programs apply what one could call a principle of locality. The global state of the computation is distributed over many local environments, one for each clause being executed. In the framework of [1], a local environment keeps track of the state of the variables of the clause it annotates. A correct approximation of an operation which is part of a clause, e.g. unification of a variable X_i with a constant, correctly describes the new state of the local variables after execution of the concrete operation, e.g. the variable X_i becomes *ground* and all other local variables X_j which possibly share their value with X_i cannot retain a mode *free*. To allow such a safe approximation, the abstract state must be sufficiently detailed. In case one includes a mode such as free which is not closed under substitution (instantiating the variable makes the mode incorrect), then sufficiently detailed information about possible sharing must be included in the abstract state, otherwise all modes *free* should be turned into *any* as soon as a single variable is involved in a unification. Notice that the unification between X_i and the constant can also affect non-local variables e.g. a variable Y_j in the caller's environment of the current clause. That environment no longer reflects the current state of the variable Y_j. However, there is no interest in the current state of non-local variables. It is the role of procedure exit to update the state of the caller's environment at the point where its state becomes again relevant.

The same observations remain valid when considering abstract interpretation of constraint logic programs. [9] extend the abstract state with a constraint component listing the concrete constraints between local variables. Although the effect of bindings between non-local variables on the values of local variables is rather easily captured in a *possible sharing* component between local variables, it is not that easy for numeric constraints. A binding to a local variable can propagate through a chain of non-local constraints and finally bind another local variable. [9] take a very rough approach : variables involved in a non-local constraint are always in a state which is instantiation closed. So one can ignore non-local constraints without compromising the safety of the analysis. In the current paper we develop a different schema and consider a distributed computation where one aims at correctly approximating the global state corresponding to the current program point. When approximating the effect of an operation, say unification of X_i with a constant, the change to the local state, i.e. that X_i becomes ground, is propagated to the neighbouring environments and the computations already done in that environment are reexecuted. Eventual updates are in turn propagated. For example, assume that X_i occurs in the head and that the corresponding variable in the caller's environment is Y_i. That variable becomes also ground. Suppose a part of the executed computation in the caller's environment was unifying Y_i with Y_j, that computation is reexecuted and Y_j becomes also ground. These changes in the state of Y_i and Y_j are in turn propagated. Eventually a stable state is reached where every variable in every environment safely approximates the concrete value in the program point of the concrete execution. To a certain extent, the idea is a generalisation of the repeat-previous-call strategy mentioned in [1]. Due to the transparency of logic programs, one can "repeat" the execution of an already executed part to obtain a better approximation of the state. For some applications (e.g. [4]) it is necessary to make a distinction between the first execution and the reexecutions for precision reasons.

Apparently, we have a more complex algorithm for abstract interpretation. So, what is gained ? The advantage is that one can use simpler abstract domains and simpler abstract operations, which, when only applied locally, would yield incorrect results. The global propagation mechanism turns the result in a safe approximation. For example, one can perform mode analysis with a very simple abstract domain not including any information about sharing and obtain results which appear as precise as with the complex PROP domain of [10] or the EXP domain of [3].

Finally, notice that it is unlikely that the same effect can be achieved by extending the abstract domain with a "global store" containing all operations already executed. Firstly, there is the problem that a procedure can be defined by different clauses, each performing different operations, on exiting from such a procedure, one would have to keep alternative states for the global store. Secondly, recursion causes an unlimited growth of the global store.

2 Sketch of the Framework

Our framework is an extension of the framework of [1] for logic programs which constructs an abstract AND-OR graph adorned with abstract substitutions. The construction of the graph involves three application specific operations : procedure entry, procedure exit and abstract interpretation of built-ins. Plugging in these three operations which satisfy the correctness conditions stated in the framework guarantees that the final graph describes all concrete AND trees (and hence all derivations) that can occur during execution of the intended class of queries.

We assume programs are in so called normal form i.e. arguments of predicates occurring in the heads and bodies of clauses are distinct variables. This simplifies considerably the procedure entry and procedure exit operations as all effects of unifications are concentrated in built-ins of the form $X_i = X_j$ and $X_i = f(X_{j_1}, \ldots X_{j_n})(n \geq 0)$. We first discuss abstract interpretation of built-ins. In [1] a built-in has an abstract call substitution β_{in} which is assumed to be correct. It means that any concrete computation which reaches the program point where the built-in has to be performed has an accumulated substitution whose projection upon the variables of the clause the built-in is part of (upon the local environment of the clause), say τ_{in}, is described by β_{in}, i.e. $\tau_{in} \in \gamma(\beta_{in})$ with γ the concretisation function. Executing the built-in under substitution τ_{in} yields a new substitution $\tau_{in}\theta$. The corresponding abstract operation computes an abstract success substitution β_{out} from β_{in}. Correctness requires that $\tau_{in}\theta \in \gamma(\beta_{out})$, i.e. any concrete computation which reaches the program point immediately after the built-in has an accumulated substitution whose projection upon the local environment of the clause is described by β_{out}.

Consider mode analysis where the abstract substitution gives for each local variable the mode which is one of *free*, *ground* and *any*. When unifying two variables say X_i and X_j, it is not difficult to define an abstract operation which correctly computes the new modes for X_i and X_j from the old ones. However, for the other local variables, there is a problem. In case the mode of X_i or X_j has changed from *free* to *any* or *ground*, one has to change the mode of all *free* variables to *any*. Indeed, one cannot exclude that other variables share (have been unified) with one of X_i or X_j. This is a very rough approximation. In order to obtain more precision, works on mode analysis include in one or another form a sharing component in the abstract substitutions.

Our novel approach consists of a **propagation** step which follows the abstract interpretation of the built-in. Only after this propagation step has updated the abstract substitution produced by the abstract interpretation of the built-in, do we require that $\tau_{in}\theta$ is described by it. Consequently, the abstract interpretation of the built-in can be a much weaker (simpler) operation. Let us try to make our ideas more precise. One can say that the state reached by concrete execution is the result of a sequence of operations $O_1, \ldots, O_n$. Let CS be the concrete state corresponding to it (e.g. the accumulated substitution). Performing another operation, say O_{n+1}, yields a new state CS'. In the abstract interpretation framework, the part of CS local to the clause on which O_{n+1} is performed is abstracted as AS. With AO_{n+1} the abstract counterpart of O_{n+1}, we have $AS' = AO_{n+1}(AS)$ and correctness requires that AS' describes the local part of CS'. For example, in Fig. 1 AS could be the abstract call substitution of B_j and AS' its abstract success state (assuming B_j is a call to a built-in).

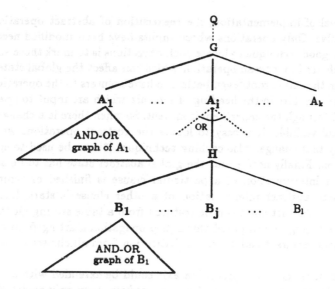

Fig. 1. Partial AND-OR-graph.

In the new framework, we assume that clauses are not only adorned with abstract call and success substitutions but also with a current state AS_c which describes correctly the part of CS local to the clause in question. In Fig. 1, assuming B_j is to be executed, that means that all clauses on the path between B_j and the root ($H \leftarrow B_1, \ldots, B_l, G \leftarrow A_1, \ldots, A_k, \ldots$) are adorned with an AS_i but also the clauses which can be involved in solving calls to the left of calls on the path to the root, i.e. clauses in the AND-OR subgraph for $B_1, \ldots$, in the AND-OR subgraph for $A_1, \ldots$. All these local abstract states AS_i together make up a global state AS. In the new framework we have $AS' = Prop(AS)$ where both AS' and AS are global states and $Prop$ is a function which reexecutes all operations affecting AS until a stable state is reached. In particular, if $O_1, \ldots, O_n, O_{n+1}$ is the sequence of operations leading to CS', then their abstract counterparts, say $AO_1, \ldots, AO_n, AO_{n+1}$, are reexecuted until a stable state is reached. All these operations AO_i are weak operations, without propagation, which by themselves need not to be safe. The point is that when the global propagation ceases, a safe approximation is reached, i.e. all local states AS'_i correctly describe state CS' and this for any CS represented by AS. The operations which need to be considered for reexecution are :

- All operations involved in reexecuting calls to the left of the path from the

current program point to the root $(B_1, \ldots, B_{j-1}, A_1, \ldots, A_{i-1}, \ldots$ in Fig. 1)
 – The unifications between calls and headings on the path to the root (cfr. Fig.
1, unifications between H and A_i, G and $\ldots, \ldots$). Notice that the other OR-
branches evolving from calls on the path to the root, are not reexecuted. (cfr.
Fig. 1 : the other clauses defining $A_i, \ldots$).

At the level of implementation, the reexecution of abstract operations can
be very selective. Only operations whose inputs have been modified need to be
reexecuted. A good technique to locate such operations is to mark those variables
in a call which are input to an operation which can affect the global state in the
subgraph with the call as root (even better, to have pointers to the operation) and
to mark those variables in the heading of a clause which are input to operations
to be reached through the caller's environment. So when there is a change to the
state of a local variable, it is easy to locate the abstract operations which can
be affected by that change. Also caching techniques [8] can be used to speed up
the reexecution. Finally notice that the global abstract state has to be adjusted
when abstract interpretation of a particular clause is finished or temporarily
interrupted and abstract interpretation of another clause is started/resumed.
The new global abstract state can be restored from a table storing global states
on context switching or computed through propagation starting from a known
state : the latest abstract call/success substitution from all clauses contributing
to the state.

   Also procedure entry and procedure exit could be extended with a propagation step. With programs in normal form, procedure entry only copies the state
of the variables in the call into the state of the variables in the head. So it is
hard to imagine an application where propagation could be needed to obtain a
safe procedure entry. Procedure exit copies information the other way around
but also has to update the state of variables not participating in the call. Similar
as for built-ins, it is feasible to have a weak procedure exit which needs to be
followed by propagation to obtain a safe approximation.

   As described in [1] the AND-OR tree is folded into a graph in case of recursion. E.g. in Fig. 2, the call $P(Y_1, \ldots, Y_n)$ refers back to its ancestor call
$P(X_1, \ldots, X_n)$. We have that the projection of β_1 respectively β_2 upon $X_1, \ldots, X_n$
is equivalent to the projection upon $Y_1, \ldots, Y_n$ of β_3 respectively β_4. Propagation through recursion requires a special mentioning. Reexecuting $P(Y_1, \ldots, Y_n)$
means that one has to reexecute $P(X_1, \ldots, X_n)$. Reexecuting the unification between $P(X_1, \ldots, X_n)$ and $P(Z_1, \ldots, Z_n)$ means that one has also to update the
abstract substitution in terms of $Y_1, \ldots, Y_n$.

3   Mode Analysis

We illustrate our novel framework with a mode analysis application. An abstract
substitution over a set of variables $X_1 \ldots X_n$ consists of a n-tuple of modes, the
ith element being the mode for X_i. The possible modes are *free*, *ground* and *any*
with their obvious meaning. No information about possible sharing is included.

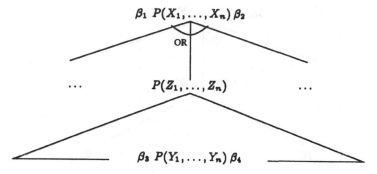

Fig. 2. AND-OR graph with a recursive call $P(Y_1, \ldots, Y_n)$.

Within existing frameworks, such an abstraction leads to very imprecise approximations. There is a very simple partial order over the modes of one variable *free* $<$ *any* and *ground* $<$ *any*. This partial order can easily be extended to n-tuples of modes : $(m_1, \ldots, m_n) \leq (m'_1, \ldots, m'_n)$ iff $m_1 \leq m'_1$ and ...and $m_n \leq m'_n$. We obviously have an abstract domain satisfying all requirements in [1]. Assuming that the top level query consists of a single call Q/n, the abstract AND-OR graph is initialised with a root consisting of the call $Q(X_1, \ldots, X_n)$ and an abstract call substitution $(m_1, \ldots, m_n)$ specifying the mode of the class of queries being analysed. It is assumed that there is no sharing between the arguments of the concrete query. If sharing is desired, it has to be created explicitly in the code for Q.

Procedure entry (Fig. 3), an operation applied on a call $P(Y_{i_1}, \ldots, Y_{i_m})$, performs abstract unification between the call and the heads of the defining procedures. With $P(X_1, \ldots, X_m)$ such a head, the effect is to pass the mode of Y_{i_j} to X_j. In addition, the modes of the variables $X_{m+1}, \ldots, X_n$ local to the body of the defining procedure are initialised as *free*. Proving that this operation (without propagation) satisfies the condition imposed by the framework of [1] is straightforward.

Procedure exit (Fig. 3) considers the defining clauses one by one. Assume $P(X_1, \ldots X_m)$ is the head of a defining clause and $(m_1, \ldots, m_m, \ldots, m_n)$ the final abstract success substitution of the body. Assume $P(Y_{i_1}, \ldots Y_{i_m})$ is the call. The modes $m_1, \ldots, m_m$ are passed on to the variables $Y_{i_1}, \ldots Y_{i_m}$. This yields a tuple $(m_1, \ldots m_m)$ of modes. Having such a tuple for every defining clause, the least upper bound is taken. Assuming that the call is part of a clause with variables $Y_1, \ldots, Y_n$, the variables in $\{Y_1, \ldots, Y_n\} \setminus \{Y_{i_1}, \ldots, Y_{i_m}\}$ are given the mode they had in the abstract call substitution for the call $P(Y_{i_1}, \ldots, Y_{i_m})$. It is easy to prove that the modes for the variables Y_{i_j} satisfy the condition of [1], however, the modes of the other variables do not satisfy the condition because the effects of sharing of variables from $\{Y_1, \ldots, Y_n\} \setminus \{Y_{i_1}, \ldots, Y_{i_m}\}$ and from $\{Y_{i_1}, \ldots, Y_{i_m}\}$ are ignored. It is claimed that the condition is satisfied after propagation. Propagation, as explained in Sect. 2, consists of :

- reexecuting (without propagations) calls to the left of the path from the

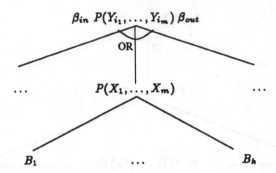

Fig. 3. Fragment of AND-OR graph with call $P(Y_{i_1}, \ldots, Y_{i_m})$ and clause $P(X_1, \ldots, X_m) \leftarrow B_1, \ldots, B_k$.

current program point the call $P(Y_{i_1}, \ldots Y_{i_m})$ to the root.

- reexecuting the unifications between calls and heads on the path from the current program point to the root. This is slightly simpler than procedure entry. Unification of a call $P(U_1, \ldots, U_n)$ with a head $P(V_1, \ldots, V_n)$ updates the mode of U_i and V_i as described below for the abstract interpretation of $X_i = X_j$ (for $1 \leq i \leq n$). Currently we have no formal proof of the claim, we discuss the claim in more detail after describing abstract interpretation of built-ins.

This propagation continues until a stable state is reached.

Abstract interpretation of built-ins is concerned with the effect of unification. Let $X_1, \ldots X_n$ be the variables of the clause. The abstract interpretation of the built-in $X_i = X_j$ is defined as follows : if either X_i or X_j has mode *ground*, both variables get mode *ground*; else if X_i and X_j have mode *free*, the modes are unchanged; else mode *any* is given to X_i and X_j. For a unification $X_i = f(X_{j_1}, \ldots, X_{j_m})$ one has similar rules : if either X_i has mode *ground* or all of X_{j_k} have mode *ground*, then all of $X_i, X_{j_1}, \ldots, X_{j_m}$ are given mode *ground*; else if X_i has mode *free*, then its mode is changed to *any* while the modes of the X_{j_k} are unchanged; else all modes become *any*. Again it is obvious that the modes of $X_i, X_{j_1}, \ldots, X_{j_m}$ satisfy the conditions of [1], but that propagation is needed to obtain correct modes for the other variables.

Claim *Abstract interpretation of built-ins terminates and yields an abstract success substitution which satisfies the condition of [1], i.e. the abstract success substitution describes all concrete substitutions which can be observed in this program point.*
Although we do not have a formal proof, we can mention some crucial elements.

Lemma. *Propagation terminates.*
The proof is based on the observation that the changes to the global state are monotone with regard to the instantiation order *free* < *any* < *ground*.

Conjecture. *Let AS be an abstract substitution over the variables* $X_1, \ldots, X_n$. *Let* $U_1, \ldots, U_m$ *be a set of unifications involving variables from* $X_1, \ldots, X_n$. *Let* θ *be the mgu of* $U_1, \ldots, U_m$. *Let* $WAU_i(S)$ *be the abstract substitution resulting from the weak (without propagation) abstract interpretation of* U_i *with abstract call substitution S. If* $\sigma \in \gamma(AS)$ *then* $\sigma\theta \in \gamma(AS')$ *with* AS' *the abstract substitution obtained by applying* $WAU_1, \ldots, WAU_m$ *starting in the state AS until no more changes are observed.*

The conjecture claims that the abstract interpretation of the U_i can be done in isolation, ignoring the interaction with the other unifications and that propagation will recover the effects of the interaction. It is interesting to point out an example operation not satisfying the conjecture. Consider as operations U_i numerical constraints of the form $a_1 X_1 + \ldots + a_n X_n = b$. Consider an abstract operation assigning *ground* to the mode of X_i only if $X_1, \ldots, X_{i-1}, X_{i+1}, \ldots, X_n$ have mode *ground* and assigning *any* to the mode of X_i only if $X_1, \ldots, X_{i-1}, X_{i+1}, \ldots, X_n$ have mode *ground* or *any* and at least one of them has mode *any*. Consider $X + Y = 5, X - Y = 2$ with modes *free* for X and Y. Abstract interpretation of the individual statements will not change the modes of X and Y, so also propagation will not modify them. However, a constraint solver for linear equations will, given the two equations, ground X and Y. In other words the proposed abstraction (domain and operations) is not giving a correct approximation of the behaviour of the constraint solver.

4 Some Examples

We start with a simple artificial example :
$P(X, Y) :- Q(X, Y), R(X), S(Y).$
$Q(U, V) :- U = V.$
$Q(U, V) :- V = b.$
$R(U) :- U = b.$
$S(U) :- \ldots$
Assume the abstract state for the call to P is $\{X = f, Y = f\}$. Q has the same call state. Abstract interpretation of the first clause for Q ends with a current state $\{U = f, V = f\}$; for the second clause, one obtains $\{U = f, V = g\}$. After taking the upper bound and renaming, one obtains $\{X = f, Y = a\}$ as abstract success state of Q and both X and Y are marked as having non-local operations eligible for reexecution ($U = V$ in the first clause of Q). Because V has mode *ground* in the local context of the second clause, and *ground* is the most instantiated state, one need not to add $V = b$ to the list of unifications to be redone upon modifying the mode of Y. We have applied this optimisation throughout our examples. After entering the clause defining R (the variable U in the head literal is marked as having a non-local unification), the unification $U = b$ is abstractly executed. It changes the mode of U from f to g. As there is a non-local operation, propagation is initiated. Back to the top level clause, the non-local operation on X is found to be hidden in the call to Q. The execution of Q is repeated, this time with abstract call state $\{X = g, Y = a\}$. Abstract inter-

pretation of the first clause ends with $\{U = g, V = g\}$, for the second clause the result $\{U = g, V = g\}$ is already available, procedure exit yields $\{X = g, Y = g\}$ and one obtains the correct and precise result that Y is *ground* when calling S (also X and Y in Q are no longer marked as having non-local operations). A simple mode analysis in a logic programming setting, keeping track of possible sharing will return from Q with $\{X = f, Y = a\}$ and possible sharing between X and Y. Executing R will turn the mode of X into g but will leave $Y = a$ (if Y were f it would become a due to the possible sharing with X) and one ends with the imprecise result that Y has a mode a when S is called. It requires more sophisticated abstract domains such as the sharing domains in [5] and [11], the PROP domain of [10] and the EXP domain of [3] to obtain the more accurate result.

The next example involves recursion :
$Q(X, Y) :\text{-} X = Y.$
$Q(X, Y) :\text{-} X = f(X'), Q(X', Y).$
Consider the query $? - Q(A, B), B = c.$ with $\{A = f, B = f\}$. First, the call $Q(A, B)$ is executed. The first clause ends with state $\{X = f, Y = f\}$ while $X = Y$ is a candidate for reexecution (to reduce the amount of non-local propagation, such unifications between variables of the head could be returned as part of the abstract substitution). The second clause calls Q recursively with the same pattern. Using the result of the first clause as an approximation, one obtains mode a for X, f for Y and X' and a non-local operation on both X' and Y. Taking the upper bound of both clauses results in a new success state $\{X = a, Y = f\}$ with non-local operations on X and Y, which is also the final one. Next, the call $B = c$ is considered : it grounds B and triggers reexecution of $Q(A, B)$ with state $\{A = a, B = g\}$. The corresponding success state $\{A = g, B = g\}$ is computed by again considering the two clauses for Q and using the result of the first one as an approximation for the recursive call. By reexecution $\{A = g, B = g\}$ is obtained as success state of the query.

Using an instantiation closed domain, e.g. modes a and g, the propagation becomes optional. It was easy to adapt our existing abstract interpreter such that it only performed limited propagation : not doing propagation through the head into the caller's environment and only repeating the unifications amongst the previous calls in the body. We tested several programs. For a lot of simple programs, one obtains already good results as illustrated by Table 1 which contains for each program all the call states and the corresponding success states of the involved predicates. The code of the programs is found in Appendix A.

For other programs local propagation is not able to derive precise results. A good example is quicksort. $qsort(L, S) :\text{-} H = nil, qs(L, S, H).$

Table 1. Call and success states of simple programs with limited local propagation.

program	call state	success state
naive-reverse(X,Y)	$\{X = g, Y = a\}$	$\{X = g, Y = g\}$
append(Rt,H,Y)	$\{Rt = g, H = g, Y = a\}$	$\{Rt = g, H = g, Y = g\}$
append(X,Y,Z)	$\{X = g, Y = g, Z = a\}$	$\{X = g, Y = g, Z = g\}$
append(X,Y,Z)	$\{X = a, Y = a, Z = g\}$	$\{X = g, Y = g, Z = g\}$
reverse(X,Y)	$\{X = g, Y = a\}$	$\{X = g, Y = g\}$
reverse(X,Y,Z)	$\{X = g, Y = a, Z = g\}$	$\{X = g, Y = g, Z = g\}$
insert(E,OT,NT)	$\{E = g, OT = g, NT = a\}$	$\{E = g, OT = g, NT = g\}$
insert(E,T)	$\{E = g, T = a\}$	$\{E = g, T = a\}$

$qs(U, R, T) \text{ :- } U = nil, R = T.$
$qs(U, R, T) \text{ :- } U = E.V, T_1 = E.T_2, partition(E, V, Sm, Gr),$
$qs(Sm, R, T_1), qs(Gr, T_2, T).$
$partition(E, L, Sm, Gr) \text{ :- } L = nil, Sm = nil, Gr = nil.$
$partition(E, L, Sm, Gr) \text{ :- } L = F.T, Gr = F.Gr1, F > E,$
$\qquad partition(E, T, Sm, Gr1),$
$partition(E, L, Sm, Gr) \text{ :- } L = F.T, Sm = F.Sm1, F \leq E,$
$\qquad partition(E, T, Sm1, Gr),$

Abstract interpretation computes for the call state $\{X = g, Y = a\}$, the success state $\{X = g, Y = a\}$ with limited local propagation and is summarised by Table 2 which gives for each call state of qs/3 the call states of the recursive calls in the second clause and the success state.

Table 2. Call and success states for qs/3 with limited local propagation

Call state	Call state 1st recursive call Call state 2nd recursive call	Success state
$\{U = g, R = a, T = g\}$	$\{Sm = g, R = a, T_1 = a\}$ $\{Gr = g, T_2 = a, T = g\}$	$\{U = g, R = a, T = g\}$
$\{U = g, R = a, T = a\}$	$\{Sm = g, R = a, T_1 = a\}$ $\{Gr = g, T_2 = a, T = a\}$	$\{U = g, R = a, T = a\}$

Full propagation is needed to obtain the success state $\{X = g, Y = g\}$. For the sake of simplicity we use the modes a and g, although including f leads to similar results. With call state $S = \{U = g, R = a, T = g\}$ the abstract interpretation of the first clause computes a first approximation of the success state $\{U = g, R = g, T = g\}$. On procedure entry for the 2nd clause with call state S, the unification $T_1 = E.T_2$ remains eligible for reexecution and T_1 and T_2 have mode a. The abstract interpretation of the first recursive call marks the variables R and T_1 - originating from the unification $R = T$ with R and T

having mode a in the first clause. The second recursive call uses the approximation and grounds T_2. Then $T_1 = E.T_2$ is reexecuted and grounds T_1. This forces the reexecution of the first recursive call as it is marked on T_1 with call state $\{S_m = g, R = a, T_1 = g\}$. This call state is identical to the call state of its ancestor.

Finally, the abstract interpretation of a PROLOG III program [2] is worked out : mode analysis with modes f, g and a of the following lengthlist program which has in addition to unifications (represented as constraints) some typical numeric constraints :

$lengthlist(L, N) : -\{L = nil, N = 0\}.$
$lengthlist(L, N) : -\{L = E.T, N > 0, N = Nt + 1\}lengthlist(T, Nt).$

and the query $? - lengthlist(X, Y)$ with call state $S_0 = \{X = g, Y = f\}$ (see Fig. 4). Constraint propagation has to deal with two typical numeric constraints.

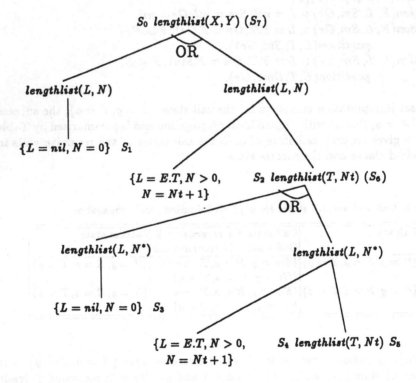

Fig. 4.
AND-OR graph for the call $lengthlist(X, Y)$ with call state $S_0 = \{X = g, Y = f\}$ at the moment when propagation is started for N^*.

The constraint $N > 0$ changes the mode of N from *free* to *any* (there could be another constraint $N \leq 0$) but it does not need to be reexecuted. For $N = Nt + 1$

there are two possible cases :

1. If N or Nt has mode g, then abstract interpretation turns both into g.
2. Otherwise, N and Nt are given mode *any* while the constraint is eligible for reexecution.

Note that if both N and Nt have mode f before constraint propagation, constraint propagation imposes mode a to them because they are possibly involved in more constraints : e.g. $\{N + Nt = 7, N - Nt = 3\}$ has one solution in which N and Nt are ground.

The abstract interpretation starts with call state S_0. Procedure entry computes $S_1 = \{L = g, N = g\}$ and (after interpreting the constraints) $S_2 = \{L = g, N = a, E = g, T = g, Nt = a\}$ with $N = Nt + 1$ eligible for reexecution. The projection of S_2 on the recursive call $lengthlist(T, Nt)$ gives rise to a second call state $S' = \{T = g, Nt = a\}$ for lengthlist/2. So, procedure entry extends the AND-OR graph, marks the variable N in the headings as having operations in the caller's environment eligible for reexecution and computes $S_3 = \{L = g, N = g\}$ and $S_4 = \{L = g, N = a, E = g, T = g, Nt = a\}$. During the computation of S_3 the mode of the marked variable N is changed. However propagation does not cause any changes (as N is the only marked variable and it has mode *ground*, further instantiation is impossible and propagation can be avoided). In the second clause the projection of S_4 on $lengthlist(T, Nt)$ is again S'. An iterative process is started and the renaming of S_3 is used as a first approximation of the success state corresponding to S'. Local propagation reactivates $N = Nt + 1$ (N gets mode g). This results in $S_5 = \{L = g, N = g, E = g, T = g, Nt = g\}$. Again the mode of the marked N is changed. At this point we have computed the AND-OR graph of Fig. 4 and non-local propagation is triggered. State S_5, $\{T = g, Nt = g\}$, is propagated into the caller's environment changing the mode of Nt from a into g, the reexecution of $N = Nt + 1$ is triggered and S_2 is updated into $S_2' = \{L = g, N = g, E = g, T = g, Nt = g\}$. Now the recursive clause is reentered using procedure entry : N is no longer marked, the updated state S_5' of the second clause is identical to S_5. In theory, S_5^c should be obtained by computing $S_4^c = \{L = g, N = g, E = g, T = g, Nt = g\}$ and computing the abstract interpretation of $lengthlist(T, Nt)$ with call state $\{T = g, Nt = g\}$, but the reexecution of lengthlist can again be avoided as it cannot further instantiate variables as they already have mode g in S_5. Next, we continue with procedure exit for lengthlist : S_3 and S_5' are restricted and renamed into terms of $\{T, Nt\}$; their upper bound $S'' = \{T = g, Nt = g\}$ yields the same result as the used approximation, so a fixed point has been computed. The next step in procedure exit is the extension which combines S'' with S_2'. At this point the modes of T and Nt remain unchanged, thus no propagation is triggered and $S_6 = S_2'$. Finally, procedure exit computes S_7 by combining S_0 (which is still the current state for that program point) and the upper bound computed from S_1 and S_6 : $S_7 = \{T = g, Nt = g\}$.

5 Discussion

We have sketched a novel framework for abstract interpretation of logic programming. We have illustrated the framework for the simple case of mode analysis of logic programs and have sketched how it can be extended to handle numerical constraints in a CLP language. With a very simple abstract domain, we obtain for a number of practical examples the same precision as other authors with substantially more complex domains as the sharing domain of [5], [11], the PROP domain of [10], the EXP domain of [3]. In our framework propagation is not only needed for precision, but also for correctness. Apparently we have traded simplicity of the framework against simplicity of the domain. Experimental evaluation will have to show whether our approach is advantageous. Compared to the repeat-previous-call strategy, whose practical value has been shown in [8], our propagation reactivates previous calls sooner. Reexecuting a call in the caller's environment is not delayed until normal procedure exit. Whether this has an advantageous effect upon the execution time of the analysis requires experimentation. Also more work has to be done in formalising our novel framework (including the handling of recursion) and the correctness of applications. A promising approach we are currently investigating is to reformulate the concepts in terms of the abstract OLDT framework [7]. Further work is required to find out whether other applications (in logic programming or constraint logic programming) can be simplified through the use of propagation in this novel framework. One candidate is the type analysis in [6] where the SVAL- and PSHR-components could be omitted substantially simplifying the abstract domain. However, that propagation terminates and yields safe approximations is less obvious than in the mode analysis case.

Acknowledgement
This work was funded in part by Diensten voor de Programmatie van het Wetenschapsbeleid (project RFO-AI-02) and ESPRIT project 5246 PRINCE. Maurice Bruynooghe is supported by the Belgian National Fund for Scientific Research. We are grateful to Danny De Schreye for reading the draft and discussing some technicalities and to Wim Simoens for the practical experiments.

References

1. M. Bruynooghe, "A Practical Framework for the Abstract Interpretation of Logic Programs", *Journal of Logic Programming*, Vol.10, pp. 91-124, (1991).
2. A. Colmerauer, "An Introduction to PROLOG III", *Communications of the ACM*, Vol.30(7), pp. 69-96 (1990).
3. A. Cortesi and G. File, "Abstract Interpretation of Logic Programs : an Abstract Domain for Groundness, Sharing, Freeness and Compoundness Analysis", *Proc. Symposium on Partial Evaluation and Semantics-Based Program Manipulation, PEPMA'91*, June 17-19, Connecticut, USA, pp. 52-61, (1991).
4. V. Dumortier, G. Janssens and M. Bruynooghe, "Detection of free variables in the presence of numeric constraints by means of abstract interpretation", *CW Report CW145*, (1992).

5. D. Jacobs and A. Langen, "Accurate and Efficient Approximation of Varaiable Aliasing in Logic Programs", "*Proc. NACLP89*, MIT-press, pp. 154-165, (1989).
6. G. Janssens and M. Bruynooghe, "Deriving Descriptions of Possible Values of Program Variables by Means of Abstract Interpretation : Definitions and Proofs", *to appear in the Journal of Logic Programming 1992 (also published as CW Report CW107 Computer Science Dept., K.U.Leuven, March 1990)*
7. T. Kanamori and T. Kawamura, "Abstract Interpretation Based on OLDT Resolution", *to appear in the Journal of Logic Programming 1992*
8. B. Le Charlier and P. Van Hentenryck, "Reexecution in Abstract Interpretation of Prolog", *Technical Report No. CS-92-12, also to appear in Proc. JICLP92,* (1992).
9. K. Marriott and H. Sondergaard, "Analysis of Constraint Logic Programs", *Proc. NACLP90*, ed. S. Debray and M. Hermenegildo, pp. 532-547, (1990).
10. K. Marriott and H. Sondergaard, "Notes for a Tutorial on Abstract Interpretation of Logic Programs", *Proc. NACLP89*, Cleveland, (1989).
11. K. Muthukamar and M. Hermenegildo, "Determination of Variable Dependence Information at Compile-Time through Abstract Interpretation", *Pro. NACLP89*, MIT-press, (1989).

A Code of the Simple Example Programs in Table 1

```
naive-reverse(X, Y) :- X = nil, Y = nil.
naive-reverse(X, Y) :- X = E.T, H = [E], naive-reverse(T, Rt)
             append(Rt, H, Y).


reverse(X, Y) :- Z = nil, reverse(X, Y, Z).
reverse(X, Y, Z) :- X = nil, Y = Z.
reverse(X, Y, Z) :- X = E.T, Z1 = E.Z, reverse(T, Y, Z1).


insert(E, OT, NT) :- OT = nil, NT = tree(nil, E, nil).
insert(E, OT, NT) :- OT = tree(L, F, R), NT = tree(NL, F, R), E < F,
             insert(E, L, NL).
insert(E, OT, NT) :- OT = tree(L, F, R), NT = tree(L, F, NR), E > F,
             insert(E, R, NR).


insert(E, T) :- T = tree(L, E, R), !.
insert(E, T) :- T = tree(L, F, R), E < F, insert(E, L).
insert(E, T) :- T = tree(L, F, R), E > F, insert(E, R).
```

CLP(Q) for Proving Interargument Relations

Frédéric Mesnard[1,2]   Jean-Gabriel Ganascia[2]

(1) Iremia Université de la Réunion, 15, av. René Cassin, 97489 St Denis Cedex France
(2) Laforia Université Paris VI, 4, place Jussieu, 75252 Paris Cedex 05 France
e-mail: fred@iremia.fr and ganascia@laforia.ibp.fr

Abstract. In the logic programming community, the concept of interargument relation, that is, the relation that holds between the size of the arguments of a procedure, appears in numerous works on termination proofs for logic programs. In this paper, we present a method for proving linear interargument inequalities. Our technique relies on the notion of abstract procedures and on CLP(Q). We prove its correctness and fully describe its implementation in Prolog III. The applications we present go beyond termination proofs and demonstrate its usefulness.

1   Introduction

Termination is an important research area in Logic Programming [Kowalski, 74] [Lloyd, 87] and *a fortiori* in Constraint Logic Programming [Jaffar & Lassez, 87] [Cohen, 90]. This problem is in general undecidable, but it would be useful to have means which may insure termination. The various approaches pursued toward this end can be roughly divided into three classes:

    • loop checking mechanisms: at run-time, goals are compared with previously generated goals, and if they are "sufficiently similar", the derivation may be pruned [Brough & Walker, 84] [Bol, 90];

    • static-syntactic analysis: at compile-time, the program text is analyzed and terminating and non-terminating queries may be characterized [Vasak & Potter, 86] [De Schreye *et al.*, 89];

    • static-semantic analysis: at compile-time, given a description of a class of queries and a computation rule, from properties of an abstract version of the program, termination proofs may be achieved [Ullman & Van Gelder, 88] [Plümer, 90] [Versaechtse & De Schreye, 91].

    The main drawback of the first approach seems to be its run-time cost. Although the second approach provides useful tools for studying the recursive behavior of simple logic programs, only the third approach seems to be powerful enough to cope with the *problem of local variables*. We illustrate this point by the following

Prolog III [Colmerauer, 90] program:

```
q(<X>.Xs,X) {Xs::L,L=<2}.
q(Xs,Z):- p(Xs,Ys),q(Ys.Ys,Z) {Xs::L,L>=4}.

p(<X,Y,Z>.Xs,<X>) {Xs::L, L=<2}.
p(<X,Y,Z>.Xs,<X>.Zs):- p(Xs,Zs).
```

Let us explain the first clause defining $p/2$. The term $<X,Y,Z>$ denotes the list of length 3 whose the elements are X, Y and Z in that order. The dot means concatenation of *lists*. Bracketed terms, e.g., {Xs::L, L=<2}, represent constraints. The constraint $Xs::L$ means that the list Xs has length L. So the clause means: for every pair of lists $<X,Y,Z>.Xs$ where the length of Xs is less or equal than 2 and $<X>$, p holds. The remaining clauses should be clear. Now suppose we are interested in proving the termination of the class of queries $:- q(Xs,Y)$. where Xs is a list whose length is known, under the Prolog computation rule. Only the second clause defining $q/2$ may cause a problem. Notice that we can not syntactically compare the recursive call $q(Ys.Ys,Z)$ with the head of the clause $q(Xs,Z)$. Using abstract interpretation [Janssens & Bruynooghe, 90], we know that $q/2$ calls $p/2$ with as its first argument the list Xs and its second argument a free variable. Methods from the static-syntactic approach may infer that such calls terminate. Using abstract interpretation one more time, we know that once the call to $p/2$ succeeds, the second argument of $p/2$ is a list. To conclude the termination proof, we have to show that the size, say P_1, of the first argument of $p/2$ is greater than twice the size, say P_2, of the second argument when $p/2$ succeeds. The relation $P_1 > 2*P_2$ is called an *interargument relation* for $p/2$.

The concept of linear interargument relation was introduced by [Karr, 76] in the context of imperative programming languages. [Ullman & Van Gelder, 88] explain how to generate and prove relations of the form $P_i + c \geq P_j$, where c is an integer, for logic programs. This work was first extended by [Plümer, 90] for relations of the form $P_1 + ... + P_k + c \geq P_{k+1} + ... + P_n$, then [Versaechtse & De Schreye, 91] propose to use relations of the form $c_0 + c_1 P_1 + ... + c_n P_n = 0$ (c_i integers). Let us point out that *none* of these approaches would be helpful in our case.

In this paper, we propose a technique for proving interargument relations that take the following form: $c_0 + c_1 P_1 + ... + c_n P_n \geq 0$ (c_i given integers). The rest of the paper is organized as follows. First, we briefly recall the transformation from a logic program to an abstract logic program. Then we explain our technique and prove its correctness. Examples, ranging from termination proof to the improvement of the run-time behavior of logic programs, follow a complete description of the implementation. At last we conclude, comparing our method with related work and summarizing our contribution.

2   From Logic Procedures to Abstract Procedures

In [Versaechtse & De Schreye, 91] the authors show how, given correct type norms and call or success descriptions, one may abstract concrete terms from a logic procedure, resulting in a 'related' abstract procedure where each argument P_i denotes the size of the argument A_i of the original procedure. For example, if we agree to measure the size of a list by its length and if we know the success description of $p/2$: $p(\text{List},\text{List})$ then let $p'/2$ reference the abstract version of $p/2$:

```
p'(X+3,1) {X>=0,X=<2}.
p'(X+3,Y+1):-p'(X,Y) {X>=0,Y>=0}.
```

The following property specifies the link between the logic procedure $p/2$ and its abstract version $p'/2$:

for all ground lists L1 and L2, if $p(\text{L1},\ \text{L2})$ holds then $p'(\text{L1'},\text{L2'})$ holds, where L1' (resp. L2') denotes the length of L1 (resp. L2).

We refer the interested reader to [Versaechtse & De Schreye, 91] for a complete and formal discussion about this topic. Let us go back to our problem of section 1. If we are able to prove:

$\forall$ X, Y natural numbers $p'(X,Y) \implies X \geq 2Y+1$

then we may combine the two previous properties. It gives :

for all ground lists L1 and L2, if $p(\text{L1},\text{L2})$ holds then L1' $> 2*\text{L2'}$ holds, where L1' (resp. L2') denotes the length of L1 (resp. L2).

and concludes our termination proof.

3   The Proof Method

The basic idea underlying our proof method, already mentioned in [Cohen, 90] and [Colmerauer, 90] for instance, relies on the following fact. To prove that *one* linear constraint C is a logical consequence of a set of linear constraints S, where all variables take their values in **N** (the set of natural numbers) and all coefficients in **Q** (the set of rational numbers), it suffices to prove that $S \cup \{\neg C\}$ has no solution in **Q**, which can be easily checked using a symbolic simplex-like algorithm, such as those available in CHIP [Dincbas *et al.*, 88] or Prolog III. Section 3.1. formalizes this idea and section 3.2. describes the implementation.

3.1   Theoretical Study

We assume familiarity with the main notations and results about Logic Programming and Constraint Logic Programming as described in [Lloyd, 87], [Jaffar & Lassez, 87] and [Cohen, 90]. Let V be the countable set $\{X_0,...,X_n,...\}$ of variables. var(T) denotes the set of variables of T. Let F_0 be the set $\{0,1\}$ of constant symbols, F_2 the singleton $\{+\}$ of 2-ary function symbol and $F = F_0 \cup F_2$. Let R_1 be the singleton $\{\neg\}$ of unary predicate symbols, R_i be the set $\{\geq, \leq\}$ of 2-ary predicate symbols, R_e be the singleton $\{ = \}$ of 2-ary predicate symbol and $R_2 = R_e \cup R_i$. We now define the syntax of linear terms and linear constraints:

Definition 1 (linear terms)
$TL(V,F)$ (abbreviated to TL) is the least set such that
- $V \subset TL(V,F)$, $F_0 \subset TL(V,F)$
- If $T1 \in TL(V,F)$ and $T2 \in TL(V,F)$ then $T1+T2 \in TL(V,F)$.

Definition 2 (linear constraints)
$CL(V,F,R)$, where $R \subset R_2$, is the set of atoms of the form $T1$ RelOp $T2$, where
- $T1 \in TL(V,F)$ and $T2 \in TL(V,F)$
- RelOp $\in R$.

We switch from syntax to semantics:

Definition 3 (semantics of linear terms and linear constraints)
- 0 (resp. 1) is interpreted as the natural number 0 (resp. 1),
- + is interpreted as the addition on natural numbers (this justifies why we do not introduce parenthesis when defining terms),
- $\geq$ (resp. $\leq$, =) is interpreted as the relation greater or equal than (resp. less or equal than, equal) on natural numbers,
- $\neg$ is interpreted as logical negation.

Definition 4 (N-valuation and N-interpretation)
A N-valuation is a mapping from V to N. A N-interpretation is an interpretation whose domain is N and which satisfies the conditions of definition 3.

Here is our notion of logical consequence:

Definition 5 ($\models_N$)

Let S be a set (i.e., a conjunction) of linear constraints, C a linear constraint. $S \models_N C$

means that for every θ N-valuation, for every I N-interpretation of $S\ \theta$, I is a model of $S\ \theta$ implies I is a model of $C\ \theta$.

Let us define the operator neg from $CL(V,F,R_i)$ to $CL(V,F,R_i)$:

Definition 6 (neg(C))
Let C be an element of $CL(V,F,R_i)$. neg(C) is the element of $CL(V,F,R_i)$ defined as

follows:
- if $C \equiv T1 \leq T2$ then $\quad$ neg(C) $\equiv T1 \geq T2+1$
- if $C \equiv T1 \geq T2$ then $\quad$ neg(C) $\equiv T1+1 \leq T2$.

We specify the relation between neg and logical negation with the above proposition:

Proposition 1
Let C be an element of $CL(V,F,R_i)$. Then neg(C) $\models_N \neg$ C and $\neg C \models_N$ neg(C).

Proof: obvious. $\qquad\qquad\qquad\qquad\qquad\qquad\qquad\qquad\qquad$ ◊

We formalize the basic idea of our proof method. Let us first give a new definition:

Definition 7 (Q-unsatisfiable)
Let S be a set of linear constraints.
S is Q-unsatisfiable iff S does not have any rational solution.

Proposition 2
Let S be a set of linear constraints from $CL(V,F,R_2)$, C a linear constraint from $CL(V,F,R_i)$. $S \cup \{neg(C)\}$ Q-unsatisfiable implies $S \models_N C$.

Proof: If $S \cup \{neg(C)\}$ is Q-unsatisfiable, then $S \cup \{neg(C)\}$ has no solution in N. Hence by proposition 1, $S \cup \{\neg C\}$ has no solution in N, thus $S \models_N C$.

Of course, the converse does not hold. We have:
$$\{4 \leq 3X \leq 5\} \models_N X \geq 3 \text{ and } \{4 \leq 3X \leq 5, X \leq 2\} \text{ is not Q-insatisfiable.} \qquad ◊$$

Let P be a definite constraint logic program whose terms belong to TL and whose constraints belong to $CL(V,F,R_2)$. Let p/n denote a n-ary logic procedure of P. The clauses defining p/n may be divided into 2 classes (as $\wedge$ is an associative-commutative operator, the following notations are justified):
- k clauses where p/n does not appear in the body of those clauses:
$$Cb_j : \quad p(\overline{H}_j) \leftarrow Q_j \wedge Cs_j$$
- m clauses where p/n does appear in the body of those clauses:
$$Cr_i : \quad p(\overline{H}_i) \leftarrow Q_i \wedge \left[\wedge \, p(\overline{B}_i^j)\right]_{1 \leq j \leq n_i} \wedge Cs_i$$

where
- the Q's denote conjunctions of atoms whose predicate symbols are not p/n,
- the Cs's denote conjunctions of linear constraints.

We now give the format of the properties we want to prove:

Definition 8 (allowed implication)
Let I be the formula:
$$\forall \overline{X} \in N^n \; [\, p(\overline{X}) \rightarrow c(\overline{X}) \,]$$

where
- $\overline{X}$ is a vector of n *distinct variables*,
- p is a n-ary predicate symbol,
- $c(\overline{X})$ is an element of $CL(\overline{X}, F, R_i)$.

Then I is an allowed implication.

We can now formulate the main result of this section:

Theorem (the proof method and its correctness)
Let P be a definite constraint logic program whose terms belong to TL and whose constraints belong to $CL(V, F, R_2)$, p/n denote a n-ary logic procedure of P and I the allowed implication:
$$\forall \overline{X} \in N^n \; [\, p(\overline{X}) \rightarrow c(\overline{X}) \,]$$

If
- for each non-recursive clause Cb_j, $1 \le j \le k$,

$$\{\overline{H}_j = \overline{X}\} \wedge Cs_j \wedge neg(c(\overline{X})) \quad \text{is } \mathbb{Q}\text{-unsatisfiable}$$

- for each recursive clause Cr_i, $1 \le i \le m$,

$$\left[\bigwedge\!\left(\{\overline{B}^{\,j}_i = \overline{X}^{\,j}\} \wedge c(\overline{X}^{\,j}) \right) \right]_{1 \le j \le n_i} \wedge \{\overline{H}_i = \overline{X}\} \wedge Cs_i \wedge neg(c(\overline{X}))$$

is $\mathbb{Q}$-unsatisfiable where the n_i $\overline{X}^{\,j}$'s are vectors of n distinct variables s.t.
$$var(\overline{X}^{\,j}) \cap var(\overline{X}) = \varnothing \qquad \text{for all } 1 \le j \le n_i$$
$$var(\overline{X}^{\,j}) \cap var(\overline{X}^{\,k}) = \varnothing \quad \text{if } j \ne k$$

then I is true in the least N-model of P.

Proof: We prove by induction the following property:
$$\forall \, n \in N \; \left[p(\overline{C}) \in Tp{\uparrow}n \;\Rightarrow\; \models_N c(\overline{C}) \right]$$

For n=0, the property trivially holds. Suppose it holds for each $n < n_0$. Let
$$p(\overline{C}) \in Tp{\uparrow}n_0$$

- if $\exists\, n < n_0$ such that $p(\overline{C}) \in Tp{\uparrow}n$

then by the induction hypothesis, the result holds.
- otherwise:
  - either there exists a ground instance of a non-recursive clause:
    $$p(\overline{H}_j) \leftarrow Q_j \wedge Cs_j$$

and some θ such that:
$$\overline{C} = \overline{H}_j\theta \text{ and } [\, Q_j \wedge Cs_j \,]\, \theta \in Tp{\uparrow}(n_0\text{-}1)$$

Since by the hypothesis and proposition 2:

$$\{\overline{H}_j = \overline{X}\} \wedge Cs_j \models_N c(\overline{X})$$

we have:

$$\{\overline{H}_j\theta = \overline{C}\} \wedge Cs_j\theta \models_N c(\overline{C})$$

then:

$$\models_N c(\overline{C})$$

• or there exists a ground instance of a recursive clause:

$$p(\overline{H}_i) \leftarrow Q_i \wedge \left[\wedge p(\overline{B}_i^{\,j})\right]_{1\leq j\leq n_i} \wedge Cs_i$$

and some θ such that:

$$\overline{C} = \overline{H}_i\theta \text{ and } \left\{Q_i \wedge Cs_i \wedge\left[\wedge p(\overline{B}_i^{\,j})\right]_{1\leq j\leq n_i}\right\}\theta \in Tp\!\uparrow\!(n_0\text{-}1)$$

Since by the hypothesis and proposition 2:

$$\left[\wedge\left(\{\overline{B}_i^{\,j} = \overline{X}^{\,j}\} \wedge c(\overline{X}^{\,j})\right)\right]_{1\leq j\leq n_i} \wedge \{\overline{H}_i = \overline{X}\} \wedge Cs_i \models_N c(\overline{X})$$

and by the induction hypothesis:

$$\left[p(\overline{B}_i^{\,j})\theta \models_N c(\overline{X}^{\,j})\theta\right]_{1\leq j\leq n_i}$$

then we have:

$$\models_N c(\overline{C})$$

i.e., I is true in the least N-model [Jaffar & Lassez, 87] of P. $\Diamond$

3.2 Implementation

First, we would like to point out that restricting the linear constraints to be equalities or large inequalities insures a *safe behavior* of the constraint solver. For instance, the external Prolog III procedure `particular_value/2` allows us to compute counter-examples that do belong to the satisfiability domain of the constraints. It would not be the case if we were using strict inequalities.

Let I be the formula to be proven:

$$\forall <X_1, \ldots, X_n> \in N^n\left[p(X_1, \ldots, X_n) \rightarrow c(X_1, \ldots, X_n)\right]$$

where   • $X_1, \ldots, X_n$ are *n distinct variables*,

      • $c(X_1, \ldots, X_n)$ is an element of $CL(\{X_1, \ldots, X_n\}, F, \{=, \#, \leq, \geq\})$.

The meaning of the main procedure `prove/4` (which does not check the above conditions) is:

    • `prove(<`$X_1, \ldots, X_n$`>,p,c(`$X_1, \ldots, X_n$`),true)` only if I is true,

    • `prove(<`$X_1, \ldots, X_n$`>,p,c(`$X_1, \ldots, X_n$`),dont_know)` iff

        the proof method can not give a definite answer but, if it computes
        some values for $<X_1, \ldots, X_n>$, then $c(X_1, \ldots, X_n)$ is false
        and if $p(X_1, \ldots, X_n)$ is true then I is false.

We want to stress that knowing that I is true does not imply that $p(X_1, \ldots, X_n)$ terminates. It simply proves that, *if* $p(X_1, \ldots, X_n)$ *terminates*, then $c(X_1, \ldots, X_n)$ holds. The code of the procedure prove/4 is listed below:

```
prove(Var,P,C,A):-prove'(Var,P,C,A),!.
prove(Var,P,C,dont_know).

prove'(Var,P,C,true):-
    not(notBase(Var,P,C)),not(notInductive(Var,P,C)).
prove'(Var,P,C,dont_know):-
    notBase(Var,P,C),particular_values(Var).
prove'(Var,P,C,dont_know):-
    notInductive(Var,P,C),particular_values(Var).

notBase(Var,P,C):-rule(P[Var],L),count(L,P,0),neg(C).
notInductive(Var,P,C):-
    rule(P[Var'],L),count(L,P,N),copy(N,P[Var],C,Ps,Cs),
    call(Cs),unifys(Var,Var'),unify(Ps,L),neg(C)
    {Var::Nv,Var'::Nv,  N>=1}.

count([],Pred,0).
count([Pred[_]|Ls],Pred,N+1):-count(Ls,Pred,N).
count([Prd[_]| Ls],Pred,N):-count(Ls,Pred,N) {Prd#Pred}.

copy(N,P,C,Ps,Cs):- assert(imp(P,C),[]),copy(N,Ps,Cs).
copy(1,[P],[C]):-retract(imp(P,C),[]).
copy(N+2,[P|Ps],[C|Cs]):-imp(P,C),copy(N+1,Ps,Cs) {N>=0}.

unify([],L).
unify([P[A1]|Ps],[P[A2]|Ls]):unifys(A1,A2),unify(Ps,Ls).
unify([P[As]|Ps],[Q[_]|Ls]):-unify([P[As]|Ps],Ls) {P#Q}.
unifys(<>,<>).
unifys(<A>.As,<A>.Bs):-unifys(As,Bs).

particular_values(<>).
particular_values(<X>.Xs):-
    particular_value(X,X),integer(X),
    particular_values(Xs).
```

```
'='(X,X).                         neg('='(X,Y)) {X >= Y+1}.
'#'(X,Y) {X >= Y+1}.              neg('='(X,Y)) {X =< Y-1}.
'#'(X,Y) {X =< Y-1}.              neg('#'(X,X)).
'=<'(X,Y) {X =< Y}.               neg('=<'(X,Y)) {X >= Y+1}.
'>='(X,Y) {X >= Y}.               neg('>='(X,Y)) {X =< Y-1}.
```

4 Applications

4.1 Proving Simple Properties of a Program

Consider the following program:
```
whoami(0,X,X) {X>=0}.
whoami(X+1,0,X+1) {X>=0}.
whoami(X+1,Y+1,Z+2):- whoami(X,Y,Z) {X>=0,Y>=0,Z>=0}.
```

Is it true that: $\forall$ X,Y,Z natural numbers whoami(X,Y,Z) $\Rightarrow$ Y=Z ?
```
?- prove(<X,Y,Z>,whoami, Y = Z ,A).
{X = 1, Y = 0, Z = 1, A = dont_know}
?- whoami(1,0,1).
{}
```
As we know that whoami(1,0,1) is true, the above formula is false.

Is it true that: $\forall$ X,Z natural numbers whoami(X,X,Z) $\Rightarrow$ Z $\geq$ X ?
As it is not an allowed implication, we have to generalize it:
```
?- prove(<X,Y,Z>,whoami, Z >= X ,A).
{A = true }
?- prove(<X,Y,Z>,whoami, Z >= Y ,A).
{A = true }
```
So we know that the above formula is true.

At last:
```
?- prove(<X,Y,Z>,whoami, Z = X+Y ,A).
{A = true,X!num ,Y!num }
```

4.2 Finishing the Termination Proof of Section 1

Let us recall the abstract version of p/2:
```
p'(X+3,1) {X>=0,X=<2}.
p'(X+3,Y+1):-p'(X,Y) {X>=0,Y>=0}.
```

We want to prove: $\forall$ X,Y natural numbers p'(X,Y) $\Rightarrow$ X$\geq$ 2Y+1
```
?- prove(<X,Y>,p', X >= Y+Y+1 ,A).
{A = true,Y!num }
```
But we have neither the equality nor the strict inequality:
```
?- prove(<X,Y>,p',X = Y+Y+1,A).
{X = 5, Y = 1, A = dont_know}
?- p'(5,1).
{}
?- prove(<X,Y>,p',X >= Y+Y+2,A).
{X = 3, Y = 1, A = dont_know}
```

```
?- p'(3,1).
{}
```

4.3 An Example from [Versaechtse & De Schreye, 91]

Consider the following program :
```
a(0,0).
a(X+2,Y+3):-a(X,Y) {X>=0,Y>=0}.
a(X+3,Y+2):-a(X,Y) {X>=0,Y>=0}.
```

As there is no non-trivial linear interargument relation that holds, the technique the authors describe does not give any interesting result. On the other hand we can prove that:

$$\forall\ X,Y\ \text{natural numbers}\ a(X,Y) \Rightarrow\ 9X \geq 6Y \geq 4X$$

i.e.:
```
?- prove(<X,Y>,a, Y+Y+Y >= X+X ,A).
{A = true,      X!num ,Y!num }
?- prove(<X,Y>,a, X+X+X >= Y+Y ,A).
{A = true,      X!num ,Y!num }
```

4.4 Improving the Run-Time Behavior of a Program

Here are the classical clauses defining the Fibonacci sequence:
```
fib(0,1).
fib(1,1).
fib(N+2,F+G):-fib(N+1,F),fib(N,G) {N>=0,F>=0,G>=0}.
```

We can show: $\forall\ X,F$ natural numbers $\text{fib}(X,F) \Rightarrow F>=1$
```
?- prove(<X,F>,fib, F >= 1 ,A).
{A = true }
```

The above property, which is a logical consequence of the program, allows us to refine the third clause of the previous program:
```
fib(N+2,F+G):-fib(N+1,F),fib(N,G) {N>=0,F>=1,G>=1}.
```
and fib/2 is *now* a reversible predicate:
```
?- fib(N,8) {N>=0}.
{N=5}
?-
```

4.5 Verifying Constraints

Here is the abstract version of mergeSort/2:

```
mergeSort(0,0).
mergeSort(1,1).
```

```
mergeSort(X,Y):-
    split(X,A,B),
    mergeSort(A,A'),mergeSort(B,B'),
    merge(A',B',Y) {X>=2,X=A+B,Y=A'+B'}.

split(0,0,0).
split(1,1,0).
split(Z+2,X+1,Y+1):- split(Z,X,Y){Z>=0,Y>=0,X>=0}.

merge(0,0,0).
merge(1,0,1).
merge(X+1,Y+1,Z+1):- merge(X,Y+1,Z){X>=0,Y>=0,Z>=0}.
merge(X+1,Y+1,Z+1):- merge(X+1,Y,Z){X>=0,Y>=0,Z>=0}.
```

The two following checks justify the last two constraints of the third clause of mergeSort/2:

```
?- prove(<X,Y,Z>,split, X = Y+Z ,A).
{A = true,Y!num ,Z!num }
?- prove(<X,Y,Z>,merge, Z = X+Y ,A).
{A = true,X!num ,Y!num }
```

We can also show :

```
?- prove(<X,Y>,mergeSort, X = Y ,A).
{A = true }
```

and add this constraint to obtain a reversible predicate.

5   Discussion

The proof method we presented is not a decision procedure: it is correct but not complete. Let us illustrate this remark by considering the following program:

```
r'(X,Y)  :- p'(X,Y) {X>=0,Y>=0}.
```

where p'/2 is the logic procedure defined section 2 and the allowed implication I:

$\forall$ X, Y natural numbers r'(X,Y) $\Rightarrow$ X>=3

Although I is true, we obtain:

```
?- prove(<X,Y>,r',X >= 3,A).
{X = 0, Y = 0, A = dont_know}
```

One may argue that the arithmetic with + as the only function symbol is decidable [Presburger, 30]. Actually this approach has been implemented [Cooper, 72] [Lane, 88] but it seems to be computationally far too expensive. This remark leads us to talk about the complexity of our method. In theory, solving linear rational constraints is a polynomial problem. But the numerical module of Prolog III consists in an incremental implementation of the simplex algorithm, which is an exponential algorithm in the worst case, so the complexity of our method is exponential.

We owe a lot to the work described in [Versaechtse & De Schreye, 91] since our proposal relies on the elegant notion of abstract procedures the authors introduce. Although our method seems simpler and more efficient, its main drawback comes from the fact that the user has to provide the linear interargument relation to be proven. On the other hand, their method computes the coefficients of linear interargument equalities. As we can easily handle linear interargument inequalities, we are currently working on a top-down approach to conjecture the coefficients of the inequalities. Related ideas may be found in [Cousot & Halbwachs, 78] and [Van Gelder, 90]. On the other hand, finding a characterisation of the formulae for which the proposed method is complete is obviously another important issue to investigate

At last, let us summarize our work. We proposed a simple yet powerful method to prove linear interargument inequalities, relying on CLP(Q) and the notion of abstract procedure. We proved its correctness and gave its full code in Prolog III. The implementation, thanks to the external `particular_value/2`, sometimes allows the computation of a counter-example if the relation is false. This technique has several applications, namely the proof of simple properties of programs including help in termination proof, improvement of run-time behavior and verification of constraints in CLP programs, which clearly show its usefulness.

Acknowledgment Frédéric Mesnard is supported by a grant from the French Ministère de la Recherche et de la Technologie.

References

[Bol, 90] R.N. Bol: Toward More Efficient Loop Checks. In *Proc. of NACLP'90*, pp. 465-479, 1990.

[Brough & Walker, 84] D.R. Brough, A. Walker: Some practical properties of logic programming interpreters. In *Proc. of FGCS'84*, pp. 149-156, 1984.

[Cohen, 90] J. Cohen: Constraint Logic Programming Language. In *CACM*, vol. 33, n° 7, pp. 54-68, July 1990.

[Colmerauer, 90] A. Colmerauer: An introduction to Prolog III. In *CACM*, vol. 33, n° 7, pp. 70-90, July 1990.

[Cooper, 72] D.C. Cooper: Theorem Proving in arithmetic without multiplication. In *Machine Intelligence 7*, pp. 91-99, 1972.

[Cousot & Halbwachs, 78] P. Cousot, N. Halbwachs: Automatic discovery of linear restraints among variables of a program. In *Proc. of the 5th ACM Symposium of the POPL*, pp. 84-97, 1987.

[De Schreye et al., 89] D. De Schreye, M. Bruynooghe, K. Versaechtse: On the Existence of Nonterminating Queries for a Restricted Class of PROLOG-Clauses. In *Artificial Intelligence 41*, pp. 237-248, 1989.

[Dincbas *et al.*, 88] M. Dincbas, P. Van Hentenryck, H. Simonis, A. Aggoun, T. Graf, F. Berthier: The Constraint Logic Programming Language CHIP. In *Proc. of FGCS'88*, 1988.

[Jaffar & Lassez, 87] J. Jaffar, J-L. Lassez: Constraint logic programming. In *Proc. of the 14th ACM Symposium of the POPL*, pp. 111-119, 1987.

[Janssens & Bruynooghe, 90] G. Janssens, M. Bruynooghe: Deriving descriptions of possible values of program variables by means of abstract interpretation. Report CW 107, *Dpt of Computer Science, K.U. Leuven*, March 1990.

[Karr, 76] M. Karr: Affine Relationships Among Variables of a Program. In *Acta Informatica* 6, pp.133-151, 1976.

[Kowalski, 74] R.A. Kowalski: Predicate Logic as a Programming Language. In *IFIP*, pp. 569-574, 1974.

[Lane, 88] A. Lane: Trilogy: a New Approach to Logic Programming. In *Byte*, March 1988.

[Lloyd, 87] J.W. Lloyd: *Foundations of Logic Programming*; Springer-Verlag, 1987.

[Presburger, 30] M. Presburger: Über die Vollständigkeit eines gewissen Systems der Arithmetik ganzer Zahlen in welchem die Addition als einzige Operation hervortritt. In *Comptes-rendus du Ier Congrès des Mathématiciens des Pays Slaves*, Warsaw, pp. 92-101, 1930.

[Plümer, 90] L. Plümer: *Termination Proofs for Logic Programs*, LNAI 446, Springer-Verlag, 1990.

[Ullman & Van Gelder, 88] J.D. Ullman & A. Van Gelder: Efficient tests for top-down termination of logical rules. In *JACM*, vol. 35, n° 2, pp. 345-373, April 1988.

[Van Gelder, 90] A. Van Gelder: Deriving Constraints Among Arguments Sizes on Logic Programs. In *Proc. of the 9th Symp. of the PODS*, pp. 47-60, 1990.

[Vasak & Potter, 86] T. Vasak & J. Potter: Characterisation of terminating logic programs. In *Proc. of SLP'86*, pp. 140-147, 1986

[Verschaetse & De Schreye, 91] K. Verschaetse & D. De Schreye: Deriving Termination Proofs for Logic Programs, using Abstract Procedures. In *Proc. of the 8th ICLP* , pp. 301-315, 1991.

Representation of Fragmentary Multilayered Knowledge

Andreas Hamfelt and Åke Hansson

Uppsala Programming Methodology and Artificial Intelligence Laboratory
Computing Science Dept., Uppsala University
Box 520, S-751 20 Uppsala, Sweden

Abstract. Formalization presupposes 'precisification'. A formal representation, therefore, cannot account for all relevant aspects of imprecise domain knowledge. In this paper we present a methodology for dealing with this problem. In an imprecise domain, part of the expertise is to know the realm within which knowledge may be faithfully specialized. In a computer reasoning system, such expert knowledge can be reproduced as a metatheory for proposing and reasoning with formal object theories, each representing one particular specialization. Metaknowledge of this kind will however most often also be imprecise and the expertise on how it may be specialized resides then at the metametalevel, etc. We show that logic provability and upward reflection are adequate means for representing such hierarchical domain knowledge and the dependencies in it between adjacent levels.

1 Background

An important topic for knowledge representation is to explain how imprecise knowledge is dealt with in reasoning and to reproduce the obtained model in a computer reasoning system. Desirable to attain is a model which does not refer to things without obvious counterparts in the domain knowledge, such as fuzzy numbers and the like. A hierarchical model proposed in legal philosophy is close to this objective. In this paper we show a representation of this model in metalogic programming with reflection. Furthermore, we describe the underlying formalization methodology of our system and illustrate its inferencing and user interaction.

The considered model—which we have regimented into an informal theory '$\mathcal{IT}$'—is a multilayered hierarchy of fragmentary rule descriptions. The *multilayering* is caused by 'rules of legal interpretation' which are "inessential in principle, in the sense that, although they are necessitated in practice by the imperfections and the dynamic character of the existing systems, they would not be needed in a perfect, unambiguously formulated, consistent, and complete legal system, conformable to a stable social reality. The actual function of rules of legal interpretation is to direct the identification of the existing system and its continuous construction and readjustment." (Horovitz, [9], p. 94). Since (meta)rules of legal interpretation are imperfect as well, metametarules also are necessary, etc., yielding a whole multilayered hierarchy whose formalization would be impracticable in a single level language whereas a metalogic program may provide the needed multilevel theory structure. Because the content of $\mathcal{IT}$ is imprecise, all that is available for its formalization are schematic descriptions of sentences, which we, analogous with axiom schemata (for logical axioms), will call *schemata* (for non-logical axioms) [4, 6].

In $\mathcal{IT}$ the basic idea is that a theory T_{i+1} completely determines the content of the lower adjacent theory T_i. Thus, the structure of $\mathcal{IT}$ could be understood as an open hierarchy of 'theories' $T_1, T_2, \ldots, T_i, T_{i+1}, \ldots$ where rules of *meaningfulness* and *acceptance* in each T_{i+1} decide, respectively, which specializations of schematic descriptions of sentences for level i are meaningful, thus belonging to the language of T_i, and which are legally acceptable, thus belonging to T_i. If a theorem can be deduced from theory T_{i+1} expressing that a sentence A_i is provable from theory T_i, then A_i belongs to theory T_i. The same holds for theory T_{i+2} with respect to theory T_{i+1}, etc. No level has rules that do not require interpretation so, in principle, this proceeds *ad infinitum*. This endlessness can be dealt with, e.g., by choosing some level n to be the 'topmost' at which discretion is used for specifying schemata, thus making the validity of all rules on each level i, $1 \leq i < n$, ultimately depend on discretion. A characterization is needed of the *provability relationship* between two adjacent levels $i+1$ and i ranging over n distinct levels (theories), and forming a hierarchy of dependent relations directed from the highest to the lowest level. On each level i this provability relation $T_i \vdash A_i$ should coincide with the rules of logic. The hierarchical dependence of the provability relationship may be characterized $T_i \vdash A_i$ *iff* $T_{i+1} \vdash {}'T_i \vdash A_i{}'$, where ${}'T_i \vdash A_i{}'$ names $T_i \vdash A_i$ and $1 \leq i < n$. With theory T_n specified, the hierarchy decides the content of the object level as well as of all the other levels. That is to say, T_n decides the contents of all the theories $T_{n-1}, \ldots, T_1$. This characterization constitutes the *rules of acceptance* of $\mathcal{IT}$, whose role thus is to determine whether or not a certain sentence really is a legal rule.

Rules of legal interpretation give legal reasoning an informal counterpart to upward reflection. If a legal rule is proposed for solving a legal case its structure and content must be shown to accord with the (meta)rules of legal interpretation, otherwise the rule is legally invalid. Likewise in automatized legal reasoning, a formula A representing a legal rule can be assumed included in an object level theory O representing legally valid rules if its inclusion accords with the metalevel theory M of formulas representing rules of legal interpretation, i.e., assuming *Demo* defines provability we have

$$(UR) \qquad Demo(O, name(A)) \leftarrow Demo(M, name(Demo(O, name(A))))$$

where *Demo* holds for formulas belonging to or deducible from a theory, cf. Kowalski [11]. This corresponds to upward reflection, where something proved on an upper level is forced upon a lower level. In contrast, downward reflection is unsound since something 'proved' from imperfect knowledge on a lower level cannot be forced upon an upper level thereby perhaps contradicting rules accepted by the legal principles on this level: the knowledge of each T_i is always imperfect and applying the rules in T_{i+1} is necessary for assessing, accepting or rejecting the rule proposed for T_i. Thus a predicate *Demo* founded on representability [1] (equivalence between object language and metalanguage proofs) and requiring both reflection rules is inadequate for characterizing the provability relation of $\mathcal{IT}$. $\mathcal{IT}$ requires *definition* of the provability relation for an object language instead of *representation* of an existing such relation. Metalevel definitions allowing upward reflection to enforce proofs on the object level which the object level theorem prover itself cannot carry out are examples of *Demo* predicates that deviate from the representability notion and give *definitional* extensions of theories, cf. Kowalski [11]. It should be noted that in our system no theorem prover exists in advance for a theory T_i. Therefore, *Demo* in theory T_{i+1} is used both for defining an object level theorem prover for theory T_i as well as in the representation of the upward reflection rule.

That (UR) is part of the provability characterization helps to reduce the complexity caused by $\mathcal{IT}$'s hierarchical structure since it allows each T_i to be characterized 'locally' towards its immediate object of study, i.e., each meta/object language relation is represented separately. The *Demo* predicate defines the provability relation on a lower level thus providing a link between adjacent levels. For example, the formula $Demo(name(T_1), name(A))$ says (on level 2) that the rule A is provable from, and thus included in, the object theory T_1. Stated as a goal the formula reads 'is A provable from the theory T_1?' the proof of which corresponds to a line of arguments to the effect that the inclusion of A in T_1 should be regarded as in accordance with the metarules of legal interpretation in T_2, thus showing that the rule is legally acceptable. (UR) makes this inclusion dependent on these metarules whose inclusions in T_2 in turn depend on theories of higher levels, whose provability definitions are characterized in a similar way yielding a whole hierarchy of interdepending provability definitions. Still, however, each T_i may be described and considered as a separate theory. Moreover, we have proved [3] that (UR) is a derived inference rule in our system, i.e., though (UR) is part of our provability definition, it does not extend the notion of logic provability.

The declarative reading of $Demo(name(T_1), name(A))$ is rendered by taking T_1 as a static theory implicitly consisting of all rules fulfilling the conditions for inclusion. Only the boundary between rules shown to satisfy the conditions and those not yet shown to satisfy the conditions moves just as in Sergot's 'query the user' [12] the boundary between the facts, given to the computer by the user and those not yet given, moves.

Part of our methodology for formalizing $\mathcal{IT}$ is thus to specify the theory T_1 metatheoretically in terms of what can be proved from T_1. The topmost theory T_n gives an axiomatic definition of provability of theory T_{n-1} and indirectly of all theories T_i, $i < n$, hence, embracing all the non-logical axioms of these theories.

This methodology must be elaborated, though, to account for the *assimilation of external knowledge*. Three separate and distinct 'theories' are involved in the process of a formalization, cf. Kleene ([10], pp. 65, 69): $\mathcal{IT}$ the informal theory of which the formal system constitutes a formalization, $\mathcal{OT}$ the formal system or object theory, and $\mathcal{MT}$, the metatheory, in which the formal system $\mathcal{OT}$ is described and studied. Theories $\mathcal{IT}$ and $\mathcal{MT}$, which are informal, do not have an exactly determined structure, as does $\mathcal{OT}$. Consider the following two approaches for studying $\mathcal{OT}$: (i) the formal theory $\mathcal{OT}$ is "introduced at once in its full-fledged complexity" and investigated by methods without making use of an interpretation. (This is known as the metamathematical approach if the methods are finitary.) (ii) the formal theory $\mathcal{OT}$ is studied by recognizing an interpretation of the theory under which it constitutes a formalization of $\mathcal{IT}$, i.e., we analyse existing informal theories $\mathcal{IT}$, "selecting and stereotyping fundamental concepts, presuppositions and deductive connections, and thus eventually arrive at a formal system", i.e., at the formal theory OT. Approach (i) presupposes that the complexity of the informal theory is fully understood. This does not hold in our domain where a realistic system can only have a partial axiomatization of the formal object theory. This axiomatization can gradually be extended, though, by consulting the user both for supply of metalinguistic entities representing objects of the formal object theory and for completing formal proofs in it. Thus, in a practical system we must adhere to approach (ii).

Although the non-logical axioms of the formal object theory OT cannot be enumerated in advance its possible content can nevertheless be discussed in a metatheory MT. To this end we have devised a 'semiformal' metalanguage—whose object language is the whole composite n-level language of OT—for a theory MT which axiomatizes the 'available' part of the formal object theory OT and encodes rules for the assimilation into it of externally supplied knowledge fragments. Knowledge assimilation is dependent on the deductive structure of OT, which can be accounted for in MT since its objects of discourse include formal proofs, i.e., sequences of formulas of OT. In MT prover(demo(name(t(1)),name($\langle A \rangle$)),...,...,Proof) expresses that a formalization of a rule A is included in the formal theory $t(1)$ of OT, which represents the informal theory T_1 of IT. In the metalanguage this inclusion is verified by a sequence Proof of statements each of which names an 'object/object'-inference or a 'meta/object'-inference of the object language, thus constituting a metaproof in MT of a formal proof in OT. MT has a formal part and an informal part and it is in this sense it is a 'semiformal' theory. MT's formal part consists of the assertions the program makes about the content of OT in unit clauses for prover, the informal part simply consists of those assertions about OT's content that the user would approve of.

2 The Semiformal Metatheory

We now briefly describe our metatheory MT, mainly as a metaprogram partially characterizing OT whose intended interpretation is IT. Our metalogic consists of Horn clauses and the inference mechanism of Prolog.

To the *formation rules* and *rules of inference* (including axiom schemata) of proof theory correspond the *rules of meaningfulness* and the *rules of acceptance*, respectively, both of which have a more vague character than their proof theory counterparts. The *rules of meaningfulness* in MT can only partially characterize sentences of the object language, i.e., the language of OT. In the representation in the metalanguage, we assume a fixed structure for designating a class of rules, i.e., a schema. Within this structure local differences must be met, i.e., different specializations of the schema have to give different representations of sentences (rules) of the object language. These local differences are expressed by metavariables which have to be filled in by a user and satisfy certain interactively investigated typing conditions. Let us illustrate this with the program clause that characterizes a provision and its 'open texture'. This provision has the linguistic wording of the metavariable Text. But this provision may, e.g., be used to regulate a case of 'a hire of goods', although its linguistic wording does not say so. Thus the phrase 'a sale of goods' may be open for different interpretations. This understanding of a provision is the basis for our representation of a provision as a schema. Moreover, all the possible specializations of this schema that give meaningful and acceptable legal rules cannot be decided in advance.

In the clauses below n(...) is a shorthand for name(...) for which we postulate an inverse law of naming, i.e., $n(A)=n(B){\rightarrow}A=B$. As to the problem of naming in metaprogramming, observe that all variables are metavariables; there are no object variables, the object theory consists of propositional sentences only.

```
meaningful_sent(t(1),RuleProp1,[ModAt1,unspec],LegSet1,Text):-
  RuleProp1 = (legal_cons(pay,X,Y,goods,price):-
            and(actor1(X,goods),and(actor2(Y,goods),and(unsettled_price(goods),
            and(demands(Y,price),reasonable(price,goods))))))),
  ModAt1 = [X/vendee,Y/vendor], Types = [actor(X),actor(Y)],
  LegSet1 = [[provision_no(sga(5))|_],LegSet0],
  Text = 'If a sale of goods has been made but no price settled then
        the vendee should pay what the vendor demands if reasonable.'
  proper_typing(t(1),RuleProp1,ModAt1,Types,Text).
```

In our representation of $\mathcal{IT}$ we assume this provision open with respect to the concepts 'vendee' and 'vendor'. (Note that a provision can be open also to its 'logical' structure.) So, the assumed fixed structure of this provision is represented in the metalanguage as the term specified for the metavariable RuleProp1 with open places expressed by the metavariables X and Y. These variables have to be specialized interactively with the user. The predicate proper_typing is defined for this interaction. The metavariable ModAt1 expresses the relation between the concepts of $\mathcal{IT}$, i.e., the text of Text and its open parts 'vendee' and 'vendor', and its formal counterpart in $\mathcal{OT}$ partly specified in RuleProp1. A proper typing carried out by the user gives a meaningful rule of the language of theory t(1) represented in the metalanguage by the specialized term of RuleProp1. The metavariable LegSet1 identifies what part of level 1 in $\mathcal{IT}$ is relevant for a particular case.

The *rules of acceptance* may also only be partially characterized in the metalanguage. However, a user can interactively add interpretation data, thereby extending the partial characterization of $\mathcal{OT}$ in metatheory $\mathcal{MT}$. Whether or not a meaningful rule belongs to a theory of $\mathcal{IT}$—i.e., is legally acceptable, and thus should have a formal counterpart in $\mathcal{OT}$—is determined by the prover clauses below which belong to $\mathcal{MT}$ and take as object theory the whole multilayered $\mathcal{OT}$. Their first demo argument defines the formalization in $\mathcal{OT}$ of logic provability between a theory T_i of $\mathcal{IT}$ and a sentence of $\mathcal{IT}$ but though, e.g., the fourth proof term argument has a counterpart in $\mathcal{OT}$—a formal proof extending over the whole hierarchy of $\mathcal{OT}$—it includes expressions solely of $\mathcal{MT}$ as well.

```
prover(demo(n(t(I)),n(SentPropl)),ModI,LegSetI,ProofI):-                [UP]
  propose_sent(t(I),SentPropl,ModI,LegSetI),
  J is I + 1,
  ground([SentPropl,ModI,LegSetI]),
  permissible(t(I),SentPropl),
  prover(demo(n(t(J)),n(demo(n(t(I)),n(SentPropl)))),
```

```
       [ModAtJ,ModI],[LegSetAtJ,LegSetI],ProofJ),
Proofl = (sentence_of(theory(I),SentPropI):-
           proof_of(theory(J),proved(theory(I),SentPropI),ProofJ)).
```

```
permissible(t(I),SentPropI):-I = 1.
permissible(t(I),SentPropI):-I ≥ 2,\+ SentPropI = (Head:-Body).
```

Clause [UP] encodes in $\mathcal{MT}$ upward reflection between two theories T_i and T_j of arbitrary adjacent levels in $\mathcal{IT}$, with formal counterparts t(I) and t(J) in $\mathcal{OT}$. A sentence is assumed to belong to a theory T_i if this accords with the rules of theory T_j of the higher adjacent level. In $\mathcal{MT}$, LegSetI and ModI identify and modify formula schemata corresponding to known fragments of sentences of the theory T_i. The predicate propose_sent is defined to specialize interactively with a user such meaningful_sent schemata and see to it that variables range over correct domains. Proofl is a metaproof in $\mathcal{MT}$ of the existence of a sequence of formulas in $\mathcal{OT}$'s formalization of $\mathcal{IT}$ constituting a formal proof of the proposed sentence.

The permissible subgoal constrains upward reflection. If each sentence were upward reflected directly when proposed, the reasoning process would ascend directly to the topmost level since the metarule proposed for assessing the sentence would itself directly be upward reflected, etc. Therefore, at levels i, $i \geq 2$, only sentence proposals which are ground facts may be upward reflected, postponing the assessment of rules, which may only be proposed as non-ground conditional sentences, till facts are activated by their premises. Under this reasoning scheme the content of all sentences involved in the reasoning process will eventually be assessed.

Clause [ANDI] handles $\wedge$-introduction. In $\mathcal{MT}$ a theory T_i of $\mathcal{IT}$, with t(I) as formal counterpart in $\mathcal{OT}$, is assumed to include a sentence which is a conjunction if both its conjuncts may be assumed included in T_i.

[ANDI]

```
prover(demo(n(t(I)),n(and(G1,G2))),[[ModG1,ModG2],ModsBelow],LegSetI,Proofl):-
  I ≥ 2,
   prover(demo(n(t(I)),n(G1)),[ModG1,ModsBelow],LegSetI,ProofG1),
   prover(demo(n(t(I)),n(G2)),[ModG2,ModsBelow],LegSetI,ProofG2),
   Proofl =(sentence_of(theory(I),and(G1,G2)):-
             and(proof_of(theory(I),G1,ProofG1),proof_of(theory(I),G2,ProofG2))).
```

Clause [MP] encodes modus ponens. In $\mathcal{MT}$ a theory T_i of $\mathcal{IT}$, with t(I) as formal counterpart in $\mathcal{OT}$, is assumed to include a sentence which is the consequence of a proposed implication of T_i whose antecedent can be assumed included in T_i.

```
prover(demo(n(t(I)),n(HeadI)),ModI,LegSetI,Proofl):-                          [MP]
  I ≥ 2,
   propose_sent(t(I),(HeadI:-BodyI),ModI,LegSetI),
   prover(demo(n(t(I)),n(BodyI)),ModI,LegSetI,ProofBodyI),
   Proofl = (sentence_of(theory(I),HeadI):-and(rule_of(theory(I),(HeadI:-BodyI)),
                                         proof_of(theory(I),BodyI,ProofBodyI))).
```

The knowledge of rules in $\mathcal{IT}$ for assessing sentence proposals for the adjacent lower level theory T_i will at some level j be too rudimentary for composing a theory T_j.

At this level, T_j is considered to be the user's opinion of the sentences proposed for T_i. This is encoded in $\mathcal{MT}$ in the clause [TOP].

```
prover(demo(n(t(J)),n(demo(n(t(I)),n(RulePropl)))),ModJ,LegSetJ,ProofJ):-          [TOP]
J ≥ 2,
\+ propose_sent(t(J),(demo(n(t(I)),n(RuleProplI)):-BodyJ),ModJ,LegSetJ),
external_confirmation(t(I)),RuleProplI,ModJ,LegSetJ),
ProofJ = externally_confirmed(sentence_of(theory(I),RuleProplI)).
```

3 Knowledge Assimilation

The open textured character of legal knowledge entails that computerized legal reasoning must rely on the ability of the user both for supply and assessment of data. Moreover, the user interaction has to be structured and frequent. An appropriate 'query the user facility' (cf. Sergot [12]) is necessary and should promote (1) meaningful user answers and queries, (2) a construction of adequate terms for describing proofs, (3) an intelligible explanation of derived conclusions by appropriate display of proof terms, and (4) a natural order in which the system poses questions to the user. In the present study we have delved into all four of the above aspects.

The frequency of the user interaction distinguishes our case from the system SIMPLE [12] which deals with some of these aspects. In our system it is vital that questions are posed in an intuitively intelligible order. Flagging 'Ask-about' predicates as in SIMPLE is too rough an approach for obtaining a natural order of questions. We exploit the metalogic structure of our system to get a question generator independent of the inference strategy of the formal object system. The target aimed at is that questions should appear in an order which resembles an average lawyer's lines of reasoning. We have reason to believe that it is appropriate to begin by asking about lowest level rules and then gradually move up in the hierarchy. Replacing this protocol by some other is simplified by the separation of the question generator from the object system.

In contrast to the SIMPLE system we do not presuppose that a user by himself is capable of composing an adequate query. This cannot be expected in the legal domain. In our system the user query is composed interactively in order to promote meaningfulness. For example, we have typing guidelines which determine the range of acceptable concepts in a query. Also, the various levels in our system have their own strategies for accessing and using the query information data base for that level. Sergot concludes his study by sketching on a so called 'knowledge base dictionary' with similar possibilities.

Our system has a 'knowledge driven' inference strategy, i.e., resolving rules are not simply retrieved from the rule base. Both the searching for appropriate rule schemata and their specializations are influenced by the user with the support of encoded type checking knowledge. This demarcates the range of possible rules and may improve efficiency, especially for large rule bases. When the proof term is constructed, only rules that have been deemed relevant are included.

SIMPLE's query evaluator is a metaprogram implemented on top of micro-PROLOG

$$Evaluate(query) \leftarrow Demo(GlobalDB \cup User, query, result) \wedge ExtractOutput(result).$$

The *result* parameter is the proof of *query*. The program executor of micro-PROLOG does not make available the proof's structure. Therefore reflection cannot be used in SIMPLE. In our system consisting of $\mathcal{MT}$ and $\mathcal{OT}$ we *simulate*, in our semiformal metalanguage, upward reflection between a theory T_{i+1} in $\mathcal{OT}$ and a theory T_i on the adjacent lower level in $\mathcal{OT}$. Since the semiformal metalanguage takes the whole hierarchy of formal languages as its object language, proof structures may be construed for reasoning between these levels as well as within a certain level and we can use reflection without loosing the proof structure. It is to be noted that there is no reflection between $\mathcal{MT}$ and $\mathcal{OT}$; the upward reflections take place within the object theory $\mathcal{OT}$.

If the explanation facility displays every resolution step in the proof, the result often becomes too full of details to be really useful. This problem can be reduced by selectiveness either when composing the proof term, or when displaying it, or both. SIMPLE uses the first of these two approaches. Some relations are declared as 'built-in' to the effect that the individual steps in their execution are not included in the proof term. Sergot calls for a neater scheme for displaying proofs. We combine both approaches. Our knowledge driven inference strategy imposes selectiveness when the proof term is built. The proof term is presented by a piecemeal display of appropriate portions of the proof and well demarcated parts of it may be skipped if the user so wishes.

It is awkward displaying a proof which demonstrates that a query could not be proven since all failure branches in the computation are included. Why-not questions are however desirable. In our system the user is explicitly told that a certain method for legal reasoning, say *analogia legis*, cannot be used for resolving the case, and is asked whether he desires to try another method. When all possible methods are exhausted he could be told that no more method exists whereupon his query fails.

Sergot proposes an approach for making input and output declarative. With reference to the query evaluator described above, he says "The way to understand declaratively what is going on here is to realize that the combined database, the union between *GlobalDB* and *User*, remains static and fixed. Only the boundary between the two components moves, and then only to make the machine take some burdens of the user." This is paralleled in our system for each level of $\mathcal{OT}$. But, in contrast to Sergot's *Demo*-program, it is the prover clauses of $\mathcal{MT}$ that handle the user interactions for each level of $\mathcal{OT}$. The query

prover(demo(n(t(1)),n(SentProp1)),Mod1,LegSet1,Proof).

should be understood as 'is it in accordance with the adjacent higher level to assume that the rule proposal named n(SentProp1) (modified by Mod1 and belonging to the legal context LegSet1) is included in the theory T_1?'. The theory T_1 is to be understood as the set of all sentences which satisfy the constraints imposed by the higher levels. All sentences the user may successfully input belong to this set.

Below a part of the computation of this query is described supplemented by some fragments of the user interaction session involved. This will give a flavour of the 'query the user' facilities our system provides and also a good practical illustration of the aspect of knowledge assimilation by user interaction in our system but also indirectly of the aspect of knowledge processing. Below we refer to rules of theory T_1 as primary rules, of T_2 as secondary rules, etc.

Since we want to know whether a certain primary rule is included in T_1, each session begins with the completely non-ground query above about the content of that theory. As illustrated above, the content of this and other theories is fragmentary described in the meaningful_sent clauses. Thus the goal is to assimilate new knowledge into a fragmentary described theory. The assimilation must satisfy certain conditions and the resulting theory should give a reasonable suggestion for solving the current case. Knowledge to be assimilated into the theory of some level i is proposed by the user and assessed and accepted by rules of a theory at level $i+1$.

The first task for the system is to identify which part of the legal system is relevant for the case the user has in mind. The query resolves with the prover clause [UP] leading to six subgoals, the last of which builds the proof term to bind Proof1. Below, we refrain from discussing how the proof term is built during the computation. Through user interaction the first subgoal of [UP], i.e.,

propose_sent(t(1),SentProp1,Mod1,LegSet1)

selects a legal rule and modifies it for the current case. The unifying clause

```
propose_sent(Theory,RuleProp,Mod,LegSet):-
  (Theory = t(1);RuleProp=(demo(_,_):-Body)),      % ⇐ The legal context may be
  find_legal_setting(Theory,LegSet),               % assumed unknown if any of
  meaningful_sent(Theory,RuleProp,Mod,LegSet,Text). % these two conditions hold.
```

identifies the relevant part of the legal domain from which it retrieves a proposal for a rule provided it is meaningful. The latter is sorted out by meaningful_sent clauses, say, the one presented in sect. 2. In this clause the proper_typing condition is intended to promote that user proposed modifications preserve the meaningfulness of the rule. This is accomplished interactively using standard interface and data retrieval techniques, but below shown in a compressed form. Now the interaction looks like

Which of these legal fields encompasses your case?
  Commercial Law (cl) Real Estate Law (rel) Penal Law (pel) Procedural Law (prl)
cl. (user answers in italics)
Which of these legal problems corresponds to your case?
  Cancellation (canc) Completion (compl) Risk (r) Delay (d)
  Determination of Purchase Money (dpm)
dpm.
Which of these provisions seems relevant?
  sga(5) If a sale of goods has been made but no price settled then
        the vendee should pay what the vendor demands if reasonable
  sga(6) If ... etc.
sga(5).

Now a possibly relevant provision, here the sect. 5 of the Sale of Goods Act, has been identified. The next task is to determine, first whether it is possible to adapt it to the current case, and second whether the adapted provision may be assimilated into the theory.

The following part of the session shows the information the user receives and settles interactively the proposed modifications and whether these belong to the same type as the concept they replace. The last is to promote that the resulting rule is at least meaningful (obviously we should not replace 'vendee' by 'Fido', etc.).

This text describes a schematic rule which comprises some open textured legal concepts.

If a sale of goods has been made but no price settled then the <u>vendee</u> should pay what the <u>vendor</u> demands if reasonable.

In this text the open textured legal concepts are underlined. These may be replaced by similar concepts, i.e., concepts of the same type, resulting in a modified but perhaps still legally relevant rule that may be applied to resolve the current case.

The concept: 'vendee' is the rule's example of a concept of the type: 'actor'
Please modify it: *hirer*.
The concept: 'vendor' is the rule's example of a concept of the type: 'actor'
Please modify it: *letter*.

Generally, there is a need for a representation of texts which is more structured and comprehensive than underlining for showing which are the open textured legal concepts.
    The above user interaction makes the first subgoal of [UP] return with the following ground argument bindings, i.e., the schemata from sect. 5 Sale of Goods Act is adapted into a primary rule proposal regulating a case of 'hire of goods',

LegSet1 = [[provision_no(sga(5)), provision_category('Determination of Purchase
                Money'),legal_field('Commercial Law')],unspec],   call it $\langle$leg_set_1$\rangle$
Mod1 = [[hirer/vendee,letter/vendor],unspec],   call it $\langle$mod_1$\rangle$
RuleProp1 = (legal_cons(pay,hirer,letter,goods,price):-
            and(actor1(hirer,goods),and(actor2(letter,goods),and(unsettled_price(goods),
            and(demands(letter,price),reasonable(price,goods))))))), call it $\langle$rule_prop_1$\rangle$

where $\langle$*name*$\rangle$ is shorthand for an occurrence of the term named by *name*. Now it must be established whether it accords with the higher adjacent level, i.e., the theory T_2, to assume a primary rule with this proposed content is included in the theory T_1. This is accomplished through 'upward reflection'. Before a formula with content information is upward reflected it must be checked for groundness (a hack) and permissibleness. These are the tasks of the third and fourth subgoals of [UP] which permit a conditional rule on level 1 to be upward reflected. The fifth, 'upward reflection', subgoal of [UP]

prover(demo(n(t(2)),n(demo(n(t(1)),n($\langle$rule_prop_1$\rangle$)))),
    [ModAt2,$\langle$mod_1$\rangle$],[LegSetAt2,$\langle$leg_set_1$\rangle$],Proof2),

resolves with the prover clause [MP] leading to four subgoals (the first and last of which controls the index of the current level and builds the proof term, respectively). Now a secondary rule must be proposed for assessing the lower level expression. The second subgoal of [MP] is

propose_sent(t(2),(demo(n(t(1)),n($\langle$rule_prop_1$\rangle$))):-Body2),$\langle$mod_2$\rangle$,$\langle$leg_set_2$\rangle$), where $\langle$mod_2$\rangle$ is [ModAt2,$\langle$mod_1$\rangle$], $\langle$mod_1$\rangle$ is [$\langle$mod_at_1$\rangle$,unspec], $\langle$mod_at_1$\rangle$ is [hirer/vendee, letter/vendor] and $\langle$leg_set_2$\rangle$ is [LegSetAt2,$\langle$leg_set_1$\rangle$].

Suppose the adapted provision is not already known to be included in the theory T_1. Then it may nevertheless be assumed included, provided its inclusion accords with the legal interpretation principles for reducing *lacunae* at level 2, e.g., *analogia legis*. The user interaction looks like

A textual reading of available legal sources does not indicate any solution to your case. However, various accepted methods exist for interpreting legal sources and their possible adaptions. Applying such a method might result in a proposal for a solution.

Three methods are 'analogia legis' (al), 'e contrario' (ec), and 'extensive' (e) interpretation. Specify your choice: *al*.

Analogia legis has been chosen and the relevant rules involved in reasoning by analogy are presented, and the user may either accept or reject them. The first rule encodes the relation between primary rules of theory T_1 and secondary rules for analogia legis of theory T_2.

```
meaningful_sent(t(2),RuleProp2,Mod2,LegSet2,Text):-
  RuleProp2 =(demo(n(t(1)),n(RuleProp1)):-
                analogia_legis(n(RuleProp1),n(ModAt1),LegSet1)),
  Mod2 = [_,[ModAt1,_]],
  LegSet2 = [[interpretation_theory('analogia legis')|_],LegSet1],
  Text = "A primary rule proposal is legally valid (i.e., belongs to the
          theory t1 of valid primary rules) if its inclusion accords with
          the secondary rule for analogia legis.'...',
  proper_typing(t(2),RuleProp2,[],[],Text).
```

The user interaction now looks like

A primary rule proposal is legally valid (i.e., belongs to the theory t1 of valid primary rules) if its inclusion accords with the secondary rule for analogia legis.

(If you think analogia legis is inappropriate for your case you may reject this rule whereupon other possible secondary rules will be suggested.)

This rule description does not contain any open textured legal concepts which at this point can be further particularized. Wherever such concepts occur in its premises they will be handled instead as the rules matching these premises are activated. Do you accept the rule? *yes*.

This answer verifies that the secondary rule for analogia legis should be attempted and the second subgoal of [MP] returns with its second argument bound to

```
(demo(n(t(1)),n(⟨rule_prop_1⟩)):-
   analogia_legis(n(⟨rule_prop_1⟩),n(⟨mod_at_1⟩),⟨leg_set_1⟩)))
```

and LegSetAt2 bound to [interpretation_theory('analogia legis')|_]. The third subgoal of [MP]

```
prover(demo(n(t(2)),n(analogia_legis(n(⟨rule_prop_1⟩),n((mod_at_1)),⟨leg_set_1⟩)))),
    ⟨mod_2⟩,⟨leg_set_2⟩,ProofBody2),
```

recursively calls [MP]. Now a meaningful proposal for an actual analogia legis secondary rule will, by the second propose_sent subgoal of [MP], be retrieved from the clause below

```
meaningful_sent(t(2),RuleProp2,_,LegSet2,Text):-
  RuleProp2 = (analogia_legis(n((Cons:-Ante)),n(ModAt1),LegSet1):-
                and(not(casuistical_interp(LegalField,n((not(Cons):-Ante))))),
                and(intended_for(ProvisionNo,n(TypeCase)),
                and(substantial_similarity(n(TypeCase),n(Ante),n(ModAt1)),
                and(intended_to_meet(ProvisionNo,Interests,LegalField),
                and(supports(ProvisionNo,n(ModAt1),ProInt,Interests),
                and(recommend_rejection(ProvisionNo,n(ModAt1),ContraInt,Interests),
                outweigh(ProInt,ContraInt))))))))),
  LegSet2 = [[interpretation_theory('analogia legis')|_],LegSet1],
  LegSet1 = [[provision_no(ProvisionNo),_,legal_field(LegalField)],_],
  Text = See below.
  proper_typing(t(2),RuleProp2,[],[],Text).
```

The rule is presented:

A certain rule may be applied to a case not subsumed, or at least not with certainty subsumed, under the rule's linguistic wording, if the case is not the object of a particular explicit rule, if the case has a substantial similarity to those the rule is intended for, if interests of some importance, which the rule is intended to meet, support such an application, and if no contrary interests exist recommending the rejection of such an application.

Do you accept the rule? *yes.*

The propose_sent subgoal of [MP] returns with these bindings (where ⟨rule_prop_1⟩ is ((cons_rule_1):- ⟨ante_rule_1⟩))

```
analogia_legis(n(((cons_rule_1):-(ante_rule_1))),n((mod_at_1)),⟨leg_set_1⟩):-
  and(not(casuistical_interp('Commercial Law',n((not(cons_rule_1):-(ante_rule_1))))),
  and(intended_for(sga(5),n(TypeCase)),
  and(substantial_similarity(n(TypeCase),n((ante_rule_1)),n((mod_at_1))),
  and(intended_to_meet(sga(5),Interests,'Commercial Law'),
  and(supports(sga(5),n((mod_at_1)),ProInt,Interests),
  and(recommend_rejection(sga(5),n((mod_at_1)),ContraInt,Interests),
  outweigh(ProInt,ContraInt))))))).
```

The user's acceptance implies that the system eventually assumes the completely specialized rule of *analogia legis* as a non-logical axiom. Now it must be proved that with the proposed content the antecedent of the *analogia legis* rule (call it ⟨al_body⟩)

is included in T_2. The third subgoal of [MP] is

prover(demo(n(t(2)),n(⟨al_body⟩))),¬
    [[interpretation_theory('analogia legis')],⟨leg_set_1⟩],_),

and each of the conjuncts in ⟨al_body⟩ will be demonstrated in turn by the prover clauses [ANDI], [MP], and [UP]. The conditions, under which *analogia legis* may be applied, vary from field to field. Therefore, when facts are eventually activated by the rule's premises or the premises of matching rules, the content of these facts must be assessed with respect to the current legal field and other pertinent aspects. To illustrate how a user proposed content for a sentence is accepted (or rejected) at higher levels let us focus on the fourth premise which gives rise to the goal

prover(demo(n(t(2)),n(intended_to_meet(sga(5),Interests,'Commercial Law'))),
    ¬[[interpretation_theory('analogia legis')|_],⟨leg_set_1⟩]).

A sentence that may satisfy this premise is presented

> The provision sect. 5 Sale of Goods Act is to be interpreted as were it intended to **protect consumers and similar groups**.

and the user is invited to adapt it to fit his case.

The concept: 'consumer protection'
is the rule's example of a concept belonging to the type: 'protection of weaker party'
Please modify it: *hirer protection*.

Now the user has proposed a content which is meaningful in the sense that the modified concepts belong to the same type as those they replace. If the result is an unconditional sentence its inclusion in the theory T_2 must be accepted by the rules of theory T_3. If it is a conditional sentence it is assumed included in T_2 directly after the user's acceptance. The resolving clauses in the respective cases are [UP] and [MP]. Thus, in the first case upward reflection occurs immediately whereas in the second case it is postponed until backward inferencing by modus ponens at the current level leads to the proposal of a fact. Since prompted by lack of resolving rules, this is a kind of implicit reflection cf. [2].

Suppose a fact is proposed. The goal will resolve with the prover clause [UP], whose recursive fifth subgoal resolves with the prover clause [MP] leading to the application of tertiary rules for assessing the proposed (secondary) fact. Reasons of space force us to remove a part of the trace and interaction session here. The inferencing at the tertiary level is similar to that just described for the secondary level. We conclude this section with a fragment of the trace in which a tertiary fact is proposed but no quaternary rules exist for assessing it. The upward reflected goal looks like

prover(demo(n(t(4)),
    n(demo(n(t(3)),
        n(adequate_to_equalize('actors with similar economical positions',
                'consumer protection'/'hirer protection',
                'Commercial Law'))))),Mod4,LegSet4,_).

For the theory T_4 propose_sent fails however to return any quaternary rules which may assess the adequate_to_equalize fact. The goal resolves with the prover clause [TOP] and the user may or may not accept the content of the 'adequate to equalize' rule.

Provided the rule is accepted this completes the computation of the fourth conjunct in the antecedent of the analogia legis rule. The following three conjuncts in the antecedent of the analogia legis rule are computed likewise which completes the computation of the initial query. A conclusion is not considered as final before the line of arguments leading up to it has been considered and accepted by the user. To this end the user needs a comprehensible presentation of the proof term. We illustrate elsewhere [3] how derivations of goals can be entrusted to the user's acceptance or rejection by an interactive piecemeal display of a term representing the proof of the goal.

Above, user answers have been either in the affirmative or in the negative. It is not expected that the user be left without help though. Precedent legal cases provide examples of the application of many legal concepts and user interaction systems for proper presentation of precedent cases is a subject dealt with in our previous studies [5].

4 Conclusions and Further Work

Above we have proposed a *novel approach* for representing fragmentary, multilayered, not fully formalizable knowledge. The informal metatheory of the usual formalization approach is replaced by a semiformal metalogic program which interactively composes formal object theories to be accepted or rejected by the user as formalizations of the knowledge. Our representation is structure preserving and easily copes with changes in the represented legal knowledge.

Imprecise knowledge requires advanced *user interaction* that promotes meaningful user answers and queries, constructs and intelligibly displays proof terms explaining derived conclusions, and makes the system pose its questions in a natural order. These aspects have been considered and to some extent solved in our program [7, 8].

Multiple semantic interpretations of provisions is realised by allowing the user to fill schemata with meaningful content referring to his fact situation whereupon the system accepts or rejects the thus proposed rule. Including multiple structural interpretations, e.g., adding premises, should raise no real obstacles provided rules of acceptance for such alteration can be established.

In *case law* rules of legal interpretation are as important as in statute law and apart from the difficult problem of inducing schemata from precedent cases, we hypothesize that our framework needs only minor adaptations to catch the problem of case-based reasoning.

Proof terms should, since the notion of being a persuasive line of arguments is vague, not only be displayed for user communication but also reasoned about.

Acknowledgments

We would like to thank Keith Clark and Leon Sterling for valuable comments. The research reported herein was supported by the National Swedish Board for Technical Development (STU).

References

1. K. A. Bowen, R. A. Kowalski: Amalgamating Language and Metalanguage in Logic Programming. In: K. Clark and S.-Å. Tärnlund (eds.): Logic Programming. London: Academic Press 1982, pp. 153–172
2. S. Costantini, G. A. Lanzarone: A Metalogic Programming Language. In: G. Levi, M. Martelli (eds.): Proc. Sixth Intl. Conf. on Logic Programming. Cambridge: MIT Press 1989, pp. 218–233
3. A. Hamfelt: Metalogic Representation of Multilayered Knowledge. Uppsala Theses in Computing Science 15/92. Uppsala: Uppsala University 1992
4. A. Hamfelt: The Multilevel Structure of Legal Knowledge and its Representation. Uppsala Theses in Computing Science 8/90. Uppsala: Uppsala University 1990
5. A. Hamfelt, J. Barklund: An Intelligent Interface to Legal Data Bases—Combining Logic Programming and Hypertext. In: A. M. Tjoa, R. Wagner (eds.): Proc. Database and Expert Systems Applications. Vienna: Springer 1990, pp. 56–61
6. A. Hamfelt, J. Barklund: Metaprogramming for Representation of Legal Principles. In: M. Bruynooghe (ed.): Proc. Second Workshop on Metaprogramming in Logic. Leuven: Katholieke Universiteit Leuven 1990, pp. 105–122
7. A. Hamfelt, Å. Hansson: A Semiformal Metatheory for Fragmentary and Multilayered Knowledge as an Interactive Metalogic Program. In: Proc. Intl. Conf. on Fifth Generation Computer Systems. Tokyo: OHMSHA 1992. pp. 1107–1114
8. A. Hamfelt, Å. Hansson: Representation of Fragmentary and Multilayered Knowledge: A Semiformal Metatheory as an Interactive Metalogic Program. UPMAIL TR 68. Uppsala: Comp. Sci. dept., Uppsala University 1991
9. J. Horovitz: Law and Logic. Vienna: Springer 1972
10. S. C. Kleene: Introduction to Metamathematics. New York: North Holland 1980
11. R. A. Kowalski: Problems and Promises of Computational Logic. In: J. W. Lloyd (ed.): Computational Logic. Berlin: Springer 1990, pp. 1–36.
12. M. J. Sergot: A Query-the-User Facility for Logic Programming. In: P. Degano and E. Sandewall (eds.): Integrated Interactive Computer Systems. Amsterdam: Noth-Holland 1983, pp. 27–41.

Metaprograms for Change, Assumptions, Objects, and Inheritance

Jan Grabowski

Institut für Statistik und Informatik
Universität Wien, Liebiggasse 4, A-1010 Wien
jcg@ifs.univie.ac.at

Abstract

The effects of database updates, context switches, and the introduction of reusable components to declarative programming are given concise explanations in terms of many-sorted logical metaprograms, in a way that supports concurrent execution of goals. The details of this approach can be realized in many ways; we present one of them for discussion. Possible applications are: reasoning with assumptions, self-modifying programs, object oriented structures in declarative programming, distributed knowledge bases. The method is presented for logic programming but not inherently restricted to this field.

1 Introduction

1.1 The Dilemma of Declarativity

Our starting point is the general paradigm of *declarative programming*, or programming with *executable specifications*. In declarative programming, correct execution consists in computing exactly all *solutions* of a *goal* or *query* wrt a *knowledge base* or *database*. The way in which the solution set depends on the goal and the database, the *semantics*, can be defined *declaratively*, i.e. without specifying how the solution set is computed.

In this sense, any relational database query language is an example of a declarative programming language, and a highly practical, though not very powerful one. Other, more powerful examples are logic programming, query languages of deductive databases, constraint programming, programming with term rewriting systems, and programming with subsumption inference. We assume that the reader is familiar with some of them.

Declarative programming has the advantage of a clear and "mathematically distinguished" semantical standard; it identifies programming with specification and often supports an elegant and transparent programming style. (Unfortunately the results computed by existing interpreters are not always correct under the semantical standard. We call this problem the *minor dilemma of declarativity* and neglect it in the present paper.) In contrast, the semantics of non-declarative programming languages are very difficult to specify, contain a lot of arbitrary decisions, and frequently cause misunderstandings and implementation incompatibilities.

On the other hand, no practical programming is at all purely declarative:

- As the world changes, databases must be updated. However, even a minor change of the database is a modification of the program. The declarative viewpoint demands that the program be constant during execution, and that changes be limited to periods when the program is idle. This is in many cases an impractical demand. We do not want to exclude updates from the programming capabilities of our systems. Current database programming needs unified languages that can express queries, updates, and transaction control. Moreover, self-modifying programs are a common issue in Artificial Intelligence, in particular, with systems capable of acquiring or synthesizing algorithmic knowledge.

- A running declarative program is in practice always embedded into a non-declarative environment. To be useful, it must accept an input or produce an output at least from time to time. Again, the classical approach consists in limiting input and output to the times when the program is idle, i.e. when it performs no deductions. Apart from the uncomfortably low degree of interactivity of such programs, the approach fails to give any support for distributed programs.

- It is common knowledge that large systems should be developed in a modular way and that modularity should be supported by the programming language and programming environment. It is widespread belief that the object-oriented paradigm fulfils this requirement in an adequate way. Implementations of declarative programming languages tend to provide facilities for constructing modules and/or objects and for reusing components, often without giving declarative semantics for them (cf., for example, the object-oriented extensions of BIM_PROLOG and MACProlog).

We call this situation the *major dilemma of declarativity*.

1.2 Our Approach

The major dilemma of declarativity can be given at least four possible answers:

1. Declarative programs are a just a nice exercise. Their simplicity means inadequacy to real world. Their further development is ill-spent effort. – This answer is rather implicit in the attitude of many software engineers to declarative programming.

2. The advantages of declarativity justify its limitations. All concessions made to overcome the limitations have to be revised, and non-declarative features to be replaced by declarative ones. – This is the viewpoint favored in the communities of Logic and Algebraic Programming.

3. Declarativity is fine in some cases, non-declarativity in other ones. Both should be jointly supported in hybrid programming systems. – This is the viewpoint of commercial tool developers, who emphasize versatility and do not care of theoretical clarity. It enjoys some acceptance in the Artificial

Intelligence community which likes to use everything that might contribute to intelligent behavior.

4. It is the clear semantics from where declarative programming draws its power. Whatever extensions to be made should result in an easy-to-describe extension to the original declarative semantics.

In this paper we take the last viewpoint. We will define extensions to declarative programming in terms of *metaprograms*. This idea is not new. Metaprograms have been extensively studied, in particular in Logic Programming, for several purposes.

There are at least two types of metaprogram applications:

- to specify a non-standard execution mode for declarative programs (with or without preserving the semantics of the program to which the metaprogram is applied [the *object program*]).

- to interpret object programs that contain calls to predicates which cannot be defined within the declarative semantics. These are then called *meta-level predicates*. In this case, the object program is not declarative, while the metaprogram might be.

We are interested in the second type of application of metaprograms. We want to define several *non-declarative* extensions to declarative programming in terms of *declarative* metaprograms.

A well-known problem of metaprogramming is the loss of efficiency resulting from interpreting a metaprogram instead of directly interpreting the object program. Methods of diminishing this loss are, e.g., partial evaluation, or the use of special interpreters that can switch from meta-level to object-level interpretation. We largely ignore this problem in the present paper. Though executable, our metaprograms are not designed for execution at all. Rather they serve as *specifications* to regulate the functioning of any interpreters of the respective extensions of declarative programming languages.

Our metalanguage L_m is Logic Programming with several sorts. The definitions we give in terms of L_m all consist of *definite* Horn clauses.

The object language L'_o is a specific extension to some declarative programming language L_o. L_o may, for example, be one-sorted Logic Programming with negation under SLDNF semantics, or some kind of non-Horn Logic Programming language, or a term rewriting language, or a constraint programming language, or a deductive database query language etc. Our *examples* take logic programs as object-level programs. We emphasize, however, that the method is largely independent of the L_o chosen.

The non-declarative extensions we consider are:

operations to change the database,
operations to define *objects* (named databases) and to make objects and their contents available to other objects,
operations to switch the context of execution from one object to another.

We introduce them step by step, in the above order.

2 The Metaprograms

2.1 Sorts

Our meta-level language is *many-sorted*. (Cf. Hill and Lloyd [1] for a detailed analysis of many-sorted vs. one-sorted approaches to first-order logical metaprogramming, which shows that many-sorted approaches are mandatory if one wants to model objectlevel variables by meta-level ones.) We need at least the sorts

formula	for objectlevel formulae (items recorded in the database)
goal	for objectlevel goals (tasks to be performed on the database)
answer	for answers of objectlevel goals
name	for names of objects
individual	for other object-level things ('data')
database	for databases

The user is free to classify object-level data by more than just one sort *individual*. Anyway these sorts will not appear in our meta-level clauses.

A term of sort *database* represents a database. These terms will *never* appear on the object-level.

A term of sort *answer* represents an answer of a goal. What is meant here is a kind of answer which is *not* processed explicitly on the object-level. It is similar to the variable substitution induced by a goal in Logic Programming, or a reduced constraint set in Constraint Logic Programming. We will assume that there is a constant **yes** of sort *answer*, whose intended meaning is that the goal was successful and produced no specific answer.

2.2 Database Updates

For the moment, we think of a database as a finite *set of formulae* $\{f_1, \ldots, f_k\}$. The history of a database can be thought of as a finite series of operations that entered or removed formulae to/from it, beginning with the empty database. Hence the database can be represented as a term of sort *database* formed from terms of sort *formula* with the aid of the following constructors:

$$\begin{aligned} \texttt{root} &: \to database \\ \texttt{asserted} &: database \times formula \to database \\ \texttt{retracted} &: database \times formula \to database \end{aligned}$$

having in mind the following meanings:

root	represents the empty database.
asserted(d, f)	represents the database obtained from the one represented by d by entering f.
retracted(d, f)	represents the database obtained from the one represented by d by removing f (leaving the database unchanged if it doesn't contain f).

We say that two terms of sort *database* are equivalent if they represent the same set. For example, retracted(asserted(d, f), f) and d represent the same set of formulae.

To say it in Artificial Intelligence terms, we adopt the state concept of Ginsberg and Smith [2] (states are sets of formulae) and the state representation of situation calculus, cf. [3] or [4] (states are represented by constructor terms). The latter is, however, a technical decision; we might have better chosen logic programming with sets to formulate our metaprograms, but we wanted to stay on well-accepted grounds.

Entering and removing formulae are operations that can be invoked on the object-level with respect to the *current* database. This means, there exist two functors

$$\text{assert} \quad : formula \rightarrow goal$$
$$\text{retract} \quad : formula \rightarrow goal$$

As **assert** and **retract** change the database, they are non-declarative. Their meanings have to be specified in the metaprogram. The metapredicate **demo** is responsible for the interpretation of an object-level goal. It is therefore often called "the metainterpreter". Besides the object-level goal, the **demo** predicate should have three other arguments: the database *before execution*, the database *after execution*, and the answer. Thus, **demo** is of type

$$\text{demo} : database \times goal \times answer \times database$$

The following clauses for **demo** provide a definition for **assert** and **retract**:

```
demo(D, assert(F), yes, asserted(D, F)) <-
demo(D, retract(F), yes, retracted(D, F)) <-
```

An important requirement to be imposed on our definitions is that the effects of object-level goals are independent of the representation of a database. This means that if terms d_1 and d_2 are equivalent (represent the same database), then $\text{asserted}(d_1, f)$ and $\text{asserted}(d_2, f)$ should be equivalent, as well as $\text{retracted}(d_1, f)$ and $\text{retracted}(d_2, f)$. This condition is met by our intended meaning of **asserted** and **retracted**.

Hill and Lloyd [5] discuss a reduced variant of **demo**:

$$\text{demo} : database \times goal$$

where the first argument is the current database which is assumed to remain unchanged under the execution of the goal. This means that there are no object-level goals corresponding to our $\text{assert}(f)$ and $\text{retract}(f)$. Changes to the database do not occur within a call to **demo**. This is exactly the restriction discussed earlier that updates are performed on idle programs only. For the purpose of updates they use two special meta-level predicates **insert** : *database* × *formula* × *database* and **delete** : *database* × *formula* × *database* with obvious intended meanings. We do allow self-updates of running object-level programs and want to specify the effects by our metaprograms.

It may turn out to be useful to have an object-level operation

$$\text{empty} : \rightarrow goal$$

whose intended meaning is that it empties the current database. The corresponding clause for **demo** is

```
demo(D, empty, yes, root) <-
```

2.3 Goal Composition

We introduce just one operation to form compound goals: *sequential composition*. In accordance with Prolog notation we denote it by the comma (,):

$$\_ \, , \, \_ \ : goal \times goal \to goal$$

Unlike the corresponding operation in pure Logic Programming, this operation does not generally have a declarative meaning, since the overall answer can depend on side-effects produced by the component goals. In particular, the order of arguments matters. The database *before execution* of the second goal is the database *after execution* of the first goal.

```
demo(D1, (G1, G2), A, D) <-  demo(D1, G1, A1, D2) &
                             demo(D2, G2, A2, D) &
                             combine_answers(A1, A2, A)
```

The metapredicate

$$\text{combine\_answers} \ : answer \times answer \times answer$$

describes the way the answers of two successively executed goals are combined. We do not suggest a specific definition of `combine_answers` here; it is natural to require that it be an associative function in its first two arguments. We also require that `yes` be a neutral element of this function:

```
combine_answers(yes, A, A) <-
combine_answers(A, yes, A) <-
```

2.4 Objects

Operations of a declarative programming language always refer implicitly to the current database, which is anyway constant during the whole computation. Once we add operations like `assert` and `retract` that change the database, it makes sense to wish to refer to databases other than the current one. Therefore we introduce *names* for databases. Once a database has been given a name, it can be referred to explicitly by its name, provided the name is *known* (recorded) in the current database. This requires that the interpreter keeps all named databases available for later use.

Named databases are also called *objects*, to reflect an intended connection to the concept "object" in object-oriented programming. This approach to objects in logic programming is related to Chen's [6] module concept.

Our initial definition of a database is now reformulated in the following recursive way:

Definition

1. An *object* is an ordered pair $n : D$, where n is a name and D a database.

2. A *database* is the union of a finite set of formulae and a finite set of objects, the latter being a partial function from names to databases.

Thus a database has the form $\{f_1, \ldots, f_k, n_1 : D_1, \ldots, n_m : D_m\}$, with pairwise distinct n_i's.

Interesting special cases of objects are

- $n : \{\,\}$, where $\{\,\}$ denotes the empty set: just a name;

- $n : \{v : \{\,\}\}$: a name carrying another name as "value".

Names are represented on the meta-level by ground terms of sort *name*. (As names are assigned at runtime, one may wish to have a built-in object-level operation that generates fresh names at runtime, too. We do not propose such an operation here.)

Names are assigned, deassigned, and checked by the following object-level operations:

$$\mathtt{new} \ : name \rightarrow goal$$

$$\_ := \_ \ : name \times name \rightarrow goal$$

$$\mathtt{delete} \ : name \rightarrow goal$$

$$\mathtt{exists} \ : name \rightarrow goal$$

$$\mathtt{notexists} \ : name \rightarrow goal$$

$\mathtt{new}(n)$ assigns the name n to the *empty* database and records it in the current database. $n_2 := n_1$ takes the database assigned to the name n_1 and assigns it to n_2. The current database now "knows" at least two objects named n_1 and n_2 which, at the moment, happen to have identical contents.

To maintain the condition that the object set in a database forms a partial function from names to databases, name conflicts raised by the above operations are suppressed according to the rule that a new object overrides an old one with the same name.

Whether or not an object name is defined in the current database can be checked by the object-level operations $\mathtt{exists}$ and $\mathtt{notexists}$. ($\mathtt{notexists}$ is the negation of $\mathtt{exists}$; if the object-level language has negation, one of the two can be omitted.) Both $\mathtt{exists}$ and $\mathtt{notexists}$ need instantiated arguments to work properly. $\mathtt{delete}(n)$ deletes the object $n : D$ corresponding to the name n from the current database if there is such an object. (This need not imply that the database D is physically removed from the program; D may be referred to by another object that still exists.)

The definitions of the above operations are given on the meta-level by corresponding clauses for $\mathtt{demo}$. We will need two extra constructors

$$\mathtt{named} : database \times database \times name \rightarrow database$$

$$\mathtt{deleted} : database \times name \rightarrow database$$

for representing databases containing objects. The term $\mathtt{named}(d_1, d, n)$ represents the same database as d_1 except that the object $n : D$ (where D is the database represented by d) has been added, thereby possibly invalidating another object with the same name. $\mathtt{deleted}(d, n)$ represents the database obtained from D (the database represented by d) by deleting the object with name n if such an object exists in D.

Now we can extend $\mathtt{demo}$:

```
demo(D, new(N), yes, named(D, root, N)) <-
demo(D, N1 := N2, yes, named(D, D1, N1)) <- object(D, N2, D1)
demo(D, delete(N), yes, deleted(D, N)) <- object(D, N, D1)
demo(D, delete(N), yes, D) <- notobject(D, N)
demo(D, exists(N), yes, D) <- object(D, N, D1)
demo(D, notexists(N), yes, D) <- notobject(D, N)
```

This recurs to two other meta-level predicates

$$\text{object} : database \times name \times database$$

$$\text{notobject} : database \times name$$

Their definitions are

```
object(asserted(D, F), N, D1) <- object(D, N, D1)
object(retracted(D, F), N, D1) <- object(D, N, D1)
object(named(D, D1, N), N, D1) <-
object(named(D, D1, N1), N2, D2) <- N2\==N1 & object(D, N2, D2)
object(deleted(D, N), N1, D1) <- N\==N1 & object(D, N1, D1)
```

and

```
notobject(root, N) <-
notobject(asserted(D, F), N) <- notobject(D, N)
notobject(retracted(D, F), N) <- notobject(D, N)
notobject(named(D, D1, N1), N2) <- N2\==N1 & notobject(D, N2)
notobject(deleted(D, N), N) <-
notobject(deleted(D, N), N1) <- N\==N1 & notobject(D, N1)
```

Again, we might have defined notobject by object using negation as failure, but we wanted a definite definition.

In the above clauses we used a metapredicate

$$\_ \ \backslash== \ \_ : name \times name$$

which when applied to two terms, is successful iff the terms are different. \== is a predefined *definite* metapredicate; it is identical to the predicate defined by the set of all unit clauses n1 \== n2 <- where n1 and n2 are different ground terms of sort *name*.

2.5 Context switches

The most important use of an object consists in *making it the current database*. This allows to evaluate a goal with respect to a database other than the current one. The corresponding operation is

$$\text{with} \ \_ \ \text{do} \ \_ : name \times goal \rightarrow goal$$

A goal with n do g is successful and returns the answer a if

1. the current database contains an object $n : D$

2. g is successful and returns a if applied to D as the current database.

The prefix with n do causes a temporary *context switch*. This device is similar to dynamic context changes in Contextual Logic Programming [7]. Our contexts are more flexible since they can be modified by assert, retract etc.

We extend demo:

```
demo(D1, with N do G, A, named(D1, D2, N)) <-
    object(D1, N, D) & demo(D, G, A, D2)
```

The reader should notice that the execution of with n do g generally changes the object referred by n. If it was $n : D$ before execution, it becomes $n : D_2$ after execution, where D_2 is the database obtained from D by applying g.

Therefore a goal with n do assert(f) makes sense: It changes the database referred to by n by entering the formula f. This is how non-empty objects are constructed.

Notice that, if an object was constructed using an existing one by the copy operation $n_2 := n_1$, and one of the two objects is later changed, say, by with n_1 do assert(f), this change does not affect the database of the object with name n_2. This is the way existing objects are reused for constructing new ones.

It is important to notice that whatever goals are performed relative to an object have no effects on databases not contained (directly or indirectly) in the object. We call this principle the *absence of implicit side-effects*. It is respected by all our non-declarative object-level goals, and of course the same should hold for the declarative ones.

2.6 The formula metapredicate

Whatever the declarative semantics of L_o be, it must be reconstructed on the meta-level by the demo metapredicate. To this end we need another metapredicate which has as its solutions all formulae contained in the current database, similar to Prolog's clause. This is the metapredicate

$$\text{formula} : database \times formula$$

defined as follows:

```
formula(asserted(D, F), F) <-
formula(asserted(D, F1), F) <- formula(D, F)
formula(retracted(D, F1), F) <- F \== F1 & formula(D, F)
formula(named(D, D1, N), F) <- formula(D, F)
formula(deleted(D, N), F) <- formula(D, F)
```

formula(d, f) is satisfied if f is unifiable with a term of sort *formula* which represents a formula in the database represented by d. If terms of sort *formula* contain variables (of sort *individual*), these must participate in the unification, and we have simultaneous meta-level and object-level unification. If object-level variables are *not* represented by variables of sort *individual* on the meta-level, object-level unification must be metaprogrammed separately.

In the above clauses \== is a built-in metapredicate of sort *formula* × *formula* defined analogously to the corresponding predicate of sort *name* × *name* introduced earlier.

The third clause says that retract removes a formula only if it is identical to the argument. This version of retract involves no unification and is not resatisfiable.

2.7 Upward and downward passing. cover

The support given to reusability by :=, new, assert and retract is too weak. Further operations are needed that allow, on the object-level, to pass objects from one database to another. They could be defined in various ways. We propose two operations

$$\_ := \_ \setminus \_ : name \times name \times name \to goal \qquad \text{(upward passing)}$$

$$\_ \setminus \_ := \_ : name \times name \times name \to goal \qquad \text{(downward passing)}$$

Like every object-level operation, these two operate relatively to the current database. $n_2 := n_1 \setminus n_3$ assumes that the current database contains some object named n_1, and that this object in turn contains some object named n_3. This latter object is copied to the current database with the name n_2, thereby possibly invalidating another object with the same name. $- n_2 \setminus n_3 := n_1$ assume that the current database contains two objects with names n_1 and n_2. The first one is copied into the second one and there assigned the name n_3.

```
demo(D, N2 := N1 \ N3, yes, named(D, D3, N2)) <-
     object(D, N1, D1) & object(D1, N3, D3)
demo(D, N2 \ N3 := N1, yes, named(D, named(D2, D1, N3), N2))
     <- object(D, N1, D1) & object(D, N2, D2)
```

As everything we define, the two operations are governed by the principle that implicit side effects are impossible. For example, $n_2 \setminus n_3 := n_1$ adds an object to the object referred to by n_2 in the current database. If the same object had been referred to by other names in the current database or elsewhere, those names still identify the old version of the object.

To create an object without reusing an existing one we introduced the object-level operation new, which always creates an empty object. Apart from the empty database there is another database which is always available and which we may wish to be able to assign a name: the current database. As this operation

$$\textbf{cover} : name \to goal$$

is not definable by the ones introduced so far, we give it a meta-level definition:

```
demo(D, cover(N), yes, named(D, D, N)) <-
```

After cover(n), the database contains a new object $n : D$, where D is what was the current database before. A previously existing object with the same name n is removed from the current database. (Note that it keeps existing in the database D which is now identified by n.)

2.8 Inheritance: Uncovering an object

So far we have learnt how to reuse objects by copying them and then entering single formulae or objects. Now we want to introduce a more powerful method which allows to conjoin several databases into a new one. According to common terminology we might call this method "import" or "inheritance". We know already how an object $n : D$ can be made available to another object $n' : D'$ by the copying operation. What remains to do is to take the contents of $n : D$, i.e.

the database D, and unite it with the contents of $n' : D'$, i.e. with D'. This is done by

$$\text{uncover} : name \rightarrow goal$$

We need another constructor to represent databases whose histories involved an uncover operation:

$$\text{uncovered} : database \times name \rightarrow database$$

Let d' represent the database D' and let D' contain an object $n : D$. Then uncovered(d', n) will represent the database obtained from D' by the following operations:

1. Remove the object $n : D$.

2. Add all formulae contained in D.

3. Add all objects contained in D, except those for which the name is already defined in D'.

(Step 1 corresponds to the common-sense understanding that inheritance involves the testator's death.)

This is reflected in the following additional clauses for our demo, object, notobject and formula:

```
demo(D, uncover(N), yes, uncovered(D, N)) <- object(D, N, D1)
object(uncovered(D, N1), N2, D2) <-
        object(D, N1, D1) & object(D, N2, D2) & N1 \== N2
object(uncovered(D, N1), N1, D2) <-
        object(D, N1, D1) & object(D1, N1, D2)
object(uncovered(D, N1), N2, D2) <-
        object(D, N1, D1) & object(D1, N2, D2) & notobject(D, N2)
notobject(uncovered(D, N1), N2) <-
        object(D, N1, D1) & notobject(D, N2) & notobject(D1, N2)
notobject(uncovered(D, N1), N1) <-
        object(D, N1, D1) & notobject(D1, N1)
formula(uncovered(D, N1), F) <-
        object(D, N1, D1) & formula(D, F)
formula(uncovered(D, N1), F) <-
        object(D, N1, D1) & formula(D1, F)
```

This is one possible way of maintaining the uniqueness of object names upon uncovering an object. It is justified by the viewpoint that objects inherited from an uncovered object are implicitly acquired knowledge, which the user probably does not want to override explicitly acquired one.

2.9 The Rest of Object-Level: reduce

We have given meta-level definitions of 10 specific non-declarative operations which we wish to be available on object-level. Further, we gave a meta-level definition for the sequential composition of goals.

Let the rest of the object-level, which we do not want to specify here, be represented by a metapredicate reduce plus the trivial goal true:

```
demo(D, G, A, D1) <- reduce(D, G, A, D1)
demo(D, true, yes, D) <-
```

reduce is our surrogate for the metainterpreter which models the reduction of user defined goals.

For example, if we wish our metaprograms to specify a non-declarative extension to pure logic programming, we have to define **reduce** as follows:

```
reduce(D1, G, A, D2) <-              % goal reduction
    formula(D1, (H :- B)) &
    rename_and_unify(G, H, B, A1) &
    instantiate(B, A1, G1) &
    demo(D1, G1, A2, D2) &  % may invoke non-declarative goals
    combine_answers(A1, A2, A)
```

with appropriate definitions for **rename_and_unify** and **instantiate**. In this application, answers are variable substitutions, **yes** is the identical substitution, und **combine_answers** is the superposition of substitutions.

In general, we can see that the semantics of our metaprogram is parameterized by the semantics of **reduce**. What we emphasize is that the additions we made are definite clauses and thus contribute minimally to the difficulty of the semantics.

3 Analysis

In this section we use the metaprogramming approach for studying properties of object-level programs.

3.1 Programming with Assumptions

assert and **retract** support *programming with assumptions*. By this we mean that a deduction is performed under a hypothesis, in order to verify or falsify the hypothesis, or to prove an implication. For example, in an application based on propositonal calculus we might wish to prove implication p **then** q by first adding p as an assumption and then proving q. This is settled by the goal

$$\text{assert}(p), \text{prove}(q), \text{retract}(p)$$

where **prove** has the appropriate meaning. Programming with assumptions can be especially useful with inductive or abductive reasoning.

3.2 Mutual Knowledge and Self-knowledge

Assume that we want two objects $n_1 : D_1$ and $n_2 : D_2$ to share data (e.g., a subset of their formulae, or an object), and both are allowed to change these data. Each time one object has made a change, it has to notify its partner explicitly of the change. This requires that the two objects have to have knowledge of each other. However, each time $n_1 : D_1$ issues an operation to update $n_2 : D_2$, it incidentally updates itself, since it corrects its knowledge of what $n_2 : D_2$ knows. To keep $n_2 : D_2$ up-to-date of this change would require another update operation etc. This means that perfect mutual knowledge, if it ever exists, cannot

4 Example from Logic Programming

The example is an object-level program for the well-known game called "Animal", where the user has in mind a species of animal and the program has to guess it by posing binary questions on properties that animals may have or not have (*property questions*). Each time it made a wrong guess the program has to learn a new property question, thereby extending a binary tree labeled with questions. We made the program nondeterministic by allowing "don't know" as a user's answer to a property question.

It is supposed that an object named user be available in the current database. A goal relativized to user represents an interaction with the user. If the user executes the goal with success, he is expected to produce an answer, i.e. to instantiate the variables of the goal.

The following is a compound goal that constructs, within the current database, an object game which contains the procedures of the game. Notice the name-valued constructor suc : *individual* → *name*.

```
new(game),                          % define a new object "game"
game \ user := user,                % give it access to "user"
with game do                        % the rest happens within "game"
 (new(node),                        % define a new object "node"
  node \ user := user,              % give it access to "user"
  with node do                      % fill "node" with clauses
   (assert(examine :-
      property(P),      % stored property
      with user do has_property(P,A),  % property question
      continue(A)),           % continue according to user's reply
    assert(examine :-
      animal(S),         % stored species of animal
      with user do is_it(S,A),      % guess
      finish_or_update(S,A)),       % see below
    assert(continue(dont_know) :-   % may mean "yes"
      continue(y)),
    assert(continue(dont_know) :-   % as well as "no"
      continue(n)),
    assert(continue(A)) :-
      with suc(A) do examine),      % recurse to object "suc(y)"
                                    % or "suc(n)"
    assert(finish_or_update(S,y) :-
      true),                        % right guess, finish
    assert(finish_or_update(S1,n) :-  % wrong guess, update
      with user do what_is_it(S2),  % correction
      with user do
        property_to_distinguish(S1,S2,P),  % new property
      retract(animal(S1)),          % remove wrong guess
      assert(property(P)),       % become a property node
      suc(y) := node,                    % "yes" successor
      suc(y) \ node := node,       % make self-reproducible
      with suc(y) do assert(animal(S1)),  % guess when P
      suc(n) := node,                    % "no" successor
```

```
    suc(n) \ node := node,           % make self-reproducible
    with suc(n) do assert(animal(S2))   % guess when not P
    )
  ),              % "node" completed, up to self-knowledge
tree := node,                        % build a tree
tree \ node := node,                 % tell it what is a node
with tree do assert(animal(cat)),    % initial guess: a cat
assert(round :-                      % define a round of game
  with tree do examine,              % work on current tree
  with user do another_round(A),     % want once more?
  continue_or_skip(A)),
assert(continue_or_skip(y) :-
         round),                     % next round on updated tree
assert(continue_or_skip(n) :-
         true)                       % go home
)                                    % "game" completed
```

Playing Animal consists in invoking a goal with game do round.

The text should be sufficiently self-documenting. The example also shows that the language can benefit from a more flexible syntax and more powerful built-in predicates.

5 Concluding Remarks

Our aim was to show that changes, assumptions, objects, and inheritance all can be easily defined as extensions to declarative programming in a way that adds little complexity to the semantics of the underlying language. We do not claim that the way we did it is the best in any sense. Rather, it may serve as a suggestion for further discussion.

A few comments should be added here:

1. One might wonder how our approach accounts for the non-declarativity of input and output, an aspect mentioned in 1.1. Input and output seem to fit quite well if they are realized in the form of goals and answers. This is exactly what we supposed it to be in the Animal example, where the environment was represented by the object user.

   One might argue here that the real environment might behave completely randomly, unlike a programmed object. But a closer look shows that user is in fact not distinguishable from a programmed object the content of which we do not know.

2. Once a system has been specified by metaprograms, one might wish to implement it. We did not address any questions of implementation in this paper. The major problem to cope with is the amount of data that has to be kept available for possible use, which includes all databases that are referred to by object names, plus all databases that may have to be restored for backtracking.

3. In 3.3 we have seen that goals relativized to different objects can be executed independently without influencing each other's data. This suggests

an architecture in which a new object can be assigned a new process, possibly running on a different processor than the object which it is part of. Each process would have to maintain the database it started from, and the history of its operations on that database. If an object was deleted and the corresponding process has no backtracking points left, the process can be removed.

Acknowledgments

The author acknowledges the inspiration which he obtained from John Lloyd's Zurich lectures on logical metaprogramming, and from discussions with the PROPEL and HU-Prolog groups at Humboldt University on modules and objects. He is grateful for the good working conditions he had during his stay at the University of Vienna 1991/92.

Since October 1991, this work has been part of the German BMFT project GOSLER on Algorithmic Learning.

References

[1] P.M. Hill and J.W. Lloyd. Analysis of meta-programs, June 1988. (Revised January 1989.) CS-88-08, Dept. Comp. Sci., University of Bristol.

[2] M.L. Ginsberg and D.E. Smith. Reasoning about action (i+ii). *Artificial Intelligence*, 35(2+3), 1988.

[3] J. McCarthy and P.J. Hayes. Some philosophical problems from the standpoint of Artificial Intelligence. In B. Meltzer and D. Michie, editors, *Machine Intelligence 4*, pages 463–502. Edinburgh University Press, 1969.

[4] R.A. Kowalski and M.J. Sergot. A logic-based calculus of events. *New Generation Computing*, 4:67–95, 1986.

[5] P.M. Hill and J.W. Lloyd. Meta-programming for dynamic knowledge bases, December 1988. CS-88-18, Dept. Comp. Sci., University of Bristol.

[6] W. Chen. A theory of modules based on second order logic. In *Proc. IEEE Int. Symp. Logic Programming*, pages 24–33. IEEE, 1987.

[7] L. Monteiro and A. Porto. Contextual logic programming. In G. Levi and M. Martelli, editors, *Proc. Sixth International Conference on Logic Programming*, pages 284–302. MIT Press, Cambridge, 1989.

[8] Jan Grabowski and Wolfgang Müller. Introduction to PROPEL (Version 1.0), August 1990. Informatik-Preprint. Humboldt-Universität zu Berlin.

Author Index

R. Bahgat	162		Å. Hansson	321
K. Benkerimi	177		A. Hamfelt	321
P. Bonatti	220		F. van Harmelen	89
A. Bossi	265		A. Herzig	11
A. Brogi	105		P. M. Hill	177
M. Bruynooghe	294		G. Lanzarone	135
I. Cervesato	148		P. Mancarella	105
H. Christiansen	205		B. Martens	192
N. Cocco	265		F. Mesnard	308
S. Costantini	135		J. Komorowski	49
P. Dell'Acqua	135		K. Konolige	26
S. K. Debray	120		D. Pedreschi	105
D. De Schreye	70, 192		S. D. Prestwich	280
S. Etalle	265		G. F. Rossi	148
L. Fariñas Del Cerro	11		L. Serafini	235
J-G. Ganascia	308		A. Simpson	235
P. Gärdenfors	1		J. L. Träff	280
F. Giunchiglia	235		F. Turini	105
J. Grabowski	336		K. Verschaetse	70
G. Janssens	294		G. A. Wiggins	250

Lecture Notes in Computer Science

For information about Vols. 1–559
please contact your bookseller or Springer-Verlag

Vol. 560: S. Biswas, K. V. Nori (Eds.), Foundations of Software Technology and Theoretical Computer Science. Proceedings, 1991. X, 420 pages. 1991.

Vol. 561: C. Ding, G. Xiao, W. Shan, The Stability Theory of Stream Ciphers. IX, 187 pages. 1991.

Vol. 562: R. Breu, Algebraic Specification Techniques in Object Oriented Programming Environments. XI, 228 pages. 1991.

Vol. 563: A. Karshmer, J. Nehmer (Eds.), Operating Systems of the 90s and Beyond. Proceedings, 1991. X, 285 pages. 1991.

Vol. 564: I. Herman, The Use of Projective Geometry in Computer Graphics. VIII, 146 pages. 1992.

Vol. 565: J. D. Becker, I. Eisele, F. W. Mündemann (Eds.), Parallelism, Learning, Evolution. Proceedings, 1989. VIII, 525 pages. 1991. (Subseries LNAI).

Vol. 566: C. Delobel, M. Kifer, Y. Masunaga (Eds.), Deductive and Object-Oriented Databases. Proceedings, 1991. XV, 581 pages. 1991.

Vol. 567: H. Boley, M. M. Richter (Eds.), Processing Declarative Kowledge. Proceedings, 1991. XII, 427 pages. 1991. (Subseries LNAI).

Vol. 568: H.-J. Bürckert, A Resolution Principle for a Logic with Restricted Quantifiers. X, 116 pages. 1991. (Subseries LNAI).

Vol. 569: A. Beaumont, G. Gupta (Eds.), Parallel Execution of Logic Programs. Proceedings, 1991. VII, 195 pages. 1991.

Vol. 570: R. Berghammer, G. Schmidt (Eds.), Graph-Theoretic Concepts in Computer Science. Proceedings, 1991. VIII, 253 pages. 1992.

Vol. 571: J. Vytopil (Ed.), Formal Techniques in Real-Time and Fault-Tolerant Systems. Proceedings, 1992. IX, 620 pages. 1991.

Vol. 572: K. U. Schulz (Ed.), Word Equations and Related Topics. Proceedings, 1990. VII, 256 pages. 1992.

Vol. 573: G. Cohen, S. N. Litsyn, A. Lobstein, G. Zémor (Eds.), Algebraic Coding. Proceedings, 1991. X, 158 pages. 1992.

Vol. 574: J. P. Banâtre, D. Le Métayer (Eds.), Research Directions in High-Level Parallel Programming Languages. Proceedings, 1991. VIII, 387 pages. 1992.

Vol. 575: K. G. Larsen, A. Skou (Eds.), Computer Aided Verification. Proceedings, 1991. X, 487 pages. 1992.

Vol. 576: J. Feigenbaum (Ed.), Advances in Cryptology - CRYPTO '91. Proceedings. X, 485 pages. 1992.

Vol. 577: A. Finkel, M. Jantzen (Eds.), STACS 92. Proceedings, 1992. XIV, 621 pages. 1992.

Vol. 578: Th. Beth, M. Frisch, G. J. Simmons (Eds.), Public-Key Cryptography: State of the Art and Future Directions. XI, 97 pages. 1992.

Vol. 579: S. Toueg, P. G. Spirakis, L. Kirousis (Eds.), Distributed Algorithms. Proceedings, 1991. X, 319 pages. 1992.

Vol. 580: A. Pirotte, C. Delobel, G. Gottlob (Eds.), Advances in Database Technology – EDBT '92. Proceedings. XII, 551 pages. 1992.

Vol. 581: J.-C. Raoult (Ed.), CAAP '92. Proceedings. VIII, 361 pages. 1992.

Vol. 582: B. Krieg-Brückner (Ed.), ESOP '92. Proceedings. VIII, 491 pages. 1992.

Vol. 583: I. Simon (Ed.), LATIN '92. Proceedings. IX, 545 pages. 1992.

Vol. 584: R. E. Zippel (Ed.), Computer Algebra and Parallelism. Proceedings, 1990. IX, 114 pages. 1992.

Vol. 585: F. Pichler, R. Moreno Díaz (Eds.), Computer Aided System Theory – EUROCAST '91. Proceedings. X, 761 pages. 1992.

Vol. 586: A. Cheese, Parallel Execution of Parlog. IX, 184 pages. 1992.

Vol. 587: R. Dale, E. Hovy, D. Rösner, O. Stock (Eds.), Aspects of Automated Natural Language Generation. Proceedings, 1992. VIII, 311 pages. 1992. (Subseries LNAI).

Vol. 588: G. Sandini (Ed.), Computer Vision – ECCV '92. Proceedings. XV, 909 pages. 1992.

Vol. 589: U. Banerjee, D. Gelernter, A. Nicolau, D. Padua (Eds.), Languages and Compilers for Parallel Computing. Proceedings, 1991. IX, 419 pages. 1992.

Vol. 590: B. Fronhöfer, G. Wrightson (Eds.), Parallelization in Inference Systems. Proceedings, 1990. VIII, 372 pages. 1992. (Subseries LNAI).

Vol. 591: H. P. Zima (Ed.), Parallel Computation. Proceedings, 1991. IX, 451 pages. 1992.

Vol. 592: A. Voronkov (Ed.), Logic Programming. Proceedings, 1991. IX, 514 pages. 1992. (Subseries LNAI).

Vol. 593: P. Loucopoulos (Ed.), Advanced Information Systems Engineering. Proceedings. XI, 650 pages. 1992.

Vol. 594: B. Monien, Th. Ottmann (Eds.), Data Structures and Efficient Algorithms. VIII, 389 pages. 1992.

Vol. 595: M. Levene, The Nested Universal Relation Database Model. X, 177 pages. 1992.

Vol. 596: L.-H. Eriksson, L. Hallnäs, P. Schroeder-Heister (Eds.), Extensions of Logic Programming. Proceedings, 1991. VII, 369 pages. 1992. (Subseries LNAI).

Vol. 597: H. W. Guesgen, J. Hertzberg, A Perspective of Constraint-Based Reasoning. VIII, 123 pages. 1992. (Subseries LNAI).

Vol. 598: S. Brookes, M. Main, A. Melton, M. Mislove, D. Schmidt (Eds.), Mathematical Foundations of Programming Semantics. Proceedings, 1991. VIII, 506 pages. 1992.

Vol. 599: Th. Wetter, K.-D. Althoff, J. Boose, B. R. Gaines, M. Linster, F. Schmalhofer (Eds.), Current Developments in Knowledge Acquisition - EKAW '92. Proceedings. XIII, 444 pages. 1992. (Subseries LNAI).

Vol. 600: J. W. de Bakker, C. Huizing, W. P. de Roever, G. Rozenberg (Eds.), Real-Time: Theory in Practice. Proceedings, 1991. VIII, 723 pages. 1992.

Vol. 601: D. Dolev, Z. Galil, M. Rodeh (Eds.), Theory of Computing and Systems. Proceedings, 1992. VIII, 220 pages. 1992.

Vol. 602: I. Tomek (Ed.), Computer Assisted Learning. Proceedings, 1992. X, 615 pages. 1992.

Vol. 603: J. van Katwijk (Ed.), Ada: Moving Towards 2000. Proceedings, 1992. VIII, 324 pages. 1992.

Vol. 604: F. Belli, F.-J. Radermacher (Eds.), Industrial and Engineering Applications of Artificial Intelligence and Expert Systems. Proceedings, 1992. XV, 702 pages. 1992. (Subseries LNAI).

Vol. 605: D. Etiemble, J.-C. Syre (Eds.), PARLE '92. Parallel Architectures and Languages Europe. Proceedings, 1992. XVII, 984 pages. 1992.

Vol. 606: D. E. Knuth, Axioms and Hulls. IX, 109 pages. 1992.

Vol. 607: D. Kapur (Ed.), Automated Deduction – CADE-11. Proceedings, 1992. XV, 793 pages. 1992. (Subseries LNAI).

Vol. 608: C. Frasson, G. Gauthier, G. I. McCalla (Eds.), Intelligent Tutoring Systems. Proceedings, 1992. XIV, 686 pages. 1992.

Vol. 609: G. Rozenberg (Ed.), Advances in Petri Nets 1992. VIII, 472 pages. 1992.

Vol. 610: F. von Martial, Coordinating Plans of Autonomous Agents. XII, 246 pages. 1992. (Subseries LNAI).

Vol. 611: M. P. Papazoglou, J. Zeleznikow (Eds.), The Next Generation of Information Systems: From Data to Knowledge. VIII, 310 pages. 1992. (Subseries LNAI).

Vol. 612: M. Tokoro, O. Nierstrasz, P. Wegner (Eds.), Object-Based Concurrent Computing. Proceedings, 1991. X, 265 pages. 1992.

Vol. 613: J. P. Myers, Jr., M. J. O'Donnell (Eds.), Constructivity in Computer Science. Proceedings, 1991. X, 247 pages. 1992.

Vol. 614: R. G. Herrtwich (Ed.), Network and Operating System Support for Digital Audio and Video. Proceedings, 1991. XII, 403 pages. 1992.

Vol. 615: O. Lehrmann Madsen (Ed.), ECOOP '92. European Conference on Object Oriented Programming. Proceedings. X, 426 pages. 1992.

Vol. 616: K. Jensen (Ed.), Application and Theory of Petri Nets 1992. Proceedings, 1992. VIII, 398 pages. 1992.

Vol. 617: V. Mařík, O. Štěpánková, R. Trappl (Eds.), Advanced Topics in Artificial Intelligence. Proceedings, 1992. IX, 484 pages. 1992. (Subseries LNAI).

Vol. 618: P. M. D. Gray, R. J. Lucas (Eds.), Advanced Database Systems. Proceedings, 1992. X, 260 pages. 1992.

Vol. 619: D. Pearce, H. Wansing (Eds.), Nonclassical Logics and Information Proceedings. Proceedings, 1990. VII, 171 pages. 1992. (Subseries LNAI).

Vol. 620: A. Nerode, M. Taitslin (Eds.), Logical Foundations of Computer Science – Tver '92. Proceedings. IX, 514 pages. 1992.

Vol. 621: O. Nurmi, E. Ukkonen (Eds.), Algorithm Theory – SWAT '92. Proceedings. VIII, 434 pages. 1992.

Vol. 622: F. Schmalhofer, G. Strube, Th. Wetter (Eds.), Contemporary Knowledge Engineering and Cognition. Proceedings, 1991. XII, 258 pages. 1992. (Subseries LNAI).

Vol. 623: W. Kuich (Ed.), Automata, Languages and Programming. Proceedings, 1992. XII, 721 pages. 1992.

Vol. 624: A. Voronkov (Ed.), Logic Programming and Automated Reasoning. Proceedings, 1992. XIV, 509 pages. 1992. (Subseries LNAI).

Vol. 625: W. Vogler, Modular Construction and Partial Order Semantics of Petri Nets. IX, 252 pages. 1992.

Vol. 626: E. Börger, G. Jäger, H. Kleine Büning, M. M. Richter (Eds.), Computer Science Logic. Proceedings, 1991. VIII, 428 pages. 1992.

Vol. 628: G. Vosselman, Relational Matching. IX, 190 pages. 1992.

Vol. 629: I. M. Havel, V. Koubek (Eds.), Mathematical Foundations of Computer Science 1992. Proceedings. IX, 521 pages. 1992.

Vol. 630: W. R. Cleaveland (Ed.), CONCUR '92. Proceedings. X, 580 pages. 1992.

Vol. 631: M. Bruynooghe, M. Wirsing (Eds.), Programming Language Implementation and Logic Programming. Proceedings, 1992. XI, 492 pages. 1992.

Vol. 632: H. Kirchner, G. Levi (Eds.), Algebraic and Logic Programming. Proceedings, 1992. IX, 457 pages. 1992.

Vol. 633: D. Pearce, G. Wagner (Eds.), Logics in AI. Proceedings. VIII, 410 pages. 1992. (Subseries LNAI).

Vol. 634: L. Bougé, M. Cosnard, Y. Robert, D. Trystram (Eds.), Parallel Processing: CONPAR 92 – VAPP V. Proceedings. XVII, 853 pages. 1992.

Vol. 635: J. C. Derniame (Ed.), Software Process Technology. Proceedings, 1992. VIII, 253 pages. 1992.

Vol. 636: G. Comyn, N. E. Fuchs, M. J. Ratcliffe (Eds.), Logic Programming in Action. Proceedings, 1992. X, 324 pages. 1992. (Subseries LNAI).

Vol. 637: Y. Bekkers, J. Cohen (Eds.), Memory Management. Proceedings, 1992. XI, 525 pages. 1992.

Vol. 639: A. U. Frank, I. Campari, U. Formentini (Eds.), Theories and Methods of Spatio-Temporal Reasoning in Geographic Space. Proceedings, 1992. XI, 431 pages. 1992.

Vol. 640: C. Sledge (Ed.), Software Engineering Education. Proceedings, 1992. X, 451 pages. 1992.

Vol. 641: U. Kastens, P. Pfahler (Eds.), Compiler Construction. Proceedings, 1992. VIII, 320 pages. 1992.

Vol. 642: K. P. Jantke (Ed.), Analogical and Inductive Inference. Proceedings, 1992. VIII, 319 pages. 1992. (Subseries LNAI).

Vol. 643: A. Habel, Hyperedge Replacement: Grammars and Languages. X, 214 pages. 1992.

Vol. 644: A. Apostolico, M. Crochemore, Z. Galil, U. Manber (Eds.), Combinatorial Pattern Matching. Proceedings, 1992. X, 287 pages. 1992.

Vol. 645: G. Pernul, A M. Tjoa (Eds.), Entity-Relationship Approach – ER '92. Proceedings, 1992. XI, 439 pages, 1992.

Vol. 646: J. Biskup, R. Hull (Eds.), Database Theory – ICDT '92. Proceedings, 1992. IX, 449 pages. 1992.

Vol. 647: A. Segall, S. Zaks (Eds.), Distributed Algorithms. X, 380 pages. 1992.

Vol. 648: Y. Deswarte, G. Eizenberg, J.-J. Quisquater (Eds.), Computer Security – ESORICS 92. Proceedings. XI, 451 pages. 1992.

Vol. 649: A. Pettorossi (Ed.), Meta-Programming in Logic. Proceedings, 1992. XII, 353 pages. 1992.